CMP

Supercomputing In Engineering Structures

Editors:
P. Melli
C.A. Brebbia

Computational Mechanics Publications
Southampton Boston

Co-published with

Springer-Verlag Berlin Heidelberg New York
London Paris Tokyo

P.MELLI.
IBM ECSEC.
Via Georgione, 159
00144
Roma
Italia

C.A. BREBBIA
Computational Mechanics Institute
Ashurst Lodge
Ashurst, Southampton
SO4 2AA
UK

British Library Cataloguing in Publication Data

Supercomputing in engineering structures
  1. Structures. Analysis. Applications of
  supercomputer systems
  I. Melli, Piero II. Brebbia, C.A. (Carlos
  Alberto), 1938-
  624.1'71'0285411
  ISBN 1-85312-020-0

Library of Congress Catalog card number 88-63525

ISBN 1-85312-020-0 Computational Mechanics Publications Southampton
ISBN 0-945824-07-6 Computational Mechanics Publications Boston, USA
ISBN 3-540-50687-X Springer-Verlag Berlin Heidelberg New York
                   London Paris Tokyo
ISBN 0-387-50687-X Springer-Verlag New York Heidelberg Berlin
                   London Paris Tokyo

Printed in Great Britain by The Eastern Press Limited, Reading

CONTENTS

PREFACE

The advent and diffusion of general purpose mainframes with supercomputing capability and very large central storage, is rapidly increasing the use of computer codes in the manufacturing industry, so that it is apparent that these codes are going to become a cost-effective integral part of the product design and development cycle.

The chapters in this book discuss the issues connected to the use of supercomputers in a field of paramount importance for the industrial world, i.e. Engineering Structures, an area in which large computer codes for the solution of several design problems have been available since the seventies. They are the result of lectures presented at the Seminar on "Supercomputing in Engineering Structures", held in Oberlech, Austria on July 11 to 15, 1988. The Seminar was a part of the IBM Europe Institute, an event that this year was completely devoted to the theme of Supercomputing in Science and Engineering. The aim of the Seminar was to gather, for a fruitful exchange of ideas and experiences, researchers in the field of Engineering Structures and industrial code developers and users.

The chapter by A. Noor carries out an overview of the existing hardware and makes a forecast of the demands that Engineering Structures will pose in terms of computing power in the near future; it concludes by proposing a strategy for the solution of large scale structural problems on the new computing systems. The chapter by E.L. Wilson is devoted to similar problems, with special emphasis on the use of computers with multiple processors in Finite Element analysis, while the chapter by B.B. Moore concentrates on the benefits derived from the effective use of the large central storage addressability allowed by ESA/370[1], by large engineering and scientific computations.

Two chapters, by C.A. Brebbia and R.A. Adey, respectively, discuss the fundamentals of the Boundary Element Method and its computational aspects and applications on supercomputers with particular reference to the BEASY [2] code, while the chapters by G.H. Powell and C. Farhat present modeling and solution strategies for nonlinear braced frames and for large space and aerospace flexible structures. The following three chapters deal with the important and complex problems of structural impact: the first, by N. Jones, is introductory and reviews the field, while the second, by E. Haug et.al. presents some industrial car crash simulations performed using the PAM-CRASH [3] code; finally, the third one by F. Angeleri discusses the aspects of explicit Finite Element codes migration and optimization on the IBM 3090/VF, with particular reference to the PAM-CRASH experience. The chapter by A.J. Morris concentrates on the critical issue of Finite Element codes validation, while the final chapters by J.J. Connor and G.N. Vanderplaats address innovative topics such as knowledge based approaches for Boundary Element Method mesh design and optimization in Engineering Structures using supercomputers.

---

[1] ESA/370 is a trademark of International Business Machines
[2] BEASY is a trademark of Computational Mechanics International
[3] PAM- CRASH is a trademark of Engineering Systems International

The editors are indebted to the authors for their efforts in the timely preparation of the manuscripts here included, as well as to Carol Firmin and Lance Sucharov for dealing with the technical aspects of the book preparation.

P. Melli
IBM ECSEC

New Computing Systems and their Impact on Structural Analysis and Design

Ahmed K. Noor

George Washington University, NASA Langley Research Center, Hampton, Virginia 23665, USA

INTRODUCTION

Four generations of computers are generally recognized, corresponding to a rapid change in the hardware building blocks from relays and vacuum tubes (1940s-1950s), to discrete diodes and transistors (1950s-1960s), to integrated circuits (ICs) (1960-mid 70s), to large- and very-large-scale integrated devices (mid 70s-present). These generations have increased computer performance four orders of magnitude over the last thirty years while dramatically reducing the cost (see Fig. 1). A summary of the hardware and software characteristics of the four generations is given in [1].

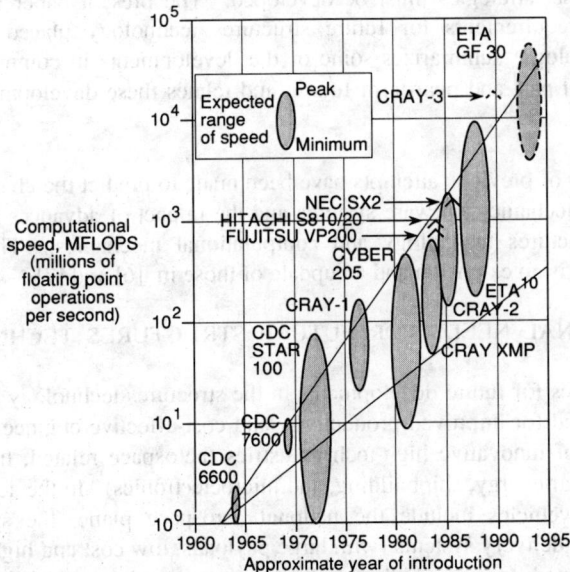

Figure 1 - Growth of computer speed.

A new (fifth) generation of computers is evolving and is likely to be available in the coming decade. The hardware building blocks for the new generation include giga-scale integrated devices [2] and [3], new transistor materials and structures, and optical components. Extensive use will be made of AI technology. Novel forms of machine architecture (e.g., new forms of parallel architecture) will be introduced and will result in a dramatic increase in computational speed. The new century will undoubtedly see more radical changes in computing technology, such as artificial neural network machines and purely optical computing. The potential of these systems is discussed in [4] and [5].

The opportunities offered by the new and projected hardware environment for structures technology are enormous. The current and evolving large supersystems, such as the CRAY-2, ETA-10, CRAY Y-MP and CRAY-3 will open the way to a vast range of new applications, and to higher levels of sophistication of current engineering problems. The small, emerging low-cost engineering workstations and personal computers will provide a high degree of interactivity and free the analysts from the constraints that are often imposed on them by large centralized computation centers. The embedded computers will aid in the control of the devices they reside in. Intelligent interfaces allowing multiple media interaction for both input and output (e.g., graphics and natural language) will facilitate the user-machine communication. Flexible high-capacity networks will allow collaborative computing by linking structural analysts and designers at different locations.

However, in order to realize the full potential of the new and emerging computing systems, the strong interrelations of numerical algorithms and software with the architecture of the systems must be understood, and special solution methodologies and computational strategies must be developed. The present paper discusses key computational requirements for future structures technology placed on evolving computer technology; summarizes some of the developments in computing systems during the recent past and near-term future; and relates these developments to structures technology.

A number of previous attempts have been made to predict the characteristics of computational mechanics software systems and the impact of advances in computing systems on structures technology and computational mechanics. The discussion presented herein is an extension and an update of those in [6] and [7].

COMPUTATIONAL NEEDS FOR FUTURE STRUCTURES TECHNOLOGY

The driving forces for future developments in the structures technology will continue to be: a) the need for improved productivity and cost-effective engineering systems; and b) support of innovative high-tech industries (aerospace related, transportation, petroleum, nuclear energy, shipbuilding, and microelectronics). In the aerospace field planned future vehicles include the national aerospace plane, the space station, improved orbital delivery systems (with large payloads, low cost and high reliability); structures subjected to very high accelerations; very high-precision shaped and controlled space structures under dynamic and thermal disturbances. The realization of the future aerospace systems requires technology advances in the structures discipline as well as in a number of other disciplines including materials, propulsion, aerodynamics, controls, avionics, optics, and acoustics. Similar advances are needed for the realization of the structures of future automotive, nuclear, and microelectronic systems. Among the technical needs for future high-performance structures are the following (see [8]):

1) Expanding the scope of engineering problems considered. This includes:

a) Examination of more complex phenomena (e.g., damage tolerance of structural components made of new material systems);

b) Study of the mechanics and failure characteristics of high-performance modern materials such as superalloys, metal-matrix composites, and high-temperature ceramic composites;

c) Study of interaction phenomena (as would be required in the thermal/control/structural coupling in space exploration, the hydrodynamic/structural coupling in deepsea mining, and the electromagnetic/thermal/structural coupling in microelectronic devices);

d) More extensive use of stochastic models to account for uncertainties associated with loads, environment, and material variability;

e) Improved representation of the details of physical systems such as damping and joints in structures (see, for example, Fig. 2); and

f) Computer simulation (modeling) of the manufacturing processes of modern materials such as solidification, interface mechanics, and superplastic forming.

2) Development of special computational strategies and effective solution methodologies for large-scale structural calculations.

3) Development of a hierarchy of models, algorithms, and procedures for the design of engineering systems. Simplified and specialized models and algorithms are appropriate for use in the preliminary and conceptual design phases and more sophisticated models are used in the detailed design phase.

4) Development of practical measures for assessing the reliability of the computational models, and estimating the errors in the predictions of the major response quantities. This also includes development of efficient adaptive improvement strategies for the computational model.

5) Continued reduction of cost and/or time for obtaining solutions to engineering design/analysis problems.

4

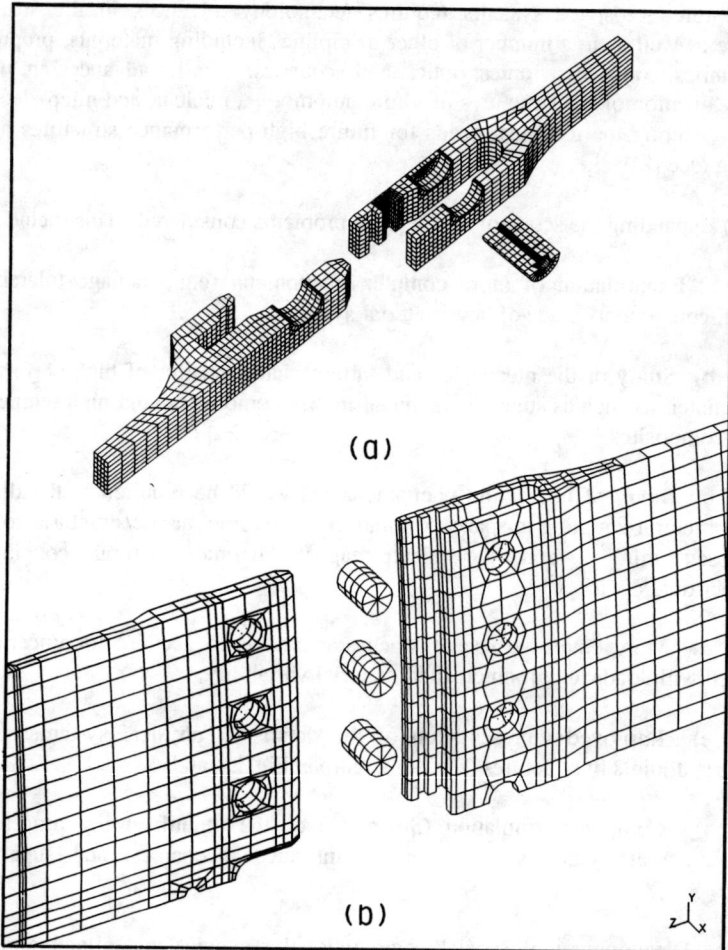

Figure 2 - Exploded views of detailed finite element models for: a) redesigned field joint of the shuttle solid rocket motor case; and b) factory joint for the shuttle motor case (courtesy of Morton Thiokol, Inc., Brigham City, Utah).

The hardware and software requirements to meet the aforementioned needs include:

1)   Distributed computing environment encompassing high-performance computers (supercomputers) for large-scale structural calculations, and a wide variety of intelligent engineering workstations for interactive user interface/control and moderate-scale calculations (e.g., interactive model generation, evaluation of results in graphic form, and generation of elemental matrices). The organization of this hardware will require:

a) Extensive facilities for local and long-range networking to make all hardware and attendant software readily available to each user.

b) Development of demand-sensitive load-sharing systems to allow programs to migrate from one hardware/operating environment to another. Local networking can facilitate cooperative multidisciplinary investigations and design projects among team members interacting through shared databases. Inexpensive long-range networking can have several significant effects. Issues of proprietary data can be reduced by sending the program to the location of the owner of the proprietary data, performing the computations there, and receiving only the processed non-proprietary results.

2) Artificial intelligence-based expert systems, incorporating the experience and expertise of practitioners, to aid in the modeling of the engineering system, the adaptive refinement of the model, and the selection of the appropriate algorithm and procedure used in the solution.

3) Intelligent hardware and software interfaces allowing multiple-media interaction. These include powerful engineering workstations; high-speed long distance communication; knowledge-based management systems to incorporate verbal (audio), visual and other natural interfaces. The interfaces allow the interactive steering and dynamic control of the computations.

4) Advanced visualization capabilities, which include local high-resolution and high-speed graphics, video and film animation capabilities.

5) Integration of analysis programs into CAD/CAM systems. With the trend of moving from software-based processing to hardware-based processing, some of the analysis modules are likely to become hardware functions. Interfaces between software and hardware functions need to be designed.

6) Computerized symbolic manipulation capability to reduce the tedium of analytic calculations and increase their reliability. This allows the analytic work to be pushed further before the numerical computations start.

7) Turnkey engineering application software systems which have advanced modeling and analysis capabilities and are easy to learn and use.

BRIEF REVIEW OF CURRENT AND PROJECTED ADVANCES IN COMPUTER HARDWARE AND NETWORKING TECHNOLOGY

The major developments in computer technology have been, and continue to be, focused on improvements of cost, size, power consumption, speed, and reliability of electrical components. The next generation of computers will be impacted by the developments in three basic areas; namely, hardware components, artificial intelligence, and computer architecture and system design methods. The major advances in hardware components are briefly reviewed in this section, and some of the new computing systems are described in the succeeding sections. The survey given here is by no means complete or exhaustive; the intention is to concentrate primarily on those developments which have had, or promise to have, the greatest impact on the structures technology. Discussion is focused on microelectronics and semiconductor technology; memory systems; secondary storage devices; user-interface facilities; and networking.

Microelectronics and Semiconductor Technology

The most notable advances in hardware components in the last three decades have occurred as a result of developments in microelectronics. Instead of connecting discrete components together by wires to produce a circuit, complete circuit patterns, components, and interconnections are placed on a small chip of semiconductor material. The predominant semiconductor material in use to date is silicon. Better understanding of this material as well as better processing, tooling and packaging techniques enabled the design of fast dense circuitry. The traditional technology used for high performance logic has been *emitter-coupled logic* (ECL), which is the fastest of the silicon logic technologies, and continues to be used in most supercomputers. The *complementary metal-oxide semiconductor* (CMOS), despite its slower speed, has low power and high component density. Therefore, more gates per chip and fewer chips are used for each logic function than for ECL.

The principal advantages of microelectronic circuits are their reliability, low cost, and low power consumption. The ever-increasing number of devices packaged on a chip has given rise to the acronyms SSI, MSI, LSI, VLSI, ULSI, and GSI, which stand for small-scale, medium-scale, large-scale, very large-scale, ultra large-scale, and giga-scale integration, respectively. Since 1960 the number of components on a chip has increased continuously. For the case when no differentiation can be made between logic and memory, the progression of development is shown in Fig. 3. The density of logic chips is projected to grow at a slower rate than the density of memory chips (a factor of 7 in five years for logic chips, compared to a factor of 10 in five years for memory chips). It is anticipated that by the year 2000 the number of components per chip will reach one billion (GSI).

Developments in microelectronics were greatly aided by two projects which started in the early 1980s: the very high-speed integrated circuits (VHSIC) project supported by the U.S. Department of Defense; and the very high performance integrated circuits (VHPIC) project supported by the United Kingdom. The full range of hardware components (computer building blocks) are now available on microelectronic chips; these include memory units, addressing units (i.e., counters and decoders), complete central processing units (CPUs) called microprocessors, and even complete microcomputers (which include the CPU, memory, and input/output functions all residing on a single chip).

The net effect of the aforementioned developments has been a sustained reduction in the cost of computing.

Current research is directed towards: 1) shrinking of conductor and device dimensions (scaling) to submicron dimensions; and 2) increasing the speed of logic circuits, to achieve a machine cycle time of the order of 1 nanosecond (1×10^{-9} second).

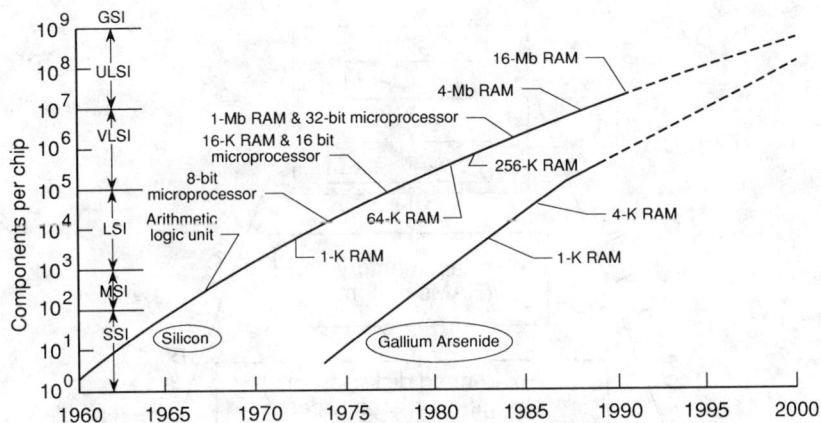

Figure 3 - Growth of number of components per ship (see [2]).

The first objective (miniaturization of electronic components) can be accomplished by using recent and improved lithography tools for etching element patterns on a chip (including optical, electron beams, ultraviolet (UV) optics, direct-wire electron beam, X-ray, and ion-beam techniques) as well as novel processing technology and fabrication techniques. Three candidate technologies are likely to achieve the second objective of ultrafast logic circuits. These technologies are based on using: 1) new material systems such as gallium arsenide (GaAs), a component semiconductor material, and super-conducting materials which do not require liquid helium temperatures (e.g., copper-oxides); 2) new transistor structures such as the quantum-coupled devices using hetero-junction-based super lattices; and 3) integrated optical circuits.

Memory Systems
Memory is the most rapidly advancing technology in microelectronics. Recent progress includes development of an entire hierarchy of addressable memories, and of high-speed, random-access memory chips with many bits of data. Each level in the hierarchy represents an order-of-magnitude decrease in access speed, and several orders-of-magnitude increase in capacity, for the same cost (see Fig. 4). The techniques of splitting and interleaving among various types of memory hierarchies in individual systems have changed some of the basic concepts of computing itself. Instead of just a few registers in the CPU and a single-level memory, a typical machine may now have:

1) a number of high-speed, general-purpose registers
2) a cache memory (or instruction buffers) for very rapid access to small amounts of data (or instructions)

3) standard central or equivalent memory

4) extended memory, directly addressable, but at a lower speed

5) hardware-implemented virtual memory, extending the amount of addressable space.

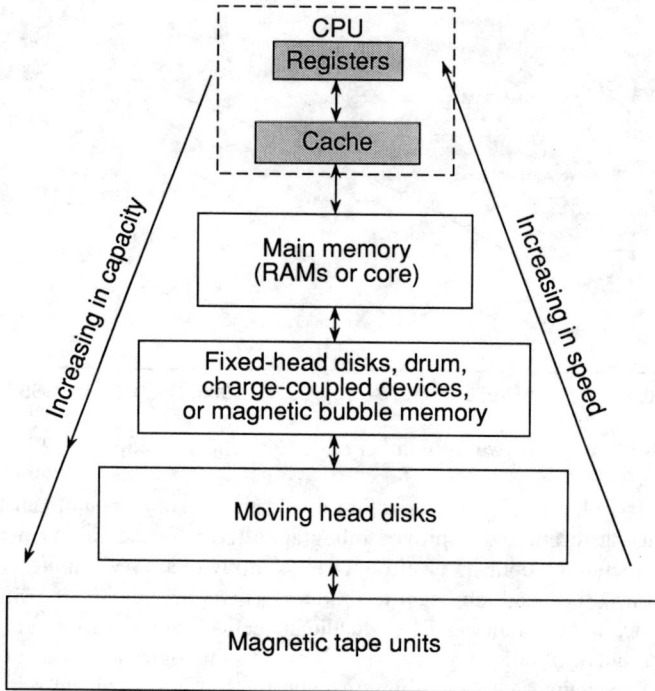

Figure 4 - Classical memory hierarchy (see [10]).

There are several types of semiconductor memories. These include random-access and read-only memories. In random-access memory (RAM), data can be written into, or read off, any storage location without regard to its physical location relative to other storage locations. Read-only memory (ROM) contains a permanent data pattern stored during the manufacture of the semiconductor chip in the form of transistors at each storage location that are either operable or inoperable. RAM can be either static or dynamic. Dynamic RAM (DRAM) requires constant refreshing to maintain its data, while static RAM (SRAM) does not. However, advances in DRAM permit double the amount of RAM at about the same cost as static RAM, so DRAM is used more frequently. The trends in the DRAM and SRAM chip capacity are depicted in Fig. 5. As can be seen from Fig. 5, the DRAM chips are typically one generation ahead of the SRAM chips.

Other types of memory include sequential access memories (SAMs) and direct-access storage devices (DASDs). In SAM information is accessed serially or sequentially. Examples of SAMs are provided by shift-register memories, charged coupled devices (CCDs), and magnetic bubble memories (MBMs). DADs are rotational devices made of magnetic materials where any block of information can be accessed directly. The relative cost-access time relationship of some memory technologies is shown in Fig. 6.

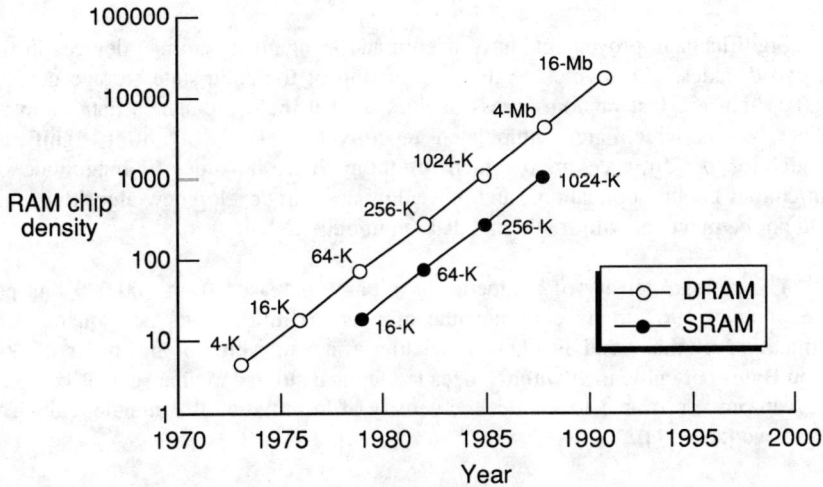

Figure 5 - Trend in RAM chip capacity.

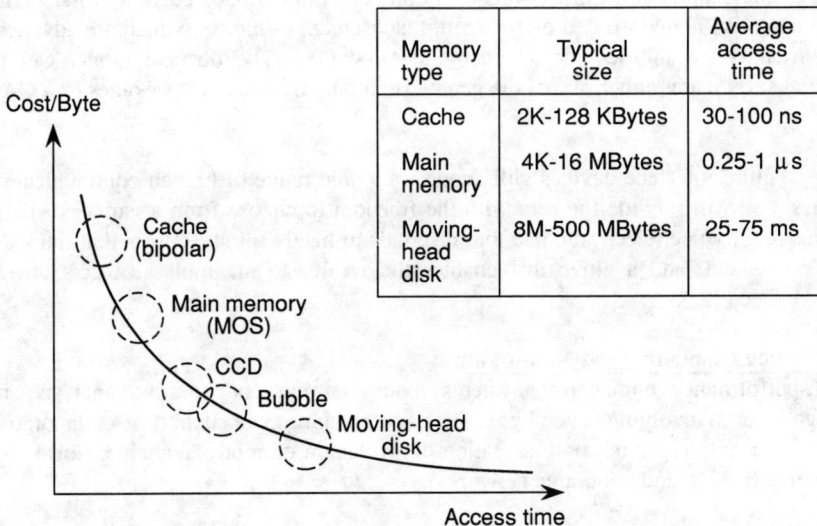

Memory type	Typical size	Average access time
Cache	2K-128 KBytes	30-100 ns
Main memory	4K-16 MBytes	0.25-1 μs
Moving-head disk	8M-500 MBytes	25-75 ms

Figure 6 - Cost and access time relationship (see [10]).

Secondary Storage Devices

Magnetic devices are widely used for data storage because they offer much greater memory capacity at a lower cost per bit of data stored than semiconductor devices. As the computational speeds increase, the computers are able to utilize and produce growing amounts of data in a given period of time. This, in turn, requires an increase in the capacity of the storage devices from which (or into which) data are drawn (or loaded).

Significant improvements have been made in magnetic storage devices in the past two decades. These include the introduction of the solid-state storage devices (SSD), which are fast random-access devices used to hold pre-staged data or intermediate results which are manipulated repetitively. The SSD offer significant potential for performance improvement of more than one order of magnitude on Input/Output-bound applications, and thus allow users to develop new algorithms that would not be practical with traditional disk input/output.

The storage density of magnetic disks has increased from 200,000 bits per square inch in 1967 to its current value of over 20 million bits per square inch. Continuation of this trend is likely to yield storage densities of the order of 300 million Bytes per cubic inch within a decade. Optical storage media, such as compact disks, can provide from 5 to 7 times the density of information that magnetic devices can achieve (see [11]).

User Interface Hardware and Software

Great efforts are now aimed at improving the productivity of the analyst and designer by developing intelligent software and hardware interfaces. More structural analysis and design software are becoming *turnkey* systems with defaults built in, and with simple menu options. Current menu options are multiwindowed (one window for each task) and are controlled by lightpen, touchscreen, or mouse (which are advanced user-friendly capabilities for accessing the system). The discrete model can be generated by using either one of the geometric modeling software packages or a CAD system.

Future interface devices will support a whole range of human communication modes, and will provide the user with the freedom to choose from a variety of these modes (e.g., voice, electronic pad that responds to handwriting, sensors that track the eye movement, and a glove that enables the wearer to manipulate objects on the screen - see [12]).

Distributed Computing and Networking

High-performance engineering systems require strong collaborative analysis and design efforts, involving several engineers and machines. In support of collaborative computing, two types of communication networks are currently available: local-area networks (LAN), and wide-area networks.

Local-area networks are designed to facilitate the interconnection of a variety of computer-based equipment within a small area. They have high transmission rates (of the order of 10 Mb/sec.), and allow different workstations to share expensive equipment and facilities. Examples of LAN are provided by Ethernet, Arcnet and token-ring network.

The first wide-area network, the ARPANET (Advanced Research Projects Area Network), was built in 1969. It demonstrated the feasibility and practicality of distributed computing, as well as of communication technology based on packet switching. A number of commercial packet networks are now available, e.g., Telenet and TYMNET. Moreover, the coupling of digital networking with the existing telephone and Digital Private Branch Exchange (PBX) systems into Integrated-Services Digital Networks (ISDNs) promises to offer access to a wide range of data and central computers via desktop workstations.

Current work is directed towards increasing the transmission rates of both local-area and wide-area networks (see [13]). Future directions include development of networking technology to support portable computing and communications. The cellular phone is likely to evolve into a portable voice and data machine that supports mobile communications.

CLASSIFICATIONS AND PERFORMANCE EVALUATION OF NEW COMPUTING SYSTEMS

Because of the rapid progress made in recent years in component technology, a number of novel forms of computer architectures have emerged. Some of the new architectures are commercially available; others are still research tools aimed at achieving high-performance and/or low-cost computations. In this section the classifications and performance evaluation of different machines is discussed, and in the succeeding sections, the major features of the new and emerging high-performance machines are described.

Classifications of Computing Systems

In an effort to identify and clarify the differences between the different machines, a number of classifications and taxonomies have been proposed. One of the earliest and most commonly-used classifications is that introduced by Flynn [14], which is based on the concurrency in instruction control and concurrency in execution. A stream is defined as a sequence of items (instructions or data) as executed or operated on by a processor.

Four broad classes can be identified according to whether the instruction or data streams are single or multiple (see Fig. 7).

1. *Single-Instruction-Stream, Single-Data-Stream (SISD) Machines.* These machines include the conventional serial computers which execute instructions sequentially, one at a time.

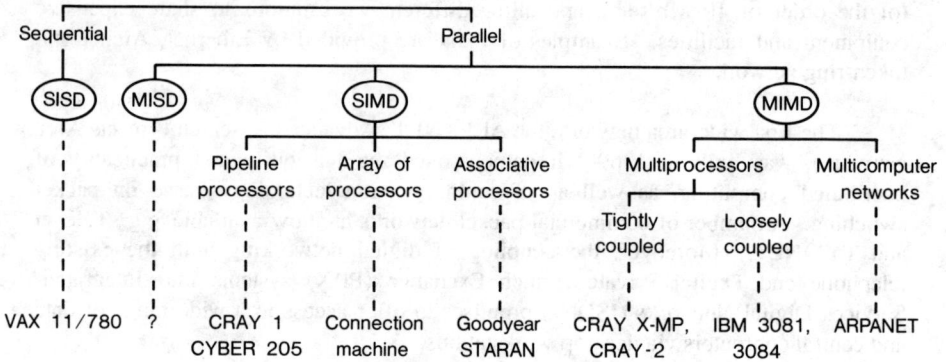

Figure 7 - Classification of computing systems

2. *Single-Instruction-Stream, Multiple-Data-Stream (SIMD) Machines.* These are computers that have a single control unit, a collection of identical processors (or processing elements), a memory or memories, and an interconnection network which allows processors to exchange data. During execution of a program the control unit fetches and decodes the instructions and then broadcasts control to the processing elements. Each processor performs the same instruction sequences, but uses different data. These operations are usually referred to as lockstep operations.

3. *Multiple-Instruction-Stream, Single-Data-Stream (MISD) Machines.* This category is included to round out the four possible combinations of single/multiple instruction and single/multiple data, but none of the current machines is classified as MISD.

4. *Multiple-Instruction-Stream, Multiple-Data-Stream (MIMD) Machines.* These are computers which contain a number of interconnected processors, each of which is programmable and can execute its own instructions. The instructions for each processor can be the same or different. The processors operate on a shared memory (or memories), generally in an asynchronous manner.

MIMD machines can generally be divided, according to the level of interaction between processors and their physical location, into multiprocessors and multicomputer networks. The latter class refers to physically dispersed and loosely-coupled computer networks. Multiprocessors can be further subdivided into tightly-coupled and loosely-coupled systems (see Fig. 6). In a tightly-coupled system the processors access a global, shared memory by means of some interconnection networks. In a loosely-coupled system, each processor has its own local memory and communicates with other processors by sending messages through the interconnection network.

Examples of computing systems which belong to each of the classes are given in Fig. 7. Also, block diagrams of the SISD, SIMD and MIMD machines are shown in Fig. 8. The architectural evolution of computing systems from sequential scalar processors to concurrent vector/scalar processors is depicted in Fig. 9. The trend has been to build more hardware and software functions into the system.. The skewed tree demonstrates that most current high-performance computers are designed with look-ahead techniques, functional parallelism, pipelining at various levels, using explicit vectors, and exploiting parallel processors in SIMD or MIMD mode. Note that some of the SISD machines have parallel processing mechanisms (due to the presence of multiple functional units and/or facilities for overlapping computations and I/O); however, the parallel processing is embedded in the hardware below the instruction level, and the appearance of sequential execution of instructions is preserved.

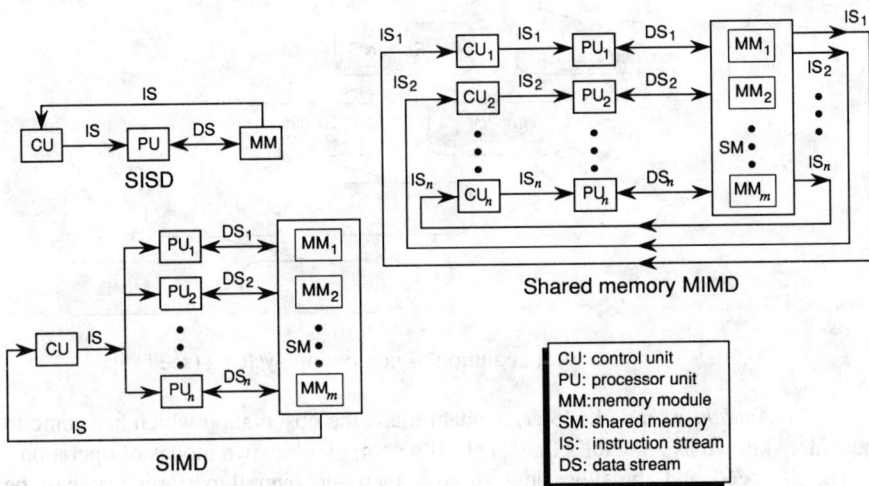

Figure 8 - Block diagrams for SISD, SIMD and MIMD machines

Performance Evaluation of Computing Systems

1) *Peak Versus Sustained Performance of Machines.* The performance of sequential computers is usually measured by a) the peak computational speed measured in millions of floating-point operations per second (MFLOPS), and b) the peak execution rate of arithmetic, logic, and program control instructions, measured in millions of instructions per second (MIPS). In the case of SIMD and MIMD machines, the *peak computational speeds* in terms of millions of floating-point operations per second can be estimated. However, the *sustained computational performance* in these machines is difficult to estimate since it varies with the level of parallelism achieved, and the overhead incurred in exploiting the parallelism in the particular application [16]. These, in turn, are functions of the formulation used, the numerical algorithm selected, the computer implementation, the compiler and the operating system used, as well as the architecture of the hardware. For vector multiprocessors the peak performance can be estimated as the sum of the vector performance of all the processors, and the lowest performance is that of a single-

scalar processor. A widely used approach for estimating the sustained performance of a machine is *benchmarking* - running a set of well-known programs on the machine. A discussion of the effectiveness and pitfalls of benchmarking is given in [17].

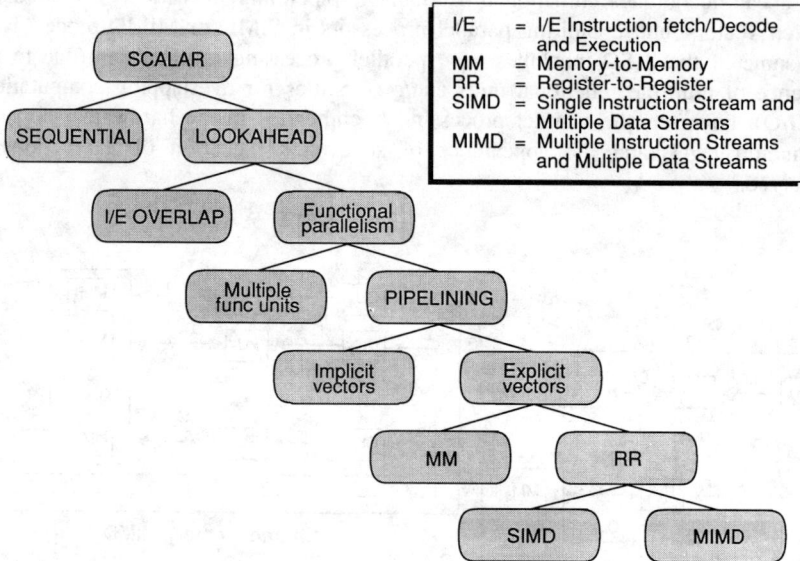

Figure 9 - Architectural evolution of computing systems (see [15])

2) *Amdahl's Law*. In 1967, Amdahl made the observation which has come to be widely known as Amdahl's Law [18]: If a computer has two modes of operation - one high speed and the other low speed - then the overall performance will be dominated by the low-speed mode. Amdahl's Law is fundamental to the understanding of computer performance. It shows that a bottleneck in a computing system is associated with a small execution rate (low-speed mode of operation) which is out of balance with the rest of the execution rates (high-speed modes of operation). In such a computing system increasing the execution rates for the high-speed modes of operation may yield only a small increase in performance unless the fraction of computation performed in the low-speed mode is essentially zero. This explains why vector computers with low-scalar speeds have not been successful. A recent discussion of the application of Amdahl's Law to massively parallel processors is given in [19].

MAJOR FEATURES OF NEW COMPUTING SYSTEMS

New computing systems cover a broad spectrum of machines ranging from the large supersystems to the small portable, embedded computers, and transputers. A description of some of these machines is given in [15, 20 and 21]. Herein, a brief discussion is given of the major features of new computing systems that are likely to have the strongest impact on the structures technology.

Most of the new and emerging machines achieve high performance through concurrent activities in the computer (or the network of computers). The exploitation of these concurrent events in the computing process is usually referred to as *parallel processing*. When parallel processing is done on physically dispersed and loosely coupled computer networks, it is usually referred to as *distributed processing*. The concurrency is used not to speed up the execution of individual jobs, but to increase the global throughput of the whole system.

Parallel processing can be applied at four distinct levels, namely: job level, program level, inter-instruction level, and intra-instruction level [10 and 22]. The hardware and software means to achieve parallelism in each case are outlined in Table 1. The hardware role increases as the parallel processing goes from high (job) level to low (intra-instruction) level. On the other hand, the role of software implementations increases from low to high levels.

Supersystems
Supersystems are a class of general-purpose computers designed for extremely high-performance throughput. Although there is no universally accepted definition or classification of supersystems, the following three classes of supersystems can be identified (based on the performance, memory system used, and cost):
  • large supersystems
  • near supers
  • minisupers
The current price ranges for the three classes in U.S. dollars are: $2M to $20M; $1M to $4M; and $100K to $1.5M, respectively.

Current *large supersystems* have the following four major characteristics: 1) high computational speeds (maximum speeds of the order of 200 MFLOPS or more); 2) high execution rates (of the order of 500 MIPS or more); 3) large main (or central) memory (with a capacity of 8 MBytes or more); and 4) fast and large secondary memory (or storage devices) with a sophisticated memory management system.

The development of large supersystems now spans three generations. The first generation included the array of processors ILLIAC IV (SIMD machine); and the pipeline (or vector) computers CDC STAR-100, and Texas Instruments Advanced Scientific Computer (ASC). The second generation supersystems included the CRAY-1, which featured the pipelined vector instructions introduced in the first generation machines, but carried out in a clever register-to-register mode. They also included the CDC Cyber 203 (an enhanced successor of STAR-100, sporting a faster scalar unit). Most of the current third-generation large supersystems use a hybrid combination of pipeline and array processors to achieve high performance. Examples of these machines are: the CDC CYBER 205, the CRAY X-MP, CRAY-2, CRAY Y-MP, ETA-10 and the Japanese machines Fujitsu VP-400, Hitachi S810/20 and S820/80; and NEC SX2-400. The top sustained speed of current supersystems is of the order of 500 MFLOPS with bursts to 1.5 GFLOPS (billions of floating point operations per second).

Table 1 - Levels of parallelism

Level	Means to achieve parallelism	Performed by
Job level	**Multiprogramming** - overlapping and interleaving the execution of more than one program (I/O and CP operations) **Multiprocessing** - running two or more CPU's concurrently on different applications or on independent job streams of the same general application	Operating system
Program level	**Multitasking** - decomposition of a program into two or more tasks (program segments) that can execute concurrently. This requires the tasks to have no data or control dependence. **Microtasking** - which permits more than one CPU to work on a program at the Do-loop level	Software
Inter-instruction level	**Concurrency among multiple instructions**. This requires an analysis of the data dependency and is accomplished by dividing each instruction into suboperations, and overlapping the different suboperations on different instructions.	Compiler
Intra-instruc-level	**Pipelining** - by dividing the instruction into a sequence of operations, each of which can be executed by a specialized hardware stage that operates concurrently with other stages in the pipeline. **Very-long instruction word (VLIW)** - performing multiple operations per instruction, each with its own address field.	Hardware

The next generation of supersystems includes the CRAY-3 and CRAY-4, the Japanese Super Speed Computer project; the Japanese fifth generation project; and the IBM TF-1 Project (an SIMD machine, with 32,768 processor nodes, capable of executing 10^{12} flops). It is anticipated that the computational speeds of the supersystems will continue to increase, and will exceed 20 GFLOPS in the next five years. The architectural characteristics of some of the new and emerging large supersystems are given in Table 2.

Near supers are high-end main frames with peak computational speeds in the range of 50-500 MFLOPS; execution rates of the order of 50 MIPS, or more; and less sophisticated memory systems than those of the large supersystems. Examples of these machines are IBM 3090/VF, CDC Cyberplus, the Connection Machine C-1 of Thinking Machine, Inc., and the BBN Butterfly.

The *minisupers* are low-cost parallel/vector computers with peak computational speeds in the range of 10 to 100 MFLOPS, and execution rates of the order of 10 MIPS, or more. Examples of these machines are the Encore/Multimax, Convex C-1 and C-2, Intel iPSC system, FPS models 300, 350 and 500 of FPS Computing (formerly Floating Point Systems, Inc.).

Supersystems can impact the structures technology in a number of ways including:

a) Increasing the level of sophistication in modeling structures to new levels which were not possible before. Examples are provided by reliability-based (stochastic) modeling of structures (to account for probabilistic aspects of geometry, boundary conditions, material properties and loading), and multidisciplinary analysis and design of structures.

b) Reducing the dependence on extensive and expensive structural and dynamic testing. This is particularly important for future large space structures (e.g., large antennas, large solar arrays, and the space station) where the reliability of testing in 1-G environment can be, at best, questionable; and,

c) Enhancing the physical understanding of some aspects of mechanical behavior which are difficult, if not impossible, to obtain by alternate approaches. An example of this is the study of the implications of various microstructural mechanisms of inelastic deformations on the macroscopic response.

Parallel Processing Machines

In the last few years there has been an explosion in the number of parallel processing machines developed. Some of these machines belong to the class of supersystems discussed in the preceding subsection. In an effort to identify and clarify the differences between these machines, a number of classifications and taxonomies have been proposed, including Flynn's classification described in the preceding subsection. However, Flynn's classification does not adequately describe several of the new multiprocessor machine architectures. More descriptive classifications of multi-

Table 2 - Architectural characteristics of some of the new and emerging large supersystems

System/ Model	Architecture/ Configuration	Maximum Number of Processors	Processor Type	Clock Cycle (ns)	Maximum Main-Memory Capacity	Peak Computational Rate (GFLOPS)
CRAY X-MP	Multiprocessor with shared memory	4	Custom	8.5	64 MW	0.84
CRAY-2		4	ECL	4.1	256 MW	1.72
CRAY Y-MP		8		6.0	32→128 MW	2.7
CRAY-3		16	Ga As/ECL	2.0	128→512 MW	16
ETA-10G	Multiprocessor with local and shared memory	8	Custom CMOS	7.0	16 MW/processor	10
IBM ES3090-600S	Multiprocessor with shared memory	6	ECL	15.0	512 MB	0.8
Fujitsu VP 400E	Single processor with multiple pipelines	1	ECL	14.0-scalar 7.0-vector	1GB	1.71
Hitachi S-810/20		1	Custom	14.0	256 MB	0.8
Hitachi S-820/80		1		4.0	512 MB	3
NEC SX2-400	Multiple processor	1	MOS	6.0	1 GB	1.3
NEC SXX		4		3.0		20

Notes: 1 MB (MW)=1,048.576 Bytes (words-each word 64-bits); 1 GB(GW)=1,073,741,824 Bytes (words-each word 64 bits); 1 GFLOP = 1 billion floating point operations per second; ECL = emitter-coupled logic; MOS = metal-oxide semiconductor, CMOS=complementary metal-oxide semiconductor.

processor systems are based on the following characteristics (see [23]): processor granularity; memory organization, connection topology between processors and memory systems; reconfigurability (to meet the performance needs in more than one environment); control mode (command flow versus data flow); extensibility (e.g., possibility of adding more processors without pronounced decrease in the efficiency of the system) and homogeneity or nonhomogeneity of the processors (for example, the CRAY X-MP has homogeneous processors, but a combination of a serial computer and an attached processor can be viewed as a heterogeneous multiprocessor). Herein, an extension of Flynn's taxonomy, proposed in [22] is described. Also, the classifications based on the first three characteristics, namely, processor granularity and speed; memory organization; and connection topology, are discussed (see [22]).

Extension of Flynn's Taxonomy. The extended taxonomy uses levels of concurrency as one principle, and instruction types as another. Four generic types of instructions and three levels of control concurrency are included in the taxonomy. The four generic types of instruction are: *scalar, vector, systolic* and *very-long instruction word* (VLIW). They are distinguished by the number of operations and number of pairs of operands as shown in Table 3.

The three levels of control concurrency include the *serial* and *parallel* types (with single and multiple instruction streams), as well as the *clustered* type (with several independent sets of multiple instruction streams). The three levels constitute a hierarchy, with parallel control being a generalization of serial control, and clustered control being a generalization of parallel control.

The extended taxonomy has twelve types of computer architecture. Examples of some of these types are given in Table 4.

Processor Granularity and Speed. Multiprocessor systems can be divided into three groups:

a) *Coarse-grain machines* consisting of a small number of very powerful processors or central processing units (CPUs) linked together. Examples of these machines are CRAY-2, CRAY Y-MP and ETA-10.

b) *Fine-grain machines.* These are massively parallel machines which combine thousands of relatively weak processors. When the processors work in concert they can form very powerful computers. Examples of machines with synchronous processors are the ICL Distributed Array Processor (DAP), with 1024 processors, at Queen Mary College in London; the Goodyear Aerospace Massively Parallel Processor (MMP), with 16,384 processors; the Connection Machine C-2 of Thinking Machine, Inc. with up to 65,536 single-bit processors; and the FPS Computing T-Series hypercube with up to 16,384 processors.

c) *Medium-grain machines.* These fall somewhere between the first two categories. They have a moderate number (e.g., tens or hundreds) of low-cost and mid-range processors (performance of the order of 2 MFLOPS for the former and

10-100 MFLOPS for the latter). Examples of the former are the Encore Multimax, Flexible Computer's FLEX/32, Sequent Balance 8000, and the Intel iPSC system. Examples of the latter are the Alliant FX/8 and Elxsi 6400 systems.

Table 3 - Classification of instruction types (see [23]).

Type of Instruction	Number of Operations	Number of Pairs of Operands	Examples
Scalar	1	1	$A = B + C$
Vector	1	N	$A_i = B_i + C_i$, i = 1,N
Systolic	M	1	Matrix operations, with data from rows and columns used repeatedly.
VLIW	M	N	Multiple operations per instruction, each with its own address field

Note: N, M > 1

Table 4 - Extended taxonomy of computer architectures (see [23]).

Level of Concurrency	Generic Instruction Types			
	SCALAR	VECTOR	SYSTOLIC	VLIW
SERIAL	CDC-7600 IBM 360/95	CRAY-1 CYBER 205	WARP	
PARALLEL	BBN-Butterfly Hypercube	CRAY Y-MP ETA[10]		
CLUSTERED	Myrias	Cedar		

Memory Organization. One of the most important classifications of multi-processor systems is that based on memory organization. According to this classification there are shared-memory machines, private- (or semiprivate) memory machines, and multilevel-memory machines.

Shared (or global) memory. This is a single monolithic main memory accessible to all processors. Examples are provided by the memories of the CRAY X-MP, the Ultracomputer of New York University, and the IBM ES3090-600S.

Distributed memory. This is a private memory connected to a single processor and requiring communication to transfer information. Examples are provided by the memories of the Connection Machine, the FPS T-Series, and the Intel iPSC.

Semidistributed memory is one which is available to a subset of processors.

Multilevel (hierarchical) memory is available in computer systems such as the ETA-10 and the Myrias 4000.

Connection Topology. Multiprocessor systems with private (or semiprivate) memories can be distinguished by the interconnection pattern (or topology) between processors, memory systems and I/O facilities. The topology affects the class of problems which can be efficiently solved on the machine. The commonly-used connection topologies are:

Bus (or ring) type connection. The various processors, memory systems, and I/O facilities reside on a common communication bus or set of buses. Most computers of this type incorporate global (or central) memory, shared by all processors, and accessed via the communication bus. Examples of machines with bus connection are the Encore Multimax, FLEX/32, and Elxsi 6400 computers.

Hypercube or n-cube connection. Each processor is directly connected to a number of its neighbors, n. The number of connections per processor results in a multidimensional cube with 2-inch nodes. This type of connection is usually used with distributed memory machines. Examples of machines with hypercube connection are the Intel iPSC and the FPS T-Series.

Switching connection. This mechanism is based on placing a switch between the processors and memory banks, thereby removing the ownership property between processors and memory. For different settings of the switch, a given processor will be connected to different memory banks.

Although most of the currently available machines for performing parallel computations belong to either SIMD or MIMD categories of Flynn's classification, there are a number of variations. Two of these variations which exploit parallelism in ways which have promise for achieving high performance are *data-flow machines* and *systolic array architectures.*

The **data-flow concept** is based on the dependency graph of a computation. The algorithm for a given computation is first written in a special programming language designed for data-flow applications (e.g., the language Id developed at the Massachusetts Institute of Technology). A compiler then translates the program into a data flow graph (which corresponds to the machine language for data flow architecture).

Systolic architectures follow from space and time representations of certain numerical algorithms which map directly onto geometrically regular VLSI/WSI (Very Large-Scale Integration/Wafer-Scale Integration) structures. The term systolic array comes from the notion of data pulsing through the processors in the network in an analogous manner to that of blood pulsing through the circulatory system in the body (see [24]). In its purest form a systolic system consists of a regular array of processing elements all doing the same calculation and passing results on to their nearest neighbors every cycle. In this way the array as a whole computes some recurrence function. The prime features which make this style attractive are the *short interconnections*, the *regularity* which gives a high packing density and simplifies the design, and the high degree of *parallelism* which, when combined with the other features, leads to high performance circuits. Potential applications of systolic arrays in finite element applications are discussed in [25, 26 and 27]. More advanced concepts of systolic arrays and systolic computing are described in [28].

Parallel processing systems can substantially expedite the multidisciplinary design process of structures by allowing the designer to carry out various analysis and design tasks in parallel. The tasks can belong to an individual discipline as well as to other disciplines (such as in multidisciplinary optimization problems).

Special-Purpose Computing Hardware

With the continued reduction in the cost of computer hardware, a number of special-purpose computers have emerged that offer increased speed for specific problems when compared with general-purpose computers. Examples are provided by the Navier-Stokes Computer (NSC) being developed at Princeton University. The NSC is a parallel processing machine comprised of a multi-dimensional array of processing nodes, each with a local memory. Its primary function is the direct numerical simulation of complex flows. A description of the NSC machine is given in [29].

There is a growing trend of moving from software-based processing to hardware-based processing, and it is likely that the classical linear finite-element analysis will become a hardware function in the future.

Small Systems

A broad spectrum of low-cost small systems exist now, including new powerful microprocessors, transputers, embedded computers, portable computers, laptop computers, desktop computers, engineering workstations, minis and superminis. Several classifications have been attempted for these systems based on word length

(8-bit, 16-bit and 32-bit machines), cost, amount of directly addressable memory, and computing speed (4, 6, 16, 20, and 25 MHz - see, for example, [30]). However, the dramatic increase in hardware capabilities of small systems coupled with the rapid reduction in cost makes these classifications of questionable value. Herein, the development in microprocessors, transputers, and engineering workstations are discussed.

Microprocessors and Chip Technology. The trend of ever-increasing the number of devices packaged on a chip has resulted in the miniaturization and increase in speed of microprocessors and minicomputers. The new powerful chips can be used as monitoring systems (e.g., embedded computers) for the detection, recording and evaluation of stochastic damage, thereby increasing the mean time between inspection for structural components.

RISC Processors and Transputers. In the 1970's the efforts directed towards creating machines with very fast clock cycle, that can execute instructions at the rate of one per cycle (like microprogrammed controllers), resulted in the development of Reduced Instruction-Set Computing (RISC) processors. These are microprocessors that provide high-speed execution of simple instructions. The implementation of RISC architecture began at IBM in the mid 1970's. RISC architecture was used in the development of high-performance superminis (e.g., PYRAMID 90X and RIDGE 32 computers). The basic notion of RISC has now evolved to encompass chips in which the chip areas formerly used for decoding, and executing complex instructions, are used for caching instructions. RISC chips offer the following advantages over the Complex Instruction-Set Computing (CISC) chips: much smaller size chips, more throughput, shorter design time, better support for high-level languages, and the ability to emulate other instruction sets.

In the last few years a new type of powerful VLSI chip (superchip) which packs RISC and a limited amount of RAM has been developed. The superchip is called a *transputer*, and is manufactured by Inmos, Ltd. of Bristol, England. The transputer was designed from the outset for parallel and distributed processing. The transputer has hardware support for parallel tasks, including local (on-chip) memory, and four high-speed communication links. These can transfer data at the rate of 10-20 Mb (Megabits) per second, thereby enabling the transputer to be interconnected in powerful arrays. A high-level programming language, Occam, has been especially developed for parallel processing on the transputer. Also, other languages (e.g., Fortran, Pascal and C) are available for use on the transputer.

A range of transputer boards and modules now exist which can be plugged into conventional and desktop computers to speed up the computations. Also, a desktop supercomputer with many transputers has been developed by Meiko, a company in Bristol, England, that grew out of Inmos. Transputer networks, formed by linking together transputers in arrays, pipelines, rings and other patterns, have been efficiently used in a wide variety of applications including solution of finite element equations and graphics processing. To date the latest transputer, the T800-20 is a 32-bit transputer with four serial links, 4K of on-chip memory, and a built-in floating point unit. It has an execution rate of 10 MIPS and a computational speed of 1.5 MFLOPS. It is equivalent to a VAX 8600 on a single chip.

Engineering Workstations with Advanced Visualization Capabilities. Single-user engineering workstations using VLSI 32-bit processor chips, and having internal graphics-processing facilities, and 50 MBytes (or more) of addressable memory have been developed. Examples of such systems are the HP-9000, series 800 and Turbo/SRX, and the Apollo computers (10000 series). A superstation with a peak performance of 40 MFLOPS has been developed by FPS Computing (Model 300). It is anticipated that in the next few years, the powerful workstations (designated 5M machines) will be desktop computers with sustained speed of 100 MFLOPS, over 1 MIP execution rate, over 100 MBytes of addressable memory, over megapixel display, and over megabit transmission rate within local area networks (LAN's). Future advanced visualization facilities include high-resolution and high-speed graphics, video and film animation capabilities.

ADVANCES IN PROGRAMMING ENVIRONMENT AND SOFTWARE TECHNOLOGY

Although considerable effort is now devoted to increasing the productivity of the analyst and designer through the development of powerful programming environment, software and programming languages remain the primary pacing items for exploiting the potential of new computing systems.

Programming Environment

This refers to the array of physical and logical means by which the analyst transmits instructions to the machine. It includes the user-interface devices; interactive programming tools (e.g., debuggers, editors, file-maintenance utilities), and tools for automatic (and semiautomatic) mapping of numerical algorithms on different machines.

The programming environments on most of the existing new computing systems are limited to a standard sequential language compiler, and extensions to support concurrency (viz. vectorization and parallelization). The extensions of currently used programming languages to support concurrency (Fortran, C, Lisp) are not the same on different machines. A description of recent work in development of software environment is given in [31, 32, 33 and 34].

Current and future interface devices have been discussed in a previous section. Future powerful programming languages should enable the user to state the mathematical and logical formulation of the problem in the expectation that the language can fill in the details. These languages should be architecture-independent high-level languages to allow the portability of the programs.

AI Knowledge-Based and Expert Systems

AI-based expert systems, incorporating the experience and expertise of practitioners, have high potential for the modeling, analysis and design of structures. These systems can aid the structural analyst in the initial selection and adaptive refinement of the model, as well as in the selection of the appropriate algorithm used in the analysis. Expert systems can also aid the structural designer by freeing him from such routine tasks as the development of process and material specifications. A review of the capabilities of some of the currently available expert systems and their limitations are given in [35 and 36]. A description of a knowledge-based system used as a modeling aid for aircraft structures is given in [37].

Large Powerful Data Management Systems and Databases

Future engineering software systems are likely to have the basic analysis software (such as data management, control, etc.) as part of the software infrastructure and the discipline specifics (such as the finite element properties of the structure) as part of application software. Conventional database management systems (DBMs) developed in the last decade such as relational database management system (RIM), provided multidisciplinary coordination, and helped in the integration of structural analysis programs into CAD/CAM systems. However, these conventional DBMs do not meet the data requirements for the current and emerging engineering/design environment (see [38]). Among the different advanced data/process modeling methodologies which have high potential for multidisciplinary analysis and design applications are: the three-level IDEF methodology developed by the Air Force's integrated computer-aided manufacturing program (see [39]); Nijssen's information analysis method based on a binary relationship model; Entity relationship model; and object-oriented data model. A description of these models is given in [38].

ADVANCES IN NUMERICAL ALGORITHMS AND COMPUTATIONAL STRATEGIES

To achieve high performance from any computing system, it is necessary either to tailor the computational strategy and numerical algorithms to suit the architecture of the computer, or to select the architecture which may effectively map the computational strategy and execute the numerical algorithm. Extensive work has been devoted to the development of vectorized and parallel numerical algorithms for new computing systems. Review of some of this work is contained in a number of monographs and survey papers.

In this section a brief review is given of the recent progress made in special parallel numerical algorithms and computational strategies that are influencing structures calculations.

Parallel Numerical Algorithms

In parallel algorithms, independent computations are performed in parallel (i.e., executed concurrently). To achieve this parallelism, the algorithm is divided into a collection of independent tasks (or task modules) which can be executed in parallel and which communicate with each other during the execution of the algorithms. Parallel algorithms can be characterized by the following three factors:

1. Maximum amount of computation performed by a typical task module

before communication with other modules.

2. Intermodule communication topology, which is the geometric layout of the network of task modules

3. Executive control to schedule, enforce the interactions among the different task modules and ensure the correctness of the parallel algorithm.

The three aforementioned factors have been used in [40] as a basis for classifying parallel algorithms on the conceptual level, and for relating each parallel algorithm to the parallel (or pipeline) architecture to which it naturally corresponds.

The design of a parallel algorithm must deal with a host of complex problems, including data manipulation, storage allocation, memory interference, and in the case of parallel processors, interprocessor communication. In general, the parallel numerical algorithms reported in the literature fall into two categories: reformulation (or restructuring) of serial algorithms into concurrent algorithms, and algorithms developed especially for parallel machines.

Most of the work on parallel numerical algorithms belongs to the first category, i.e., decomposition of familiar serial algorithms into concurrent tasks. Examples are matrix operations, direct and iterative methods for solution of algebraic equations, and eigenvalue extraction techniques (see, for example, [41, 42, 43, 44 and 45]).

The second category includes the algorithms whose development was spurred by performance criteria for parallel processing. These algorithms have been referred to as *uniquely parallel* and only a few of them have been reported in the literature [46]. Examples of uniquely parallel algorithms are provided by the parallel superconvergent multigrid method [47] and the fully parallel algorithm for symmetric eigenvalue problem [48]. In some cases the performance of uniquely parallel algorithms is superior to their serial counterparts.

Construction of Parallel Algorithms

The development of parallel numerical algorithms generally follows one or both of two related approaches: *reordering*, and *divide and conquer*. Reordering refers to restructuring the computational domain and/or the sequence of operations in order to allow concurrent computations. For example, the order in which the nodes of a finite element grid are processed may change the degree of parallelism that can be achieved in the solution of the resulting algebraic equations ([49]).

The divide and conquer approach involves breaking a task up into smaller subtasks that can be treated independently. The degree of independence of these tasks is a measure of the effectiveness of the numerical algorithm since it determines the amount and frequency of communication and synchronization. This idea pervades many of the parallel algorithms and can be extended to the overall computational strategy as described in the succeeding section.

Comments on Parallel Algorithms and Their Implementation

The following comments concerning parallel algorithms and their implementation are in order:

1. Effective parallel algorithms are not necessarily effective on sequential computers. In fact, some parallel algorithms involve additional (redundant) floating point operations which make them inefficient on sequential machines. Also, restructured serial algorithms may not be the most efficient on parallel processing machines.

2. The mathematical properties of the serial and parallel implementations of the same algorithm can be different. For example, the rate of convergence and numerical stability of serial and parallel iterative techniques can be different. In some cases parallel implementation can degrade the performance and in other cases, it improves it ([46]).

3. To achieve high performance both the numerical algorithm and its implementation must be carefully tailored to the particular machine being used. This raises the question of portability of parallel programs. It is not practical to develop algorithms and programs for each new computer. Also, it is not desirable to achieve portability at the expense of performance. A number of studies have been devoted to achieving high performance and portability of numerical algorithms on advanced computers. Two approaches have been proposed in [50] and [51]: a) restructuring of algorithms in terms of high-level modules (e.g., matrix-matrix and matrix-vector operations); and b) developing and implementing an abstraction of parallel processing that is independent of the architecture.

4. On most of the currently-available multiprocessing systems vectorization offers greater performance improvement over multitasking (i.e., parallelization). Consequently, if multitasking conflicts with efficient vectorization (e.g., multitasking results in short vector lengths), then the algorithm should be vectorized rather than parallelized.

Performance of Parallel Numerical Algorithms

Computational complexity (e.g., number of floating-point arithmetic operations) has long been used as a measure of the performance of serial algorithms. However, it is not an appropriate measure for parallel numerical algorithms. This is because parallel computers can support extra computation at no extra cost if the computation can be organized properly; and parallel computers are subject to new overhead costs (e.g., synchronization and communication) that are not reflected by computational complexity.

One of the most commonly-used measures for the performance of parallel numerical algorithms is the speedup, S, which is defined as follows:

$$S = \frac{\text{execution time using one processor}}{\text{execution time using } p \text{ processors}}$$

The measure S has the advantage that it uses the execution time and, therefore, incorporates the synchronization and communication overhead. However, it has the drawback of comparing the execution time of the same algorithm on the single and multiple processors.

Another definition of speedup, based on Amdahl's Law, was proposed by Ware [52], and is expressed by the following simple formula:

$$S(p,f) = \frac{1}{1-f\left(1 - \dfrac{1}{p}\right)} \tag{1}$$

where S is the maximum speedup achievable by using p processors; and f is the fraction of computational work done in parallel (at the high execution rate).

In Eq. 1 the execution time using a single processor has been normalized to unity. The rate of change of S with f, at $f=1$, is quadratic in p, i.e.,

$$\frac{dS}{df}\Big|_{f=1} = p^2 - p \tag{2}$$

Therefore, for massively parallel processors the fraction of parallelism must grow with the number of processors in order to achieve reasonable speedups ([45]).

The utilization rate of the multiprocessor system, U, is defined as follows:

$$U(p,f) = \frac{S}{p} \tag{3}$$

A utilization rate of 1 means that every processor is busy computing all the time.

Figure 10 shows the theoretical speedup and the utilization rate of multiprocessor systems as a function of the fraction of parallelism, f, and the number of processors, p. The figure illustrates a key issue in multiprocessor machines: as the number of processors increases, then for a given fraction of parallelism, the degree of utilization decreases.

The following comments can be made regarding Amdahl's Law and Ware's model:

1. Ware's model assumes that the parallel processing machine is a two-state machine in the sense that at any instant of time either all the processors are operating or only one of them is operating.

2. Amdahl's Law can be extended to computers with more than two modes of operation with one mode having a lower rate of execution than others. For example,

if the scalar mode is taken as the low rate and a balanced higher rate, representative of vector, memory and I/O is taken as the high rate, then Eq. 1 can be used to give the maximum speedup achievable by the system ([23]).

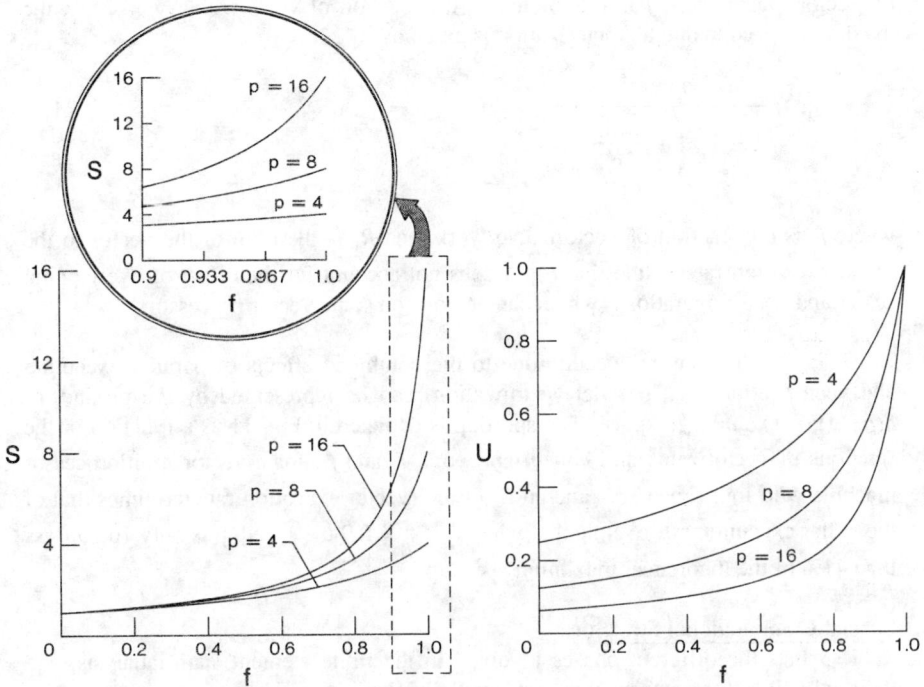

Figure 10 - Theoretical speedup, S, and utilization rate, U,
of multiprocessor systems as a function of the fraction of
parallelism, f, and the number of processors, p.

3. Ware's model does not account for the overhead associated with inter-processor communication, synchronization (for controlling data access and for program control), among others. This overhead may increase with increasing the number of processors, resulting in a speed-down behavior ([46 and 53]).

4. Eq. 1 measures the speedup relative to the implementation on a single processor of the same algorithm. It does not necessarily measure the efficiency gain due to parallelization. This will be discussed further in the succeeding sections.

5. In Ware's model the implicit assumption is made that f is independent of p, which is only true if the problem size is fixed. However, in practice the size of the problem increases with the increase in the number of available processors. The parallel part of the program scales with the problem size, but the times for the

program loading, I/O, and serial computations do not usually scale with problem size. A discussion of the effect of problem size on the performance of parallel algorithms is discussed in [54] and [55].

6. Ware's model can be used for estimating the speedup due to vectorization on vector machines, if f is interpreted as the fraction of vectorizable work, i.e., the maximum speedup due to vectorization is given by:

$$S_v = \frac{1}{1 - f_v \left(1 - \frac{1}{R_v}\right)} \tag{4}$$

where f_v is the fraction of vectorizable work, and R_v is the ratio of the vector to the scalar execution rate. Note that Eq. 4 does not account for the effect of overlapping scalar and vector operations (which can be done on some vector processors).

7. The maximum speedup due to the combined effects of parallel execution and vectorization (i.e., parallel vectorization) can be represented by the product of $S(p,f)$ (Eq. 1), and S_v (Eq. 4). The speedup is depicted in Fig. 11 as a function of the fractions of vectorizable and concurrent work, f_v and f_p, for a vector multiprocessor machine with four processors and an R_v of ten (vector execution rate ten times that of the scalar execution rate). Note that when $f_v = f_p = 0.9$ the speedup is only 16.19 (less than 41% of the theoretical maximum speedup).

Special Computational Strategies

Table 5 lists the different phases involved in the finite element static analysis, and their suitability for vectorization and parallelization. A number of special strategies can be used to increase the degree of parallelism and/or vectorization in finite element computations. These strategies are applications of the principle of *divide and conquer*, based on breaking a large (and/or complex) problem into a number of smaller (and/or simpler) subproblems which may be solved independently on distinct processors. The degree of independence of the subproblems is a measure of the effectiveness of the algorithm since it determines the amount and frequency of communication and synchronization.

Herein, three strategies are discussed: domain decomposition and substructuring; operator splitting; and element-by-element strategies.

1. *Domain decomposition and substructuring.* The basic idea of domain (or spatial) decomposition is to divide the domain into a number of (possibly overlapping) regions. The initial/boundary-value problem is decomposed into one that involves solution of initial/boundary-value problems on subdomains, thereby introducing spatial parallelism. Since the solution is not available at the interfaces between regions, it is modified iteratively as part of the solution procedure. A review of parallel domain decomposition techniques is given in [56].

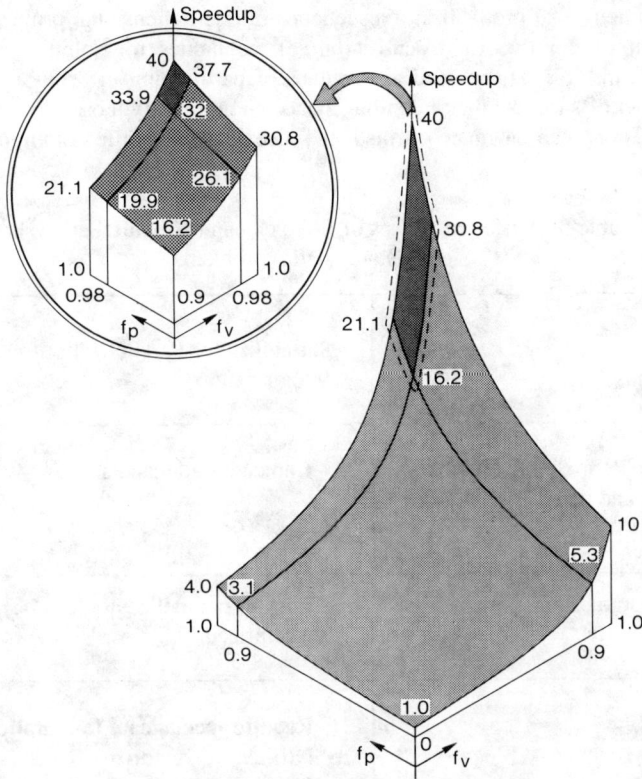

Figure 11 - Theoretical speedup on a vector multiprocessor as a function
of the fractions of parallelism and vectorization f_p and f_v

Substructuring techniques are closely related to domain decomposition. They
can also be identified at the algebraic level by partitioning the associated matrices in
an appropriate way to separate the degrees of freedom that are to be eliminated (the
internal degrees of freedom in different substructures) from those to be retained
(interface degrees of freedom). Substructuring techniques lend themselves directly to
parallel vectorization ([49] and [57]). However, the partitioning of a discretized
structure into substructures to achieve well-balanced workload distribution among the
different processors is a difficult combinatorial problem. A simulated annealing
algorithm for the approximate solution of this problem is presented in [58]. The
algorithm is analogous to a method used in statistical mechanics for simulating the
annealing process in solids. Other partitioning strategies are described in [59] and
[60].

2. *Operator splitting.* The notion of splitting has long been used to synthesize
the solution of a complicated problem from that of a simpler problem (or a sequence
of simpler problems). Among the different applications of splitting are the breaking

of a multidimensional problem into a sequence of one-dimensional problems, and the development of iterative (and semi-iterative) techniques for solution of algebraic equations. Splitting can be used as a means of partitioning the computational task into a number of subtasks that are either independent, or only loosely coupled, so that the computations can be made on distinct processors with little communication and sharing.

Table 5 - Different phases of finite element structural analysis
steady-state (static) problems

Phase	Suitability for Parallelization/ Vectorization
• Input problem characteristics, element and nodal data, and geometry	• Can be parallelized
• Evaluation of element characteristics	• Easy to parallelize and can be vectorized
• Assembly	• Require special care for parallelization • Difficult to vectorize
• Incorporation of boundary conditions	• Easy to parallelize
• Solution of algebraic equations	• Important to vectorize and parallelize
• Postprocessing	• Can be parallelized and vectorized

3. *Element-by-element solution strategies.* The modular element-by-element logic inherent in the finite element analysis procedure has been used to develop solution strategies which do not require the explicit generation of the global stiffness matrix. The frontal elimination method was originally proposed by Irons [61] to bypass the assembly process. In recent years element-by-element strategies have been developed for the solution of static and dynamic problems as well as adaptive grid generation. A review of these strategies is given in [62]. Element-by-element strategies are well-suited to vector and parallel processing and, therefore, should be

seriously considered for use in parallel processing machines.

EFFECTIVE STRATEGY FOR SOLUTION OF LARGE-SCALE STRUCTURAL PROBLEMS ON NEW COMPUTING SYSTEMS

A strategy is being developed for the efficient analysis of large complex structures on new computing systems. The strategy is designed for multiprocessor computers with either: 1) a shared memory and a small number of powerful processors such as the CRAY X-MP, CRAY-2 or CRAY-3; or 2) a hierarchical memory and a small number of clusters of medium-range processors such as the Cedar Project of the University of Illinois at Urbana-Champaign (with up to four clusters of Alliant FX/8 processors).

The strategy aims at maximizing the degree of parallelism at different levels of the finite element analysis process, including: 1) formulation level (through the use of mixed finite element models); 2) analysis level through the application of a novel spatial partitioning (or substructuring) technique based on symmetry transformations; 3) numerical algorithm level through the combined use of operator splitting techniques, iterative processes, and carefully tailored numerical algorithms for the particular hardware; and 4) implementation level through effective combination of vectorization, multitasking and microtasking, whenever available. The strategy is outlined subsequently.

Basic Idea and Key Elements of the Strategy

The foregoing strategy is based on approximating the response of the structure by a combination of symmetric and antisymmetric response vectors (or modes). The strategy can also be thought of as generating the response of the original unsymmetric structure using *large perturbations* from that of a simpler structure in which the symmetric and antisymmetric components of the response are uncoupled. The symmetric and antisymmetric modes are, therefore, generated using a reduced-size model of the modified structure (1/m the size of the original finite element model, where m is an integer which depends on the number of symmetry operations used).

The three key elements of the strategy are: 1) use of mixed (primitive variable) formulation with independent shape functions for different fields, and with all but the displacement field allowed to be discontinuous at interelement boundaries; 2) operator splitting, or restructuring of the governing discrete equations of the structure to delineate the symmetric and antisymmetric vectors constituting the response; and 3) use of preconditioned conjugate gradient (PCG) technique to generate the symmetric and antisymmetric response vectors.

Comments on the Foregoing Strategy

The following two comments regarding the strategy are in order:

1. The application of the foregoing strategy to structural problems can be divided into two phases, each of which is well-suited for parallel processing: a) preprocessing phase in which the finite element model is partitioned into substructures and reduced arrays are generated (whose dimension is 1/m times the total number of degrees of freedom of the original finite element model); b) solution phase

in which the PCG technique is used to generate the response of the original structure. It is desirable to select the symmetries such that m equals the number of processors, p. The physical interpretation of the preprocessing phase is depicted in Figs. 12 and 13 for the case m=4. The application of the foregoing strategy to linear stress analysis, free vibration, and nonlinear dynamic problems is described in [63], [64] and [65], respectively, for the case m=2. The case m=4 is described in [66].

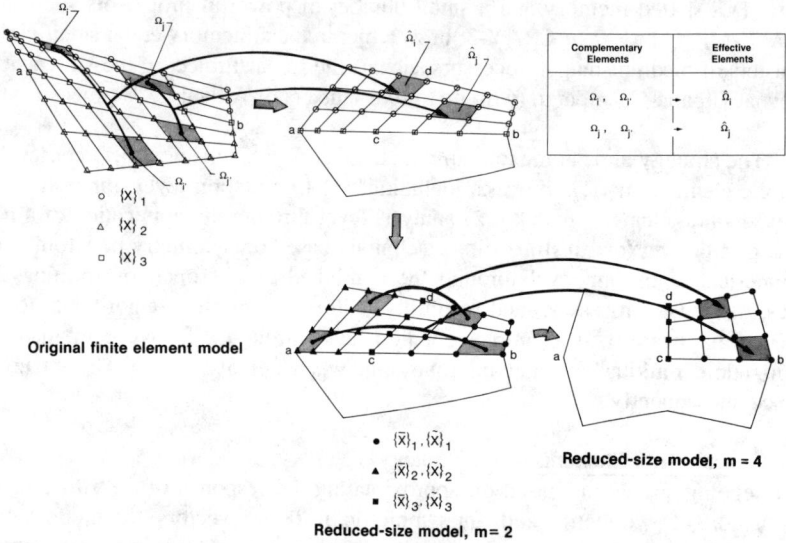

Figure 12 - Original finite element model and reduced-size models

2. The decomposition of the response vector into symmetric and antisymmetric components can be performed by means of matrix transformations. For the case m=2, the matrix transformations are given in [63] and [65].

Performance Evaluation of the Foregoing Strategy
The strategy has been implemented on the CRAY X-MP/416 at CRAY Research, Inc. in Mendota Heights, Minnesota, and the Alliant FX/8 computers at the University of Illinois at Urbana-Champaign. Herein, the efficiency gain resulting from the use of the strategy is discussed for a typical nonlinear dynamic problem of a laminated anisotropic panel with an off-center circular cutout. The loading is assumed to be uniformly distributed and normal to the panel surface and to have a step variation in time. The panel is made of graphite-epoxy material (see Fig. 14). Mixed finite element models were used for the spatial discretization, and implicit three-step method was used for the temporal integration. The details are given in [66]. Normalized contour plots for the displacement and velocity components at 3.0 msec., drawn on the undeformed middle surface of the panel, are shown in Fig. 15.* The measured CP times and processing rates obtained using the present strategy on a single CPU are shown in Fig. 16. Comparison of the wall-clock times obtained by the present strategy on one-, two- and four- CPUs, with those of the direct analysis of the panel

*See Figure 15 in colour section, page 159.

(with no partitioning) are given in Table 6. As can be seen from Table 6, the use of the present strategy on four CPU-machine reduces the total analysis time by nearly an order of magnitude, compared with that required by the direct analysis (on a single processor).

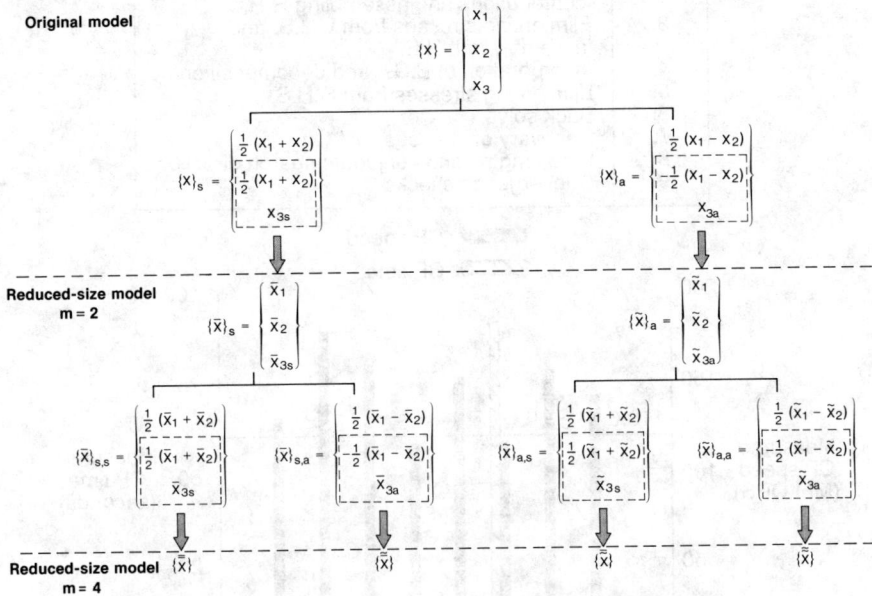

Original model

$$\{X\} = \begin{Bmatrix} x_1 \\ x_2 \\ x_3 \end{Bmatrix}$$

$$\{X\}_s = \begin{Bmatrix} \frac{1}{2}(x_1 + x_2) \\ \frac{1}{2}(x_1 + x_2) \\ x_{3s} \end{Bmatrix} \qquad \{X\}_a = \begin{Bmatrix} \frac{1}{2}(x_1 - x_2) \\ -\frac{1}{2}(x_1 - x_2) \\ x_{3a} \end{Bmatrix}$$

Reduced-size model
m = 2

$$\{\bar{X}\}_s = \begin{Bmatrix} \bar{x}_1 \\ \bar{x}_2 \\ \bar{x}_{3s} \end{Bmatrix} \qquad \{\bar{X}\}_a = \begin{Bmatrix} \bar{x}_1 \\ \bar{x}_2 \\ \bar{x}_{3a} \end{Bmatrix}$$

$$\{\bar{X}\}_{s,s} = \begin{Bmatrix} \frac{1}{2}(\bar{x}_1 + \bar{x}_2) \\ \frac{1}{2}(\bar{x}_1 + \bar{x}_2) \\ \bar{x}_{3s} \end{Bmatrix} \quad \{\bar{X}\}_{s,a} = \begin{Bmatrix} \frac{1}{2}(\bar{x}_1 - \bar{x}_2) \\ -\frac{1}{2}(\bar{x}_1 - \bar{x}_2) \\ \bar{x}_{3a} \end{Bmatrix} \quad \{\bar{X}\}_{a,s} = \begin{Bmatrix} \frac{1}{2}(\bar{x}_1 + \bar{x}_2) \\ \frac{1}{2}(\bar{x}_1 + \bar{x}_2) \\ \bar{x}_{3s} \end{Bmatrix} \quad \{\bar{X}\}_{a,a} = \begin{Bmatrix} \frac{1}{2}(\bar{x}_1 - \bar{x}_2) \\ -\frac{1}{2}(\bar{x}_1 - \bar{x}_2) \\ \bar{x}_{3a} \end{Bmatrix}$$

Reduced-size model
m = 4
$$\{\bar{\bar{X}}\} \qquad \{\bar{\bar{X}}\} \qquad \{\bar{\bar{X}}\} \qquad \{\bar{\bar{X}}\}$$

Figure 13 - Fundamental unknowns in the original and reduced-size models
(see Fig. 12)

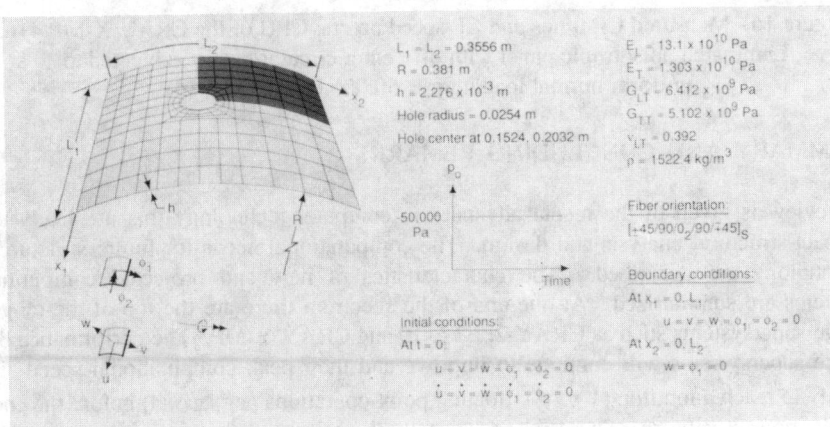

Figure 14 - Laminated anisotropic composite panel with off-center
circular cutout used in the present study

Module no.	Description
1.	Preprocessing
2.	Generation of nonlinear elemental contributions and assembling R.H.S.
3.	Eliminating stresses from L.H.S. and assembly of L.H.S.
4.	Incorporation of B.Cs. and decomposition
5.	Eliminating stresses from R.H.S.
6.	Back solve
7.	Recovery of stresses
8.	Step lengths and conjugate search directions
9.	Convergence checks

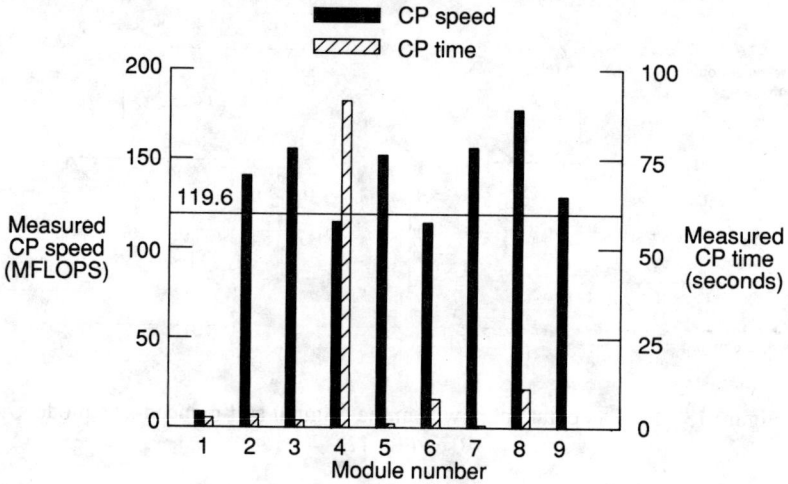

Figure 16 - Measured CP times and CP speed on one CPU of the CRAY X-MP/416. Laminated anisotropic panel with off-center circular cutout subjected to uniform normal loading p_o=-50,000 Pa (see Fig. 14).

SUMMARY AND CONCLUDING REMARKS

A review is given of the recent advances in computer technology that are likely to impact structural analysis and design. The computational needs for future structures technology are described. The characteristics of new and projected computing systems are summarized. At one end of the spectrum there are the top-of-the-range large supersystems such as CRAY-2, ETA-10 and CRAY Y-MP. The performance of large supersystems will continue to improve and their peak computational speed is likely to reach a teraflop (1 x 10^{12} floating point operations per second) before the end of the present century. These supersystems will make possible new levels of sophistication in structural modeling as well as in problem depth and scope which were not possible before. At the other end of the spectrum, embedded computers will aid in the control of the devices they reside in and will help in the realization of intelligent (smart or adaptive) structures. The emerging small low-cost engineering workstations and personal computers will provide a high degree of interactivity and free the analysts from the constraints that are often imposed on them by large centralized computation centers. Intelligent interfaces allowing multiple media interaction for both input and output (e.g., graphics and natural languages) will facilitate the user-machine communication. Flexible high-capacity networks will allow collaborative computing by linking structural analysts and designers at different locations.

Advances in programming environments, numerical algorithms, and computational strategies for new computing systems are reviewed, and a novel partitioning strategy is outlined for maximizing the degree of parallelism. The strategy is designed for computers with a shared memory and a small number of powerful processors (or a small number of clusters of medium-range processors). It is based on approximating the response of the structure by a combination of symmetric and antisymmetric response vectors, each obtained using a fraction of the degrees of freedom of the original finite element model. The strategy was implemented on the CRAY X-MP/4 and the Alliant FX/8 computers. For nonlinear dynamic problems on the CRAY X-MP with four CPUs, it resulted in an order of magnitude reduction in total analysis time, compared with the direct analysis on a single-CPU CRAY X-MP machine.

Table 6 - Performance evaluation of presented strategy on the CRAY X-MP/416 at CRAY Research, Inc. in Mendota Heights, Minnesota

	Full Structure (*Optimized Code*) (160 MFLOPS)	Partitioned Structure (*Nearly Optimized Code*) (119.6 MFLOPS)
Number of degrees of freedom	3818 displacements 6144 stresses	971 displacements 1536 stresses
Semibandwidth of equations	700	315
Wall clock time for first ten steps, (sec.)	319.7	122.6 one processor 66.7 two processors 36.3 four processors
Speedup	1.0	2.61 one processor 4.79 two processors 8.81 four processors

38

The discussion of the new computing systems presented herein is intended to give structural analysts and designers some insight into the potential of these systems for providing cost-effective solutions of complex structural problems, and to stimulate research and development of the necessary numerical algorithms, firmware and software to realize this potential.

ACKNOWLEDGMENT

The present work is supported by an Air Force Office of Scientific Research Grant No. AFOSR-88-0136, and a NASA Grant No. NAG1-730.

REFERENCES

1. Torrero, E. A. (Ed.). Next-Generation Computers, Spectrum Series, IEEE, New York, 1985.
2. Cole, B. C. Here Comes the Billion-Transistor IC, Electronics, Vol. 60, No. 7, pp. 81-85, 1987.
3. Meindl, J. D. Chips for Advanced Computing, Scientific American, Vol. 257, No. 4, pp. 78-89, 1987.
4. Sluss, J. J., Veasey, D. L., Batchman, T. E. and Parrish, E. A. An Introduction to Integrated Optics for Computing, Computer, IEEE, Vol. 20, No. 12, pp. 9-23, 1987.
5. Tank, D. W. and Hopfield, J. J. Collective Computation in Neuronlike Circuits, Scientific American, Vol. 257, No. 6, pp. 104-114, 1987.
6. Noor, A. K., Storaasli, O. O. and Fulton, R. E. Impact of New Computing Systems on Computational Mechanics and Flight-Vehicle Structures Technology, in AGARD Report No. 706, The Influence of Large-Scale Computing on Aircraft Structural Design, AGARD Structures and Materials Panel Meeting, Sienna, Italy, April 2-4, 1984.
7. Noor, A. K., Storaasli, O. O. and Fulton, R. E. Impact of New Computing Systems on Finite Element Computations. Part 4, Chapter 4, Finite Element Handbook, (Ed. by H. Kardestuncer and D. H. Norrie), pp. 4.230-4.233, McGraw-Hill, New York, 1987.
8. Noor, A. K. and Atluri, S. N., Advances and Trends in Computational Structural Mechanics, AIAA Journal, Vol. 25, No. 7, July 1987, pp. 977-995.
9. Wollard, K., Solid State, IEEE Spectrum, Vol. 25, No. 1, pp. 44-45, 1988.
10. Hwang, K. and Briggs, F. A. Computer Architecture and Parallel Processing, McGraw-Hill, New York, 1984.
11. Kryder, M. H. Data-Storage Technologies for Advanced Computing, Scientific American, Vol. 257, No. 4, pp. 117-125, 1987.
12. Foley, J. D. Interfaces for Advanced Computing, Scientific American, Vol. 257, No. 4, pp. 127-135, 1987.
13. Kahn, R. E. Networks for Advanced Computing, Scientific American, Vol. 257, No. 4, pp. 136-143, 1987.
14. Flynn, M. J. Some Computer Organizations and Their Effectiveness, IEEE Transactions on Computers, Vol. C-21, pp. 948-960, 1972.
15. Hwang, K. Advanced Parallel Processing with Supercomputer Architectures,

Proceedings of the IEEE, Vol. 75, No. 10, pp. 1348-1379, 1987.

16. Hack, J. J.   Peak vs. Sustained Performance in Highly Concurrent Vector Machines, Computer, IEEE, Vol. 19, No. 9, pp. 11-19, 1986.

17. Dongarra, J. J., Martin, J. and Worlton, J.   Computer Benchmarking: Paths and Pitfalls, IEEE Spectrum, Vol. 24, No. 7, pp. 38-43, 1987.

18. Amdahl, G.   The Validity of the Single-Processor Approach to Achieving Large-Scale Computing Capabilities, Vol. 30, pp. 483-485, Proceedings of the American Federation of Information Processing Societies, Washington, D.C., 1967.

19. Amdahl, G. M.   Limits of Expectation, International Journal of Supercomputer Applications, Vol. 2, No. 1, 1988, pp. 88-94.

20. Dongarra, J. J. and Duff, I. S.   Advanced Architecture Computers.   Technical Memorandum No. 57 (Rev. 1), Argonne National Laboratory, Argonne, Illinois, 1987.

21. McBryan, O. A.   State of the Art in Highly Parallel Computer Systems, in Parallel Computations and Their Impact on Mechanics (Ed. A. K. Noor), pp. 31-47, Proceedings of the Symposium on Parallel Computations and Their Impact on Mechanics, Boston, MA, Dec. 1987, AMD Vol. 86, American Society of Mechanical Engineers, New York, 1987.

22. Noor, A. K.   Parallel Processing in Finite Element Structural Analysis, Engineering with Computers, Vol. 3, pp. 225-241, 1988.

23. Worlton, J.   Toward a Science of Parallel Computation, in Computational Mechanics - Advances and Trends (Ed. A. K. Noor), pp. 23-35, Proceedings of the Symposium on Future Directions of Computational Mechanics, Anaheim, CA, Dec. 1986, AMD Vol. 75, American Society of Mechanical Engineers, New York, 1986.

24. Fortes, J. A. B. and Wah, B. W.   Systolic Arrays - From Concept to Implementation, Computer, IEEE, Vol. 20, No. 7, pp. 12-17, 1987.

25. Hayes, L. J.   Systolic Arrays for Finite Element Calculations, in Parallel Computations and Their Impact on Mechanics (Ed. A. K. Noor), pp. 229-238, in Proceedings of the Symposium on Parallel Computations and Their Impact on Mechanics, Boston, MA, Dec. 1987, AMD Vol. 86, American Society of Mechanical Engineers, New York, 1987.

26. Law, K. H.   Systolic Arrays for Finite Element Analysis, Computers and Structures, Vol. 20, Nos. 1-3, pp. 55-65, 1985.

27. Melhem, R. G.   On the Design of a Pipelined/Systolic Finite Element System, Computers and Structures, Vol. 20, Nos. 1-3, pp. 67-75, 1985.

28. Moore, W., McCabe, A. and Urquhart, R. (Eds.).   Systolic Arrays - Papers Presented at the First International Workshop on Systolic Arrays, Oxford, July 2-4, 1986, Adam Hilger, Bristol and Boston, 1986.

29. Hayder, M. E., Flannery, W. S., Littman, M. G., Nosenchuck, D. M. and Orszag, S. A., Large-Scale Turbulence Simulation on the Navier-Stokes Computer, Computers and Structures, Vol. 30, No. 1/2, pp. 357-364, 1988.

30. Conaway, J. H.   Structural Engineering Software on Small Computers, American Society of Mechanical Engineers, New York, Paper 80-C2/Aero-7, 1980.

31. Jordan, H.   Structuring Parallel Algorithms in an MIMD, Shared Memory Environment, Parallel Computing, Vol. 3, pp. 93-110, 1986.

40

32. Nicolau, A. A Development Environment for Scientific Parallel Programs, Applied Mathematics and Computation, Vol. 20, Nos. 1-2, pp. 175-183, 1986.

33. Pratt, T. W. The PISCES 2 Parallel Programming Environment, NASA CR-178327, 1987.

34. Gannon, D., Atapattu, D., Lee, M. H. and Shei, B. A Software Tool for Building Supercomputer Applications, in Parallel Computations and Their Impact on Mechanics (Ed. A. K. Noor), pp. 81-92, in Proceedings of the Symposium on Parallel Computations and Their Impact on Mechanics, Boston, MA, Dec. 1987, AMD Vol. 86, American Society of Mechanical Engineers, New York, 1987.

35. Dym, C. L. Expert Systems: New Approaches to Computer-Aided Engineering, Part I, pp. 99-115, in Proceedings of the AIAA/ASME/ASCE/AHS 25th Structures, Structural Dynamics and Materials Conference, Palm Springs, CA, May 1984.

36. Mackerle, J. and Orsborn, K. Expert Systems for Finite Element Analysis and Design Optimization - A Review, Engineering Computations, Vol. 5, No. 2, pp. 90-102, 1988.

37. Taig, I. C. Expert Aids to Reliable Use of Finite Element Analysis, in Reliability of Methods for Engineering Analysis (Ed. K. J. Bathe and D. R. J. Owen), pp. 457-474, Pineridge Press, Swansea, United Kingdom, 1986.

38. Fulton, R. E. and Yeh, C.-p. Managing Engineering Design Information, in Proceedings of the AIAA//AHS/ASEE Aircraft Design, Systems and Operations Conference, Atlanta, GA, Sept. 7-9, 1988.

39. Parks, C. H. Tutorial: Reading and Reviewing the Common Schema for Electrical Design and Analysis, in Proceedings of 24th ACM/IEEE Design Automation Conference, pp. 479-483, 1987.

40. Kung, H. T. The Structure of Parallel Algorithms, Advances in Computers, Vol. 19, pp. 65-112, Academic Press, 1980.

41. Schendel, U. Introduction to Numerical Methods for Parallel Computers, Ellis Horwood, Ltd., United Kingdom, 1984.

42. Dongarra, J. J. and Sorensen, D. C. Linear Algebra on High-Performance Computers, in Parallel Computing 85 (Eds. M. Feilmeier, G. Joubert and U. Schendel), pp. 3-32, Elsevier, New York, 1985.

43. Adams, L. Reordering Computations for Parallel Execution, Communications in Applied Numerical Methods, Vol. 2, pp. 263-271, 1986.

44. McBryan, O. and Van de Velde, E. Matrix and Vector Operations on Hypercube Parallel Processors, Parallel Computing, Vol. 5, pp. 117-125, 1987.

45. Ortega, J. M. Introduction to Parallel and Vector Solution of Linear Systems, Plenum Press, New York, 1988.

46. Buzbee, B. L. Uniquely Parallel Algorithms, in Parallel Computations and Their Impact on Mechanics (Ed. A. K. Noor), pp. 95-100, in Proceedings of the Symposium on Parallel Computations and Their Impact on Mechanics, Boston, MA, Dec. 1987, AMD Vol. 86, American Society of Mechanical Engineers, New York, 1987.

47. Frederickson, P. and McBryan, O. Parallel Superconvergent Multigrid, in Multigrid Methods - Theory, Applications and Supercomputing, Lecture Notes in Pure and Applied Mathematics, Vol. 110, (Ed. S. McCormick), Marcel Dekker, New York, pp. 195-210, 1988.

48. Dongarra, J. J. and Sorensen, D. C. A Fully Parallel Algorithm for the Symmetric Eigenvalue Problem, SIAM Journal of Scientific and Statistical Computing, Vol. 8, No. 2, pp. 139-154, 1987.

49. Adams, L. and Voigt, R. A Methodology for Exploiting Parallelism in the Finite Element Process, in Proceedings of the NATO Workshop on High Speed Computations, NATO ASI Series (Ed. J. Kowalik), Springer, Berlin, F-7, pp. 373-392, 1984.

50. Dongarra, J. J. and Sorensen, D. C. SCHEDULE: Tools for Developing and Analyzing Fortran Programs, MCS-TM-86, Argonne National Laboratory, Argonne, Illinois, Nov. 1986.

51. Boyle, J., Butler, R., Disz, T., Glickfeld, B., Lusk, E., Overbeek, R., Patterson, J. and Stevens, R. Portable Programs for Parallel Processors, Holt, Rinehart and Winston, New York, 1987.

52. Ware, W. H. The Ultimate Computer, IEEE Spectrum, Vol. 9, No. 3, pp. 84-91, 1972.

53. Flatt, H. and Kennedy, K. Performance of Parallel Processors, Parallel Computing (to appear).

54. Gustafson, J. L., Montry, G. R., and Benner, R. E., Development of Parallel Methods for a 1024-Processor Hypercube, SIAM Journal of Scientific and Statistical Computing, Vol. 9, No. 4, pp 609-638, July 1988.

55. Gustafson, J. L., Reevaluating Amdahl's Law, Communications of the ACM, Vol. 31, No. 4, pp. 532-533, 1988.

56. Rodrigue, G. Some Ideas for Decomposing the Domain of Elliptic Partial Differential Equations in the Schwarz Process, Communications in Applied Numerical Methods, Vol. 2, pp. 245-249, 1986.

57. Carey, G. F. Parallelism in Finite Element Modeling, Communications in Applied Numerical Methods, Vol. 2, pp. 281-287, 1986.

58. Flower, J., Otto, S. and Salama, M. Optimal Mapping of Irregular Finite Element Domains to Parallel Processors, in Parallel Computations and Their Impact on Mechanics (Ed. A. K. Noor), pp. 239-250, in Proceedings of the Symposium on Parallel Computations and Their Impact on Mechanics, Boston, MA, Dec. 1987, AMD Vol. 86, American Society of Mechanical Engineers, New York, 1987.

59. Farhat, C. H. and Wilson, E. L. A New Finite Element Concurrent Computer Program Architecture, International Journal for Numerical Methods in Engineering, Vol. 24, pp. 1771-1792, 1987.

60. Nour-Omid, B. and Park, K. C. Solving Structural Mechanics Problems on the CALTECH Hypercube Machine, Computer Methods in Applied Mechanics and Engineering, Vol. 61, No. 2, pp. 161-176, 1987.

61. Irons, B. M. A Frontal Solution Program for Finite Element Analysis, International Journal for Numerical Methods in Engineering, Vol. 2, pp. 5-32, 1970.

62. Hayes, L. J. Advances and Trends in Element-by-Element Techniques, in State-of-the-Art Surveys on Computational Mechanics (Ed. A. K. Noor and J. T. Oden), American Society of Mechanical Engineers, New York (to appear).

63. Noor, A. K. and Whitworth, S. L. Computational Strategy for Analysis of Quasi-symmetric Structures, Journal of Engineering Mechanics, ASCE, Vol. 114, No. 3, pp. 456-477, 1988.

64. Noor, A. K. and Whitworth, S. L. Vibration Analysis of Quasi-symmetric

Structures, Finite Elements in Analysis and Design, Vol. 3, No. 4, pp. 257-276, 1987.

65. Noor, A. K. and Peters, J. M. Model-Size Reduction for the Nonlinear Dynamic Analysis of Quasi-symmetric Structures, Engineering Computations, Vol. 4, pp. 178-189, 1987.

66. Noor, A. K. and Peters, J. M. A Partitioning Strategy for Efficient Nonlinear Finite Element Dynamic Analysis on Multiprocessor computers, Computers and Structures (to appear).

Finite Element Analysis on Computers with Multiple Processors

Edward L. Wilson

University of California, Berkeley, CA 94720, USA

1 Introduction

The use of multiprocessor computer systems for the dedicated solution of finite element domains was suggested in 1976 [1]. However, it has only been within the past few years that commercial multiprocessor computers have been available and experience has been obtained in their use. During the past three years research has been conducted on the automatic subdomain algorithm [2], concurrent algorithms for dynamic analysis [3], concurrent iterative solution methods [4], and direct concurrent solution algorithms [5]. Most other papers on the use of multiprocessors in computational mechanics have presented specific numerical methods to improve the speed of only one phase of a computer program, and have not proposed a major change in the finite element program architecture. Also, many papers have been presented which investigate the use of specific computer hardware for the solution of problems in computational mechanics. The purpose of this paper is to summarize a new program architecture which is not dependent on the type of computer hardware used or the class of physical problem. In addition, other ideas on the use of multiprocessor computers in the field of computational mechanics will be presented.

Summary of Computer Hardware Development

Traditionally, many large mainframe computers have used multiple processors. Over 20 years ago the first CDC six thousand series of computers used a Central Processing Unit (CPU) and several Peripheral Processing Units (PPUs). However, computers of that generation used the operating system to control the concurrent operation of the various processors. Within computationally intensive scientific programs no attempt was made to program the peripheral processors since their major functions were to minimize the time required to perform various input/output operations.

More recently multiple CPUs have been added to CRAY, IBM and other large mainframe computers. In normal use, however, the multiple processors are controlled by the operating system and are automatically assigned to different tasks in a multiuser environment. Hence, their basic purpose is to improve overall performance in the execution of traditional programs.

It is now a standard option on scientific mainframe computers to have vector or array processors. FORTRAN compilers for these modern computers recognize these standard floating point operations, and automatically produce executable programs which are significantly faster for the computationally intensive phases of a finite element program. The addition of array processors to super minicomputers has been found to reduce executable time on large scientific jobs and to improve the overall performance of multiuser systems. The major advantage of the use of vector and array processors is that a minimum of program modification is required in the source program in order to effectively utilize these computer hardware additions, which increase the speed of floating point operations only. For most linear and nonlinear finite element analyses, however, floating point operations require approximately 50 percent of execution time. Therefore, if the time for floating point operations is reduced to zero, the overall performance of the program would only be improved by a factor of two.

The use of multiple processors, where each processor may have a vector processor, has the potential for significant reductions in executable requirements for finite element analyses. Within each phase of a finite element analysis it is possible for a FORTRAN compiler to recognize parallel operations which can be assigned to different processors and executed concurrently. However, in order to fully utilize the potential of computer systems with multiple processors, it is necessary to create a new architecture for finite element programs and to develop different numerical methods.

Need For Improvement In Computational Speed

Since the introduction of inexpensive, powerful microcomputers the need for low cost computational power has virtually been eliminated for over ninety percent of linear finite element analyses. Unless a finite element analysis requires more than 15 hours (from 5 p.m. one day to 8 a.m. the next day), it is considered a short run on a microcomputer workstation. Therefore, it is possible for a design engineer to run "what if" types of problems every day without a significant increase in cost. The author has had personal contact with a large number of small structural engineering offices which are designing large (over fifty stories) buildings, subjected to dynamic earthquake loading on MS-DOS personal computer systems which cost less than $5,000. A few years ago one such iterative design would cost approximately $5,000 on a mainframe computer.

A modern microcomputer engineering workstation (20 MHZ, 180386 and 180387) is approximately 1/1000 of the speed of a large computer such as a CRAY X-MP. Therefore, problems which require 15 hours on microcomputers would require only 54 seconds on a CRAY. Hence, there is still, and always will be, a need for increases in the speed of computer systems of all sizes. The CRAY Y-MP, which has recently been released, has eight Central Processing Units and 32 million words of directly addressable central memory. There is justification for such a system for the solution of a large number of three-dimensional nonlinear, finite element analyses which are necessary in many areas of computational mechanics.

The multiprocessor mainframe computer, such as the CRAY X-MP, costs over ten million dollars therefore, only a few very large organizations can justify these expensive computer systems. Hence, their impact on the very broad field of computational mechanics will be minimal. In the author's opinion the addition of multiple processors to microcomputers, at an incremental cost of a few hundred dollars, has the potential for a significant impact on the type of computational mechanics problems which can be solved on the personal engineering workstation.

Required Research and Development

The cost of the development of computer programs has become very large compared to the cost of computer hardware. Therefore, it is very important that new programs be portable between different types of computer hardware which have multiple processors. The best method to obtain portability is to develop programs in a language which will allow tasks to be assigned to different processors and to be executed concurrently. At the present time FORTRAN 77 is the de facto standard for the development of programs in computational mechanics. Therefore, the next version of FORTRAN, 80X, should have the ability to activate operations to run concurrently. If such a language is standardized it will be possible to run the same program, which uses multiple processors, on both inexpensive microcomputers and mainframe computers such as the CRAY X-MP.

Multiple processor computers fall into two major categories - fine grain (large number of processors) and coarse grain (small number of processors). High speed memory can be shared (common) or local to each processor. Also, program storage may exist only once in common storage or it may be duplicated within the local memory of each processor. Communication between processors can take place via a common bus or by message passing to adjacent processors. At the present time, it is not apparent which hardware design is the most appropriate for the solution of problems in computational mechanics. It is the author's opinion that program developers and hardware manufacturers must work together in order to fully develop the potential of this new technology.

2   Basic Program Architecture

It is very important that a new finite element program has an architecture which will operate effectively on computers with any number of processors. At the present time supercomputers such as the CRAY have one to eight processors; whereas, other computers may have several thousand processors. In addition, some of the hardware is based on each processor having its own memory and some on computers in which the same memory is shared by all the processors. The subdomain automatic approach [2] has the advantage of being equally effective on all existing computers which have multiprocessors.

The subdomain approach requires that the finite element model be subdivided into the same number of subdomains as the number of processors, Np, which are available on the computer system. In order to obtain maximum efficiency it is essential that all processors are assigned equal computational effort. Also, for local memory processors it is very important that there is a minimum amount of data communication between processsors.

Figure 1 illustrates the concurrent solution of a finite element model on a multiprocessor computer system. After the model is subdivided into Np domains all element and subdomain calculations are completed without the need for interprocessor communication. In a computer system with local memory only the node coordinates, loads and element properties associated with each subdomain need be stored; therefore, this basic data does not need to be duplicated in the memory of the other processors. Also, the basic data is uncoupled during the stress recovery phase.

In addition, the new program stucture can be used for the solution of other types of field problems in mechanics such as fluid flow or heat transfer. Also, the automatic subdomain approach may be used in the classical numerical approach of finite difference in order to obtain improved performance on multiprocessor computer systems.

It should be emphasized that the use of the subdomain architecture does not require that existing programs for finite element analysis be completely rewritten. The existing program modules associated with pre and postprocessing, formation of element

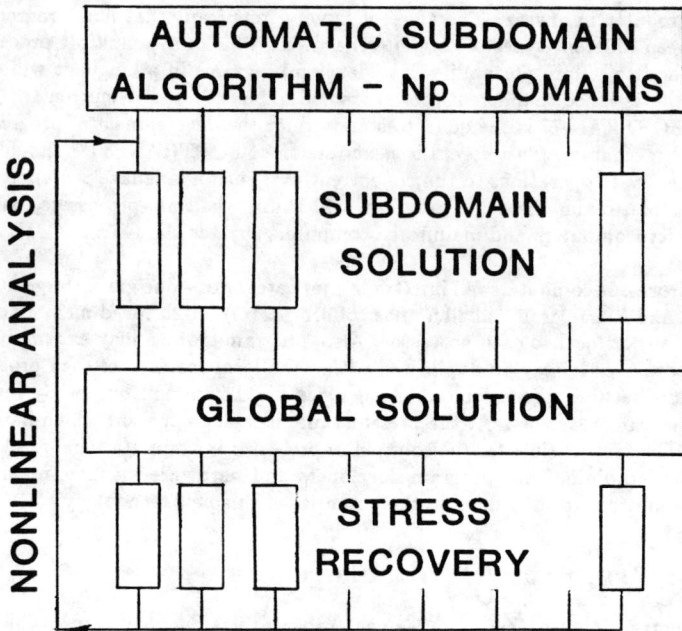

Figure 1. PROGRAM ARCHITECTURE USING MULTIPROCESSORS

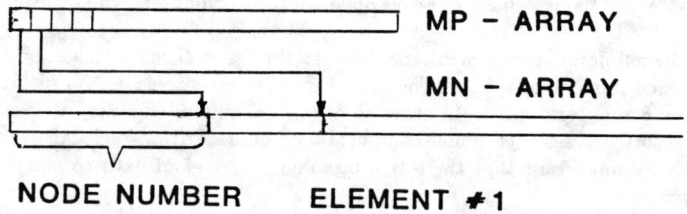

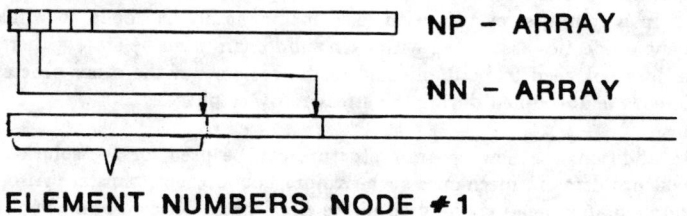

Figure 2. BASIC DATA FOR AUTOMATIC SUBDOMAIN ALGORITHM

matrices and calculation of stresses can be directly incorporated into the new subdomain architecture. Experience has indicated that the required program development for the automatic creation of subdomains, the reduction of subdomains and the solution of the global domain requires less than 1500 FORTRAN statements.

3 The Automatic Subdomain Algorithm

For static linear problems with a few load conditions, the total solution time on a sequential computer is the sum of the time required to process element information and the time required to solve the global equilibrium equations. The computational time required to form element stiffnesses, assemble the global stiffness and calculate element stresses is directly proportional to the number of elements. The computational time needed to solve the global equilibrium equations is proportional to the number of equations times the "average" band-width squared. Therefore, in the case of three-dimensional solids the solution phase tends to dominate the overall execution time.

In the case of linear dynamic response analysis in which element stresses are evaluated as a function of time the evaluation of element stresses may require the major amount of computational effort. In this case the use of concurrent processing will produce an increase in speed directly proportional to the number of processors. For nonlinear static or dynamic analysis, where the element stiffness and stresses must be computed at each load or time step, the use of concurrent computing at the element level will be very effective.

The geometry of a finite element model can be numerically defined by a list of element numbers and the node numbers associated with each element. This data can be stored in two integer arrays. The MP array is the number of elements in length in which the integer MP(i) is the location in the MN array of the last node number associated with element number "i". Therefore, the node numbers associated with element "i" are given in locations MN[MP(i-1)+1] to MN [MP(i)] where MP(0) is defined as zero. This basic data structure allows for a mixture of one, two or three-dimensional elements each with a different number of nodes.

In order for the automatic subdomain algorithm to operate at increased efficiency it is necessary to create additional arrays. The NP array is the number of nodes in length in which NP(j) is the location in the NN array of the last element number associated with node "j". Therefore, the numbers for the elements attached to joint "j" are stored in locations NN [NP(j-1)+1] to NN [NP(j)] where NP(0) is defined as zero. The data structures for these arrays are illustrated in Figure 2. The data in these basic arrays is not changed during the execution of the algorithm.

During the execution of the automatic subdomain algorithm three additional integer arrays are created and modified. The NW array is of length equal to the number of nodes and contains the number of elements attached to each node. As elements are removed from the system the numbers in the NW array are reduced. The array NF contains the number of active node numbers. An active node is one in which some of the elements have been removed. When all elements are removed from a node the node number is removed from the NF array. Finally, the ME array is of length equal to the number of elements and contains the element numbers in the sequence in which they are removed from the system.

A summary of the algorithm for the automatic creation of subdomains is given in Table 1. It should be noted that all arrays can be retained in high speed memory during the execution of the algorithm. Also, the algorithm is very fast since a minimum of array searching is required during execution.

TABLE I: THE AUTOMATIC SUBDOMAIN ALGORITHM

A.   NUMERICAL DEFINITION OF FINITE ELEMENT MODEL
       Element-Node Connectivity
           MN array containing node numbers and the
           MP element pointer array

B.   INITIAL CALCULATION OF NODE DATA
       Node-Element Connectivity
           NN array containing element numbers and the
           NP node pointer array

C.   DEFINITION OF WORKING ARRAYS
           NW array containing the number of elements
           connected to each node
           NF array containing the active node number
           ME array containing element numbers in order
           produced

D.   EVALUATION OF SUBDOMAINS OF "L" ELEMENTS EACH
   1.   Zero NF array and start at node which has a
        minimum number of elements
   2.   Remove all elements attached to node and
        update ME, NF and NW arrays
   3.   After "L" elements are removed return to
        step 1
   4.   Eliminate nodes with zero elements and
        compact NF array
   5.   Return to step 2 and use node NF(1) as next
        node

4   Direct Solution Algorithm

One of the major reasons for restricting the discussion in this section to noniterative methods is that they are easily extended to dynamic response analysis using either mode superposition or direct step-by-step integration. Further research is required in iterative methods in concurrent computation before they become as general and robust as the direct solvers.

The direct solution method suggested in this paper for the reduction of the subdomains and for the global solution of the system of subdomains is based on the profile (or skyline) storage method. The details of the algorithm and a FORTRAN listing of the subroutines are given in [6]. These subroutines have a subdomain (or substructure) reduction option. It is very important to note that all three phases (factorization, forward reduction and backsubstitution) involve vector operations and the inner DO LOOPS have been replaced with subroutine calls. If each processor has a vector processor the subdomain operations can be made very efficient.

5   Subdomain Reduction

The automatic subdomain algorithm presented is capable of subdividing any two or three-dimensional finite element model into the same number of domains as there are available

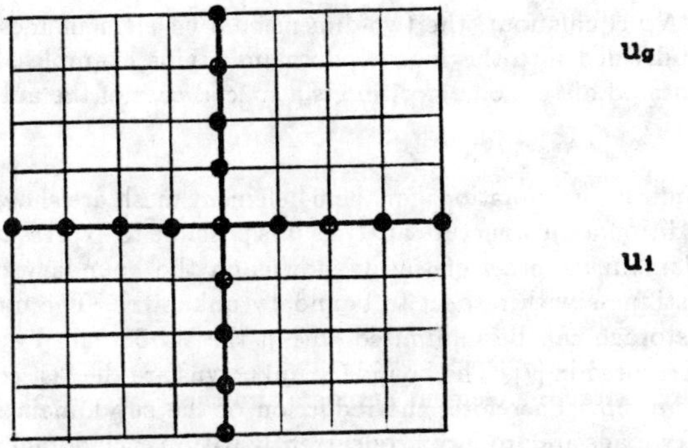

(a) FINITE ELEMENT MODEL

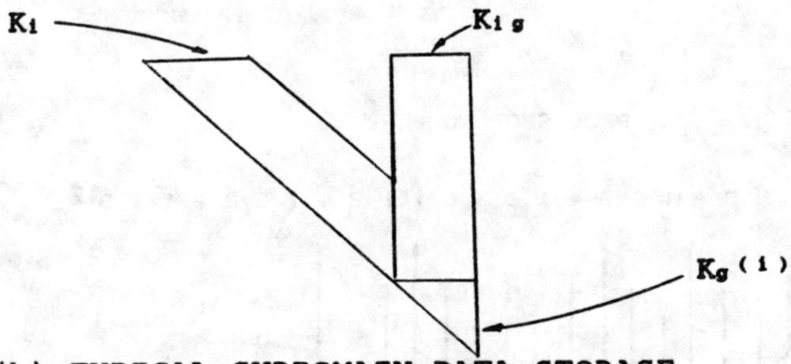

(b) TYPICAL SUBDOMAIN DATA STORAGE

$$\begin{bmatrix} K_1 & & & & K_{1g} \\ & K_2 & & & K_{2g} \\ & & K_3 & & K_{3g} \\ \text{SYMMETRIC} & & & K_4 & K_{4g} \\ & & & & K_g \end{bmatrix} \begin{bmatrix} u_1 \\ u_2 \\ u_3 \\ u_3 \\ u_g \end{bmatrix} \begin{bmatrix} F_1 \\ F_2 \\ F_3 \\ F_4 \\ F_g \end{bmatrix}$$

(c) EQUILIBRIUM EQUATIONS - GLOBAL SYSTEM

Figure 3. BASIC EQUATIONS FOR SYSTEM OF SUBDOMAINS

processors. For Np equals four, the two-dimensional 64 element mesh shown in Figure 3a would be subdivided into the four subdomains. This example also illustrates that the well-known nested dissection algorithm is a special case of the automatic subdomain algorithm.

The global equilibrium equations for the 64 element mesh are shown in Figure 3c. As in the case of traditional substructure analysis it is possible to express the basic unknowns within the subdomain in terms of the unknowns on the boundary of the domain and form a reduced stiffness with respect to boundary unknowns. The number of numerical operations and storage can be minimized within the subdomain by use of the Profile-Front Method presented in [7]. The boundary unknowns are the last equations numbered as shown in Figure 3b. Therefore, the reduction of the subdomain stiffnesses and the effective boundary loads are produced concurrently within each domain. It is immportant to note that the profile method of data storage allows the triangularization phase of solution to utilize vector processors during the reduction of each subdomain.

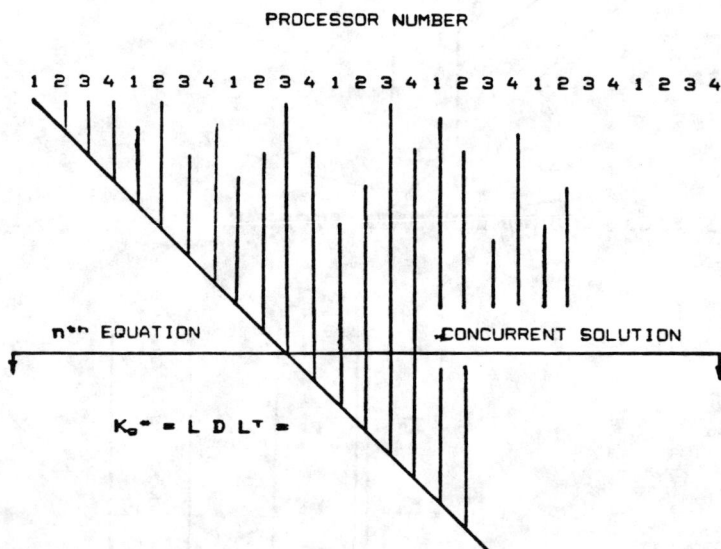

Figure 4. GLOBAL STIFFNESS MATRIX IN PROFILE FORM

6 Global Solution Of The System Of Subdomains

After all subdomains are reduced the global equations can be assembled by the application of the direct stiffness approach. In the case of multiprocessors with local storage the basic matrices can be assembled directly, if the columns of the global stiffness matrix are distributed to the different processors before the summation and the addition of the subdomain arrays can be conducted concurrently.

The global system of equations with respect to the subdomain boundary unknowns can be solved concurrently, with a minor modification of the profile equation solver previously presented. The basic topology of the global stiffness matrix is shown in Figure 4. For this example every fourth column is assigned to be reduced by one processor. An examination of the solution method indicates that after column "n" is reduced all columns greater than "n" can be reduced down to the row n concurrently in any order. Therefore, for large bandwidth problems it is possible to utilize all processors very effectively.

In the case of shared memory no communication is necessary. In the case of multiprocessors with local memory it is necessary to send the column to all other processors after it is completely reduced. In addition, it is necessary to maintain a copy of D_{ii} in the local memory of all processors. The existence of a nonzero term on the diagonal is all the information that is required to let each processor know which terms in the triangularized global stiffness matrix can be calculated.

After the global displacements are calculated it is then possible to distribute the subdomain boundary displacements to each processor for the concurrent evaluation of the displacements within each subdomain.

It is clear that the automatic subdomain and solution algorithms presented will operate on both shared memory and local memory multiprocessors. Since the local memory processors require additional time for passing data it appears that the shared memory approach is the most general. However, it should be noted that the automatic subdomain approach, as presented here, requires that all terms in the stiffness and load matrices be stored in real memory. For very large problems this may not be possible because of the address limitations of the computer. If shared storage is a limitation factor then the local storage approach within each processor may have an advantage.

For computers which automatically "page" storage, the use of multiprocessors appears to be totally ineffective. If the basic data cannot be retained in real memory the algorithm can be modified in order to create a multilevel subdomain method in which all data is moved in large blocks between real and low speed mass storage. The total number of subdomains would then be a multiple of Np.

7 Linear Dynamic Analysis

The automatic subdomain and solution algorithm, which has been presented in detail for static analysis, is easily extended to dynamic analysis of large finite element systems without a significant reduction of efficiency [3]. The basic approach is to use load-dependent Ritz vectors to reduce the size of the system [8]. The latest form of this algorithm is summarized in [5]. As in the case of static analysis, nearly all phases of the method can be carried out concurrently on multiple processors. In addition, after the modal response is evaluated, the time-dependent displacements and element stresses can be calculated concurrently within each subdomain.

This approach is fundamentally different from the multiprocessor Lanczos eigenvalue method which has been recently presented [9]. In the eigenvalue approach the complete stiffness matrix is required to be duplicated within the local memory of each processor. A different shift is used within each processor and the eigenvalues are calculated in groups near the shifts. It is clear that this method is limited to problems where the complete stiffness matrix for the finite element system can be contained in the local memory of each processor. In addition, it has been shown that the load-dependent vectors can be generated with less numerical effort and are always more accurate than if the exact eigenvectors are used [10].

8 Nonlinear Analysis

A nonlinear analysis of a finite element system is often 10 to 100 times the computational requirements for a static linear analysis. The most stable nonlinear solution algorithms involve the incremental application of the load and iteration within the load or time step in order to obtain equilibrium.

Some methods are implicit in which a direct solution of equations is not required at each load or time increment. For this approach the number of numerical operations required is directly proportional to the number of elements; therefore, the automatic subdomain approach will tend to equally divide the total computational effort between the multiple processors.

Other methods of nonlinear analysis require the formulation and the direct solution of the equilibrium equations for each increment. For this approach the methods presented in this paper for the concurrent formulation and solution of static problems on multiple processors can be used directly as indicated in Figure 1. In the case of shared storage all of the basic data must be retained in real storage for maximum efficiency. For multiprocessor computers with local storage a small amount of information is duplicated within each processor and the basic automatic subdomain appproach requires a minimum amount of messages passing between processors.

9 Numerical Examples

A large number of numerical examples were run on a hypercube multiprocessor computer using the basic numerical methods summarized in this paper [2, 3, 4, 5,]. The speed-up is defined as the ratio of the "computer time required to solve the problem using Np| processors" to the "computer time required using one processor". In general, the results indicate the approach is more efficient for larger problems. It is apparent that for very small problems the use of a large number of processors may be counter productive. Recent experience of solving problems on a shared memory computer indicates that speed-up ratios of over 90 percent can be obtained.

10 Conclusions

A considerable amount of direct and indirect experience has been obtained during the past few years on the use of multiprocessor computer systems for the solution of finite element problems using various numerical algorithms. As a result of this work the following general conclusions are made:

1. The automatic subdomain algorithm can be used as a fundamental approach to approximately divide equally the total computational work to all processors on a

multiprocessor computer system. The basic approach tends to minimize communication requirements between processors. Also, the method is effective for static, dynamic and nonlinear problems in all areas of computational mechanics.

2. The development of "smart" compilers for multiprocessor computers will speed up individual program modules of existing programs in computational mechanics. However, if the maximum potential of the hardware is to be obtained a new program architecture is required in order that all phases of the analysis can be executed concurrently.

3. The Gauss elimination method, with the basic matrices stored in profile form, is a very effective approach for the concurrent reduction of the basic equations for all subdomains. The global system of equations with respect to the subdomain boundaries can be solved directly if profile storage is used. Each Np th column is assigned to each processor and the triangularization is carried out concurrently.

4. Iteration solution methods can also be used effectively with the automatic subdomain approach at both the subdomain and the global levels. This approach appears to have the best potential for the solution of three-dimensional and nonlinear problems. The basic disadvantage of the iterative approach is that it is not directly extensible to dynamic mode superposition analysis.

5. Multiprocessor computer systems with shared memory appear to offer the most potential for the solution of problems in computational mechanics since the problem of data transfer between the local memory of the different processors is eliminated. The use of multiprocessors on computer systems which automatically "page" memory should be avoided.

6. The existence of multiprocessor computer systems is a practical reality for supercomputer systems. However, the cost is large and the development of programs requires hardware dependent programming.

7. On microcomputers the incremental cost for additional processors and memory is small and the potential increase in performance is large. However, the practical application of multiprocessor computer systems on low cost personal computers is still several years in the future. The major problem to be solved is the definition of commands within one of the standard programming languages for concurrent operations. It will then be possible to develop portable programs at a low cost.

The effective use of this new technology is an exciting challenge to researchers in computer science, numerical analysis and computational mechanics.

References

1. Wilson, E.L. Special Numerical and Computer Techniques for Finite Element Analysis, Formulation and Computational Algorithms in Finite Element Analysis, U.S. - Germany Symposium, MIT Press, pp 2-25, 1976.

2. Farhat, C.H., and Wilson E.L.; Solution of Finite Element Systems on Concurrent Processing Computers, Engineering with Computers, Vol. 2, pp 157-165, 1987.

3. Farhat C.H. and Wilson E.L.; Modal Superposition Dynamic Analysis on Concurrent Multiprocessors, Engineering Computations, 1987.

4. Farhat C.H. and Wilson E.L.; Concurrent Iterative Solution of Large Finite Element Systems, Communications in Applied Numerical Methods, Vol. 3, 1987.

5. Wilson E.L. and Farhat C.H.; Linear and Nonlinear Finite Element Analysis on Multiprocessor Computer Systems, Communications in Applied Numerical Methods, Vol. 4, pp. 425-434, 1988.

6. Wilson E.L. and Dovey H.H.; Solution or Reduction of Equilibrium Equations, Advances in Engineering Software, Vol. 1, No. 1, 1978.

7. Hoit M.I. and Wilson E.L.; An Equation Numbering Algorithm Based on Minimum Front Criteria, Computers and Structures, Vol. 16 No. 1-4, pp 225-239, 1983.

8. Wilson E.L., Yuan M.W. and J.M Dickens; Dynamic Analysis by Direct Superposition of Ritz Vectors, Earth. Engrg. Struc. Dynam. Vol. 10, pp 813-823, 1982.

9. Fulton R.; The Impact of Parallel Computing on Finite Element Computations in Reliability of Methods for Engineering Analysis, Pineridge Press, Swansea, U.K. pp 179-196, 1986.

10. Bayo E.P. and Wilson E.L.; Use of Ritz Vectors in Wave Propagation and Foundation Response, Earth. Engrg. Struc. Dynam., Vol. 12, pp 499-505, 1984.

Using ESA/370 [TM] to Speed Up Engineering and Scientific Applications

Brian B. Moore

IBM, 16, Carnelli Street, Poughkeepsie, NY 12603, USA

When machine has six vector processors, each with a peak of 120 million floating-point operations per second (MFLOPS), the upper limit of computation is 720 MFLOPS. A program that executes a large part of its instructions in scalar mode, that uses only one processor, and that spends more than half of its time waiting for DASD may thereby average only 4 MFLOPS for the time it is active in the system. This is probably far from what could achieved, and an investment in improved application software might yield substantial payback.

There are many reasons that explain why engineering and scientific programs might be inefficient. Among them are the following three. First, when using Fortran, which provides a general programming interface, the invitation is to ignore the underlying architecture of the system. Thus, for example, vector processing obstacles and opportunities are overlooked. Second, parallel processing, where several processors perform computations on shared data, is a new Fortran capability [1]. Many applications have not been upgraded to use more than one processor, and automatic exploitation of parallelism by the compiler is in the experimental stages. And third, the DASD subsystem may account for a large proportion of the time a job is resident in the system. Although I/O delays can be reduced with buffering techniques that exploit the processor storage hierarchy, this option may not be exercised because it requires an understanding of the services provided by the operating system.

There are also many reasons for seeking to improve the performance of a large-scale engineering and scientific application. The reasons may include the following:

- To run larger problems within a fixed time constraint: Gustafson [2] points out that "with a more powerful processor, the problem expands to make use of increased facilities. Users have control over such things as grid resolution, number of time steps, difference operator complexity, and other parameters that are adjusted to allow the program to be run in some desired amount of time. Hence, it may be realistic to assume that *run time*, not *problem size*, is constant."

- To convert a batch job into an interactive job: By reducing the run time for an application from, for example, 1 hour to 2 minutes, the job may be run many times during development shifts (interactively) instead of being delayed on a production queue waiting for an overnight or over-the-weekend run.

TM - Enterprise Systems Architecture/370 and ESA/370 are trademarks of the International Business Machines Corporation.

- To extend the range of problems that are effectively solvable: Buzbee and Sharp [3] observe that "although the past 40 years have seen a dramatic increase in computer performance, the number of users and the difficulty and range of applications have been increasing at an even higher rate, to the point that demands for greater performance now far outstrip the improvements in hardware." Without special programming techniques, the system time required for a very large problem may be prohibitive. If, for example, the job time can be reduced from 5 days to 4 hours, the job may be run overnight instead of being terminated prior to completion when some computing center policy is violated.

- To increase the value of an application: Developers of a licensed application may want to achieve all of the preceding results in one program that meets the needs of the broadest set of potential customers.

The common thread then is problem size. The goal is to increase the size of the problems that lie in the interactive, production and effectively-solvable regions.

The remainder of this paper discusses ways of improving the performance of large-scale engineering and scientific programs. It presents general techniques available with systems using the Enterprise Systems Architecture/370 (ESA/370) [4, 5]. The techniques are: parallel processing, I/O avoidance using large real and expanded storage, vector processing, use of stride-1 algorithms, vector register exploitation, use of compound vector operations, and locality of reference.

Problem Size

Problem size can be approached using an illustration from matrix structural analysis. In a statics problem, for example, when six displacement variables are modelled on a 1000-by-1000 grid, a system of 1000x1000x6 linear simultaneous equations is used. For this, the matrix equation $A \times X = P$ is solved for the X values. Here, A is a stiffness matrix, P is the load vector and X is the displacement vector. The matrix dimension, N in what follows, is used as the measure of problem size.

The solution matrix of a structural simulation is often sparse, with 5% or fewer of the elements having non-zero values. At some point in the simulation, a matrix multiplication process may be invoked. Matrix multiplication provides the example used throughout this paper. In the example, one of the matrices used to form the product is dense, and the other is sparse. Sparse matrices are arranged in column-major order. Only the non-zero elements of the matrix are stored (see Figure 1).

Sparse Matrix Example: The example uses matrices that are too large to fit into main storage. Parts of the matrices are read from DASD and processed, with the results written to DASD so that the cycle may be repeated. Columns of a matrix which are in main storage are said to be *active*, whereas *inactive* columns must be read from DASD before they can be processed.

Figure 2 illustrates the multiplication of the sparse matrix E by the full matrix B: $E \times B = D$. Here, D is the resulting full matrix. Real main storage buffers hold as many active columns of B and D as is possible. Successive strings of E are used in calculating contributions to the final values in the active columns of D. When the entire E matrix has been processed, the active columns of D are written to DASD and a new set of columns of B and D are activated. The cycle is repeated until all columns of D have been written to DASD.

Column number of string	C	
Row number of string	R	The string descriptor
String length	L	
Element R+1 of column C	E(R+1,C)	
Element R+2 of column C	E(R+2,C)	The element values for the string
Element R+L of column C	E(R+L,C)	

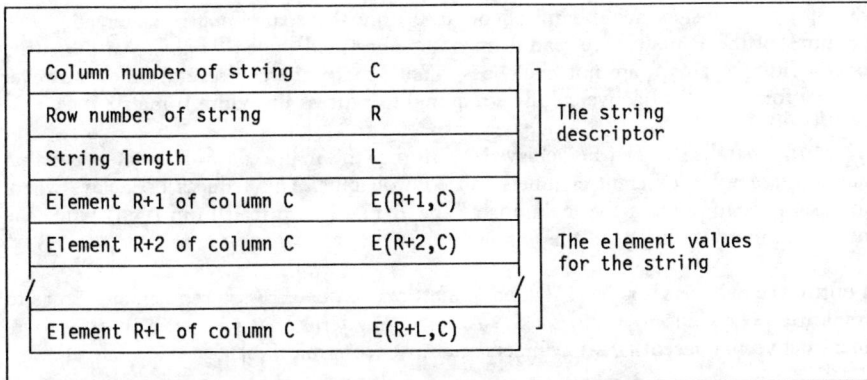

Figure 1. Column String for a Sparse Matrix

The algorithm used in computing the matrix product is indicated in Figure 2. The figure shows a string of the E matrix and S active columns of the B and D matrices. The E string contains the elements of column C starting at row $R + 1$ and ending at row $R + L$. Formulas for calculating the contribution of the E string to rows $R + 1$ through $R + L$ of the active columns of the D matrix are also shown. A vector loop for calculating contributions to column K of D multiplies $B(C,K)$ by each element of the E string. The products are added to the elements of column K of D. Final values appear in the active columns of the D matrix when this process has been followed for all of the strings in the E matrix.

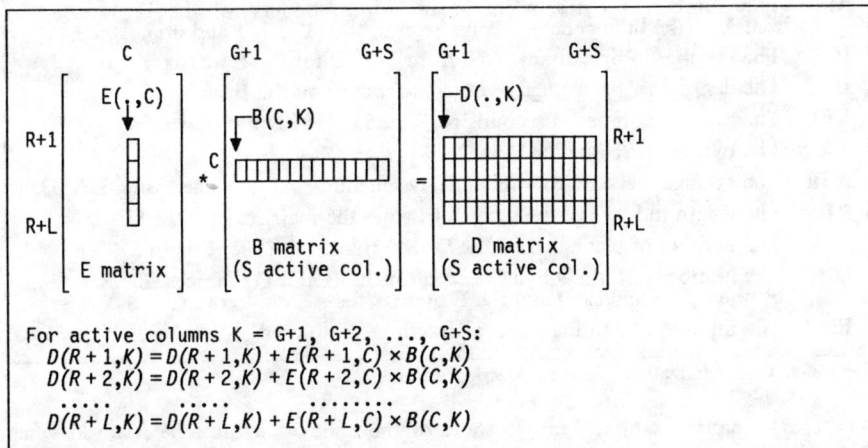

For active columns K = G+1, G+2, ..., G+S:
$$D(R+1,K) = D(R+1,K) + E(R+1,C) \times B(C,K)$$
$$D(R+2,K) = D(R+2,K) + E(R+2,C) \times B(C,K)$$
$$\cdots$$
$$D(R+L,K) = D(R+L,K) + E(R+L,C) \times B(C,K)$$

Figure 2. Sparse Matrix Multiplication Example

The matrix multiplication strategy has four characteristics worth noting. First, in a system with virtual storage and demand paging, the size of the buffers used by the program should be chosen with some care. When the buffer size exceeds the amount of real storage available to the job, paging delays will degrade the performance of the application. Thus, the buffer area set aside in virtual storage should remain within the working set parameters established by the installation policy on region size.

Second, I/O transfers are minimized because only the sparse matrix is reread. The columns of the B matrix are read from DASD once, and the columns of the D matrix, once written to DASD, are not refetched. The E matrix is read several times, however -- once for each set of active columns required to process the entire B matrix.

Third, I/O parallelism can be achieved by storing the matrices on different DASD that are accessed using different channels and control units. This reduces I/O delays when, for example, the next active columns of the B matrix are read from DASD while the newly-computed columns of the D matrix are written to DASD.

Fourth, the data sets for the B, D and E matrices can be accessed sequentially, so as to minimize delays due to DASD seeks. Accessing patterns that require, at most, movement between adjacent DASD cylinders are most economical [6].

$$RRF = (8 \times N^2)/(MSB/2.1) = (16.8 \times N^2)/MSB$$

$$IOB = (8 \times D \times N^2) \times RRF = (134 \times D \times N^4)/MSB$$

$$IOT = IOB \times XFR \quad where \quad XFR = 1/BTR + ASR/RL$$

$$DB = (8 \times 2 \times N^2) + (8 \times D \times N^2)$$

$$FPC = 2 \times D \times N^3$$

RRF The reread factor, or number of times the E matrix is read from DASD.

N The dimension of the (square) D, B and E matrices.

MSB The number of bytes in the main storage buffers for the B, D and E matrices (the buffers have relative sizes in the ratio 1, 1 and 0.1).

IOB The number of bytes transferred from DASD for the E matrix.

D The density, or percentage of non-zero values, of the E matrix.

XFR The average number of seconds per byte for DASD transfers.

TR The byte transfer rate for DASD.

ASR The average seek and rotational delays encountered when accessing DASD.

RL The length in bytes of the records to store the matrices on DASD.

IOT The number of seconds used for DASD transfers of the E matrix.

DB The number of DASD bytes needed for the B and D matrices (the first term on the right hand side) and the E matrix (the second term).

FPC The number of floating-point calculations performed.

Figure 3. Characterization of Sparse Matrix Example

The sparse matrix example exposes some of the issues that must be resolved when increasing the problem size. As a starting point for the analysis, Figure 3 estimates the number of floating point calculations made and the number of I/O bytes transferred during the matrix-multiplication process. Figure 4 shows the magnitudes associated with various problem sizes.

The number of floating-point calculations needed to form the product matrix grows with the third power of the problem size. Times for loop set-up, program loading and operating system services that make up the scalar and serial components of the execution time do not grow so rapidly. Thus, parallel and vector processing become increasingly urgent and effective as problem size grows.

Matrix size	Main Storage Buffer Size (MB)						
	17	50	100	150	200	250	
10000	99	34	17	12	9	7	**Table A**
30000	890	303	152	101	76	61	E-Matrix reread
50000	2741	840	420	280	210	168	factor (RRF)
70000	4843	1647	824	549	412	330	
10000	4.0	1.4	.68	.48	.36	.28	**Table B**
30000	320	109	55	36	27	22	I/O gigabytes
50000	2471	840	420	280	210	168	transferred
70000	9492	3228	1615	1076	808	647	for E matrix (@)
10000	1.2	.4	.2	.1	.1	.1	**Table C**
30000	96	33	16	11	8	7	I/O time for
50000	744	253	126	84	63	51	E matrix in
70000	2856	971	486	324	243	195	hours (IOT) (*)

Matrix size	Gigabytes in B or D matrix	Hours to transfer B or D matrix (*)	Gigabytes in E matrix	
10000	0.8	.24	.01	**Table D**
30000	7.2	2.2	.36	DASD loading
50000	20.0	6.0	1.00	factors
70000	39.2	11.8	1.96	

Matrix size	Billions of floating-pt. calculations	Approximate CPU hours for average MFLOPS = 20, 40, ... (#)							
		20	40	60	80	100	120	140	
10000	100	1.4	.7	.5	.3	.3	.2	.2	**Table E**
30000	2,700	38	19	12	9	8	6	5	CPU loading
50000	12,500	174	87	58	43	35	29	25	factors
70000	34,300	476	238	159	199	95	79	68	

@ 1 gigabytes = 10^9 bytes, 1 billion = 10^9.

* Transfer rate = 3 megabytes per second, record length = 32 KB, and average seek plus rotational delay = 4 milliseconds (sequential accesses) [6].

Millions of floating-point operations per second averaged by the program while CPU and vector instructions are being executed.

Figure 4. Magnitude Estimates for E-matrix Density of 5%

With matrix multiplication, the storage requirement grows with the square of the problem size. I/O overlay techniques are needed when the limits of real main storage are exceeded. Table D of Figure 4 gives examples of the amount of DASD needed as problem size increases.

The number of bytes transferred during the matrix multiplication process grows with the fourth power of the matrix dimension. This occurs because the E matrix is reread once for each set of active columns required to process the B matrix, while both the size of the E matrix and the reread factor grow with the square of the matrix dimension. In spite of the fact that E is sparse, its transfer times will come to dominate as problem size grows. Tables A and B of Figure 4 demonstrate, however, that large real storage is very effective in reducing the amount of data transferred.

Sources of Speedup

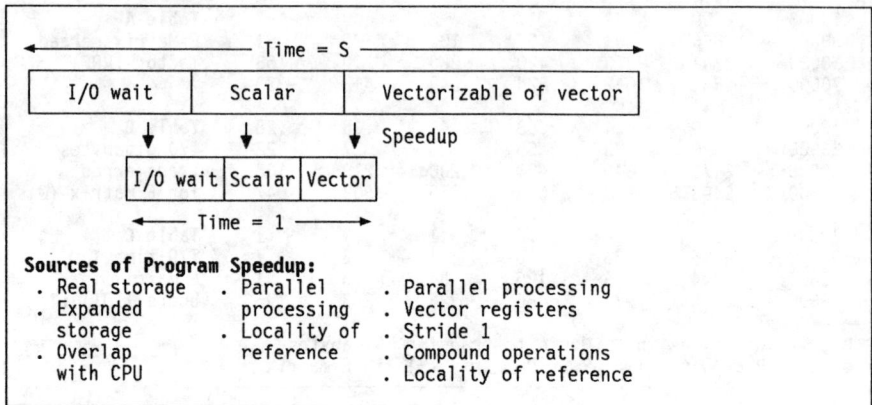

Figure 5. Job Execution Time and Job Speedup

Speedup is the ratio of the job execution times for an application that is run in two circumstances. Figure 5, for example, considers a single application in two runs made on the same machine. For the second run, program changes have been made to take advantage of the underlying architecture of the system. The first run takes S hours and the second takes 1 hour, so the speedup ratio is S.

Sources of speedup fall into categories corresponding to a division of the time used to run the job. The first segment, the *I/O wait time*, accumulates when program execution is suspended while data are exchanged with DASD. The *scalar time* accrues during the execution of CPU operations which cannot be performed using a vector facility. Scalar time includes the time needed for supervisor functions and loop set-up. Finally, the *vectorizable or vector time* accumulates during the execution of operations which can be performed using a vector facility. With systems that implement ESA/370, features such as large real storage, expanded storage, and overlap of I/O with CPU and vector computations can be used to reduce the I/O wait time. Vector and parallel processing, vector registers, compound vector operations and locality of reference can be used to shorten the scalar and vectorizable times.

ESA/370

ESA/370 defines the attributes of the system as seen by the programmer, that is the conceptual structure and functional behavior, as distinct from the organization of the data flow and controls, the logical design, and the physical implementation. ESA/370 is intended for a family of compatible computers offering a wide range of performance. The IBM 3090 provides one implementation, or model, of ESA/370. The 3090 Model 600E, for example, has six CPUs and vector facilities, with up to 256 million bytes of main storage, and up to 2,048 million bytes (2 gigabytes) of expanded storage.

The ESA/370 definitions capture everything that the programmer or compiler writer needs to know about the machines implementing it. Its instructions access data in registers, main storage and expanded storage. Execution of the instructions may at times yield undesired results -- arithmetic overflows, for example. The architecture defines the instructions, the addressing conventions and the exception conditions in a model-independent, or compatible, fashion.

Compatibility implies two things. First, compiled programs will obtain identical results when run on any model that implements the architecture -- old programs continue to run unchanged on newer models. Second, when valuable new features are identified, they are introduced in compatible fashion. Although the old programs continue to run, program modifications or recompiling may be necessary to exploit the new features. The intention of defining and evolving the architecture in this fashion is to allow continued use of application software with new families of machines.

It is useful to note that the architecture permits certain features to vary among models. The CPU data flow and machine cycle time, the size and organization of cache, the power of the vector pipelines, and the speed of the I/O channels are examples. The size of the main and expanded storage provided by a model can vary up to 2 gigabytes and 16 terabytes (16 TB, or 16,000 gigabytes) respectively. Variations in these areas are not of concern to programmers, at least from a functional point of view.

Parallel Processing

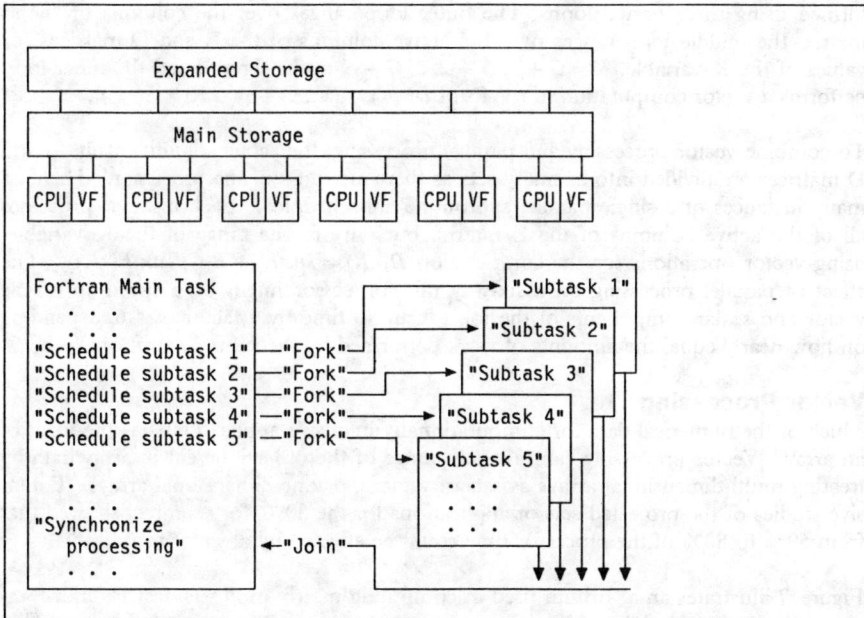

Figure 6. ESA/370 Multiprocessor and Fortran Multitasking

Figure 6 shows a multiprocessor that implements ESA/370. There are six CPUs, each of which has a vector facility installed. The CPUs share access to common main storage and expanded storage facilities, and to a common set of I/O channels, control units and I/O devices (not shown in the figure).

VS Fortran multitasking [1, 7] is available to programs which lend themselves to parallel processing. With multitasking, the job is broken up into separate tasks. In Figure 6, a Fortran job divides the work among six tasks. One will be run as the main task and five are subtasks. The "fork" and "join" operations are Fortran calls that allow the scheduling of the tasks on the six processors shown in the figure. When the tasks are finished, their results are consolidated into the main job.

Some programs may be readily suitable for parallel processing. In others, significant program restructuring may first be needed. The objectives [1] when converting an application program to parallel form should include the following:

(1) Keep the processors busy: Divide the program into subtasks that run simultaneously on different processors. Parallel subroutines should be designed to include the largest loops in terms of scope, for example, the outermost loop. They should schedule nearly equal amounts of work between synchronization points.

(2) Avoid data conflicts: Data modified by the main task or by a parallel subroutine may not be examined or modified by a parallel subroutine that might be executing simultaneously.

Sparse Matrix Example: A parallel form of the computation of Figure 2 can be performed using three nested loops. The outer loop ranges over the columns of the E matrix, the middle loop ranges over the active columns of the B and D matrices (or values of the K variable, $K = G + 1, G + 2, .., G + S$ in the figure), and the inner loop performs a vector computation of the form $D(.,K) = D(.,K) + B(C,K) \times E(.,C)$.

To combine vector processing and parallel processing, the active columns of the B and D matrices are divided into as many sets as there are CPUs in the processor. Then, as many instances of a single parallel subroutine are scheduled. Each instance processes all of the active columns of the E matrix, but only for its range of the K variable, using vector operations for the computation $D(.,K) = D(.,K) + B(C,K) \times E(.,C)$. The effect of parallel processing on a six-way multiprocessor might be a speed-up of the vector and scalar components of the job execution time by a factor of 4-6, depending on how nearly equal the amounts of work performed by the subtasks turn out to be.

Vector Processing
Much of the numerical data for computationally-intensive applications has the form of an array. Vector processing can take advantage of the order inherent in array data by treating multi-dimensional arrays as sets of vectors, or one-dimensional arrays. Extensive studies of the projected sets of applications for the 3090, for example, showed that from 50% to 80% of the processor time could be affected by a vector facility.

Figure 7 illustrates an algorithm used in computing matrix products, fast Fourier transformations, the solutions of linear equation systems, etc. The computation multiplies each element in a vector by a scalar value, adds the individual products to the elements of another vector, and places the sums in a third vector. A loop for performing the computation on an ESA/370 CPU is shown in the figure. Each iteration computes one result. The loop is sequential in nature in that each instruction is held up waiting for the result produced by the instruction preceding it. An equivalent computation can be accomplished more rapidly using a vector facility.

$$\begin{bmatrix} V1 \\ V2 \\ V3 \\ .. \\ Vn \end{bmatrix} = \begin{bmatrix} A1 \\ A2 \\ A3 \\ .. \\ An \end{bmatrix} *S + \begin{bmatrix} V1 \\ V2 \\ V3 \\ .. \\ Vn \end{bmatrix}$$

```
LD    Ai -> R2         CPU loop:
MDR   R2*FR -> R2      2 floating-point
AD    R2+Vi -> R2      calculations per
STD   R2 -> Vi         iteration
BXLE  Loop
```

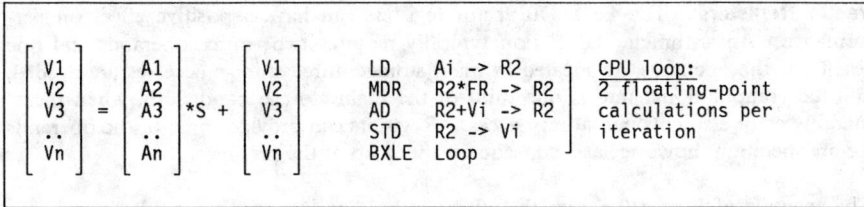

Figure 7. Vector Computations using Scalar CPU

With the ESA/370 vector architecture [5, 10, 11, 9], the scalar loop in Figure 7 is replaced by the vector loop in Figure 8. First, a VECTOR LOAD (VLD) instruction places 128 elements of the V vector in a vector register. With the 3090 vector facility, operands are shipped from main storage to the register at the rate of slightly under one per machine cycle. In a like manner, 128 results are copied from a vector register to main storage by a VECTOR STORE (VSTD) instruction, again at the rate of slightly under one per cycle for the 3090.

A vector pipeline does the computational work of the loop. The pipeline is organized like an assembly line. Operations are divided into subtasks that are executed in specialized hardware stages. Operands flow from stage to stage, and the various stages process different operands concurrently. In the figure, again using the 3090 as an example, the pipeline produces 128 products and 128 sums at the rate of about one product and sum per machine cycle. Pipeline operations are initiated by a VECTOR MULTIPLY AND ADD (VMADS) instruction. Eight individual calculations may be in process at the same time, but none holds up any of the others.

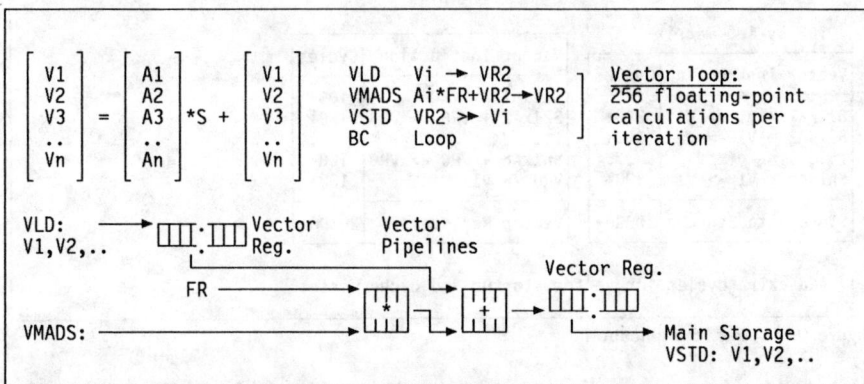

Figure 8. Vector Computations using Vector Facility

Compound Operations: ESA/370 makes available three compound vector operations that combine into one instruction the most common sequences in vector computation -- multiplication followed by addition, subtraction or summation. The VECTOR MULTIPLY AND ADD instruction, for example, fetches a vector from storage, multiplies it by a value from a register, and adds the products to another vector from a register. Compound instructions normally achieve about twice the computational rate reached for single-operation instructions such as VECTOR ADD or VECTOR MULTIPLY.

Vector Registers: The ESA/370 vector registers can have a positive effect on performance. An arithmetic calculation typically requires two source operands and one result. If the operands are located in main storage, three storage accesses are needed, and the computational rate is one third of the main storage bandwidth, when measured in vector elements per storage access. Registers can provide some of the operands for an operation, however, at a consequent speed up of the program.

The example of Figure 9 shows the effects of holding intermediate results in a vector register. There, the product of matrices A and S is placed in matrix V, where this time all of the matrices fit into main storage. The algorithm uses a sequence of N vector computations to calculate one column of the result. With the "storage to storage" algorithm, all operands are found in main storage. With the "vector register" algorithm, the intermediate results are kept in a vector register (VR 0). 100-by-100 matrices are used for the cycle comparisons at the bottom of the figure. Overhead cycles must be added to the totals given, however.

Liu and Strother [12] give examples showing the effectiveness of the IBM VS Fortran compiler in reusing values kept in the vector registers. The performance of the Fortran code approaches what can be achieved using assembly language routines.

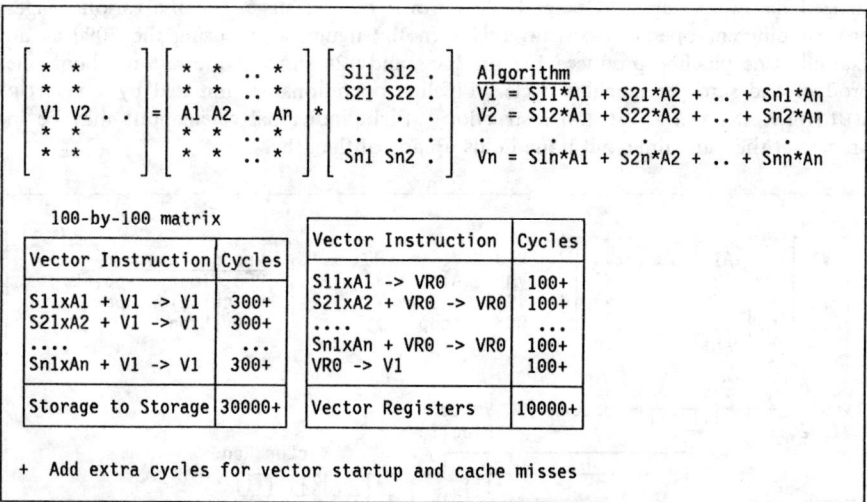

$$
\begin{bmatrix} \star & \star \\ \star & \star \\ V1 & V2 & .. \\ \star & \star \\ \star & \star \end{bmatrix} = \begin{bmatrix} \star & \star & .. & \star \\ \star & \star & .. & \star \\ A1 & A2 & .. & An \\ \star & \star & .. & \star \\ \star & \star & .. & \star \end{bmatrix} * \begin{bmatrix} S11 & S12 & . \\ S21 & S22 & . \\ . & . & . \\ . & . & . \\ Sn1 & Sn2 & . \end{bmatrix}
$$

Algorithm
V1 = S11*A1 + S21*A2 + .. + Sn1*An
V2 = S12*A1 + S22*A2 + .. + Sn2*An
Vn = S1n*A1 + S2n*A2 + .. + Snn*An

100-by-100 matrix

Vector Instruction	Cycles
S11xA1 + V1 -> V1	300+
S21xA2 + V1 -> V1	300+
....	...
Sn1xAn + V1 -> V1	300+
Storage to Storage	30000+

Vector Instruction	Cycles
S11xA1 -> VR0	100+
S21xA2 + VR0 -> VR0	100+
....	...
Sn1xAn + VR0 -> VR0	100+
VR0 -> V1	100+
Vector Registers	10000+

+ Add extra cycles for vector startup and cache misses

Figure 9. Matrix Multiplication

Cache: Main-storage accessing patterns must be considered when designing the algorithms used in an application, especially in machines with cache. The cache is a high-speed buffer that holds recently accessed information from main storage. Information fetched from the cache can be reused without access to main storage, and by loading entire "lines" of consecutive elements on any request for storage operands, the machine prefetches information for future use. Except for the effect on performance, the existence of the cache is not apparent to the program.

The elements of a vector in storage may be contiguous, or they may be separated by several element positions. The *stride* is the offset between element positions. When a stride value of 1 is used, the vector elements are found in contiguous storage locations.

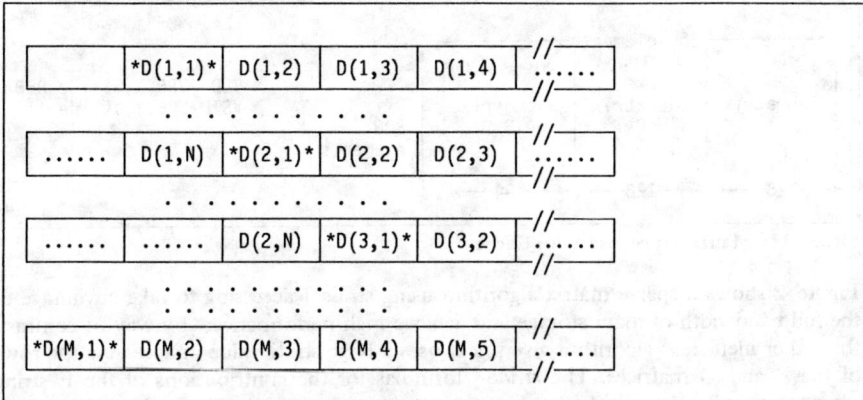

Figure 10. Cache Lines for Row 1 of a Matrix

When a different stride value is used, stride j in what follows, the elements are separated from each other.

As a specific example, an $N \times M$ Fortran array is stored so that the column elements are in consecutive storage locations. When a row of the array is accessed, the elements are not located consecutively in storage. Figure 10 shows elements $D(1,1), D(2,1), D(3,1), ...$, which form the first row of the matrix. The elements are accessed with a stride value of N. Access to each element of row one causes the cache line containing that element to be loaded. The rectangles in the figure show the set of elements that might be included in the cache lines loaded by accessing the elements of row 1. A subsequent access to the corresponding elements in row 2 of the matrix could use elements in the cache and thus would not encounter the delay associated with storage accesses.

An example illustrates nicely the value of *locality of reference* in systems with cache. When it is necessary to transpose a large matrix, restricting the processing to submatrices is beneficial. In the matrix of Figure 11, a straightforward algorithm that processes one entire row of the source matrix before going on to the next row causes the loading of 800 lines into cache before any of them is reaccessed. This is more information than fits into the cache, so lines are replaced before they are reaccessed and there is a cache miss for every element.

It is much more efficient to work, for example, with 800-by-128 element submatrices, as shown in the figure. A VECTOR LOAD (VLD) instruction with a stride of 800 collects 128 elements for a column of the transposed submatrix and a VECTOR STORE (VSTD) instruction with a stride of 1 stores them. A loop transposes the rows of the submatrix. The VLD instructions find most of the elements in the cache because only 128 lines are loaded before reaccessing begins.

Liu and Strother [12] describe cache reuse and page reuse techniques that also rely on locality or reference.

Sparse Matrix Example: Stride-j algorithms can, in many cases, be replaced by stride-1 algorithms that provide equivalent computations. When this is done, considerable speed-up can be achieved, especially in systems with cache.

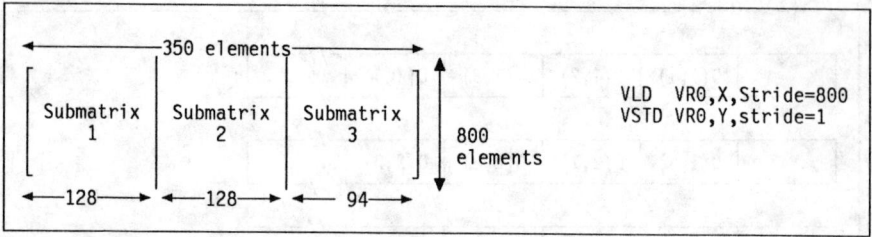

Figure 11. Matrix Transpose With Cache

Figure 2 shows a sparse matrix algorithm using stride 1 accessing to take advantage of the full bandwidth of main storage and achieve high performance. By way of contrast, the rather inefficient algorithm given next uses a large stride value in accessing the rows of the B and D matrices. The stride-j formulas for the contributions of the E string rows to rows $R + 1$ through $R + L$ of the active columns of the D matrix are:

```
For rows M = R+1, R+2, ..., R+L OF E(.,C):
```
$$D(M,G + 1) = D(M,G + 1) + E(M,C) \times B(C,G + 1)$$
$$D(M,G + 2) = D(M,G + 2) + E(M,C) \times B(C,G + 2)$$
$$..... \qquad ..... \qquad ....$$
$$D(M,G + S) = D(M,G + S) + E(M,C) \times B(C,G + S)$$

A vector loop for calculating the values for row M of D uses $E(M,C)$ as a constant that is multiplied by each element of the active columns of row C of the B matrix. The products are added to the elements of row M of the active columns of D and the results are stored back into that row. This loop uses stride-j accessing of the B and D matrices. It is executed for each of the rows $R + 1, R + 2, .., R + L$ in turn.

In a system with cache, accessing the rows of 1000 active columns of D causes the loading of as many lines into cache. Lines are usually cast out before they are reaccessed when accessing the next row. The result is a substantial performance degradation when the stride-j algorithm is compared to the stride-1 algorithm.

Processor Storage and DASD

Two types of storage are configured to an ESA/370 system -- processor storage and DASD. Retrieval time from processor storage is measured in microseconds. DASD retrieval times are measured in milliseconds [6], although data transfers may be accomplished asynchronously -- that is, concurrently with the processing of CPU programs.

There are two kinds of processor storage: real main storage and expanded storage. A 4 KB transfer between main and expanded storage takes about 75 microseconds on the 3090, including operating system overhead. The rate is about 54 megabytes per second, a substantial improvement on the DASD rate of 2-3 megabytes per second [6].

MVS [4] provides services that allow applications to place data in high-performance spaces (hiperspaces [TM]) in expanded storage. Hiperspaces are referenced from Fortran using data windowing services. Figure 12 illustrates some services that are available:

TM - Hiperspace is a trademark of the International Business Machines Corporation.

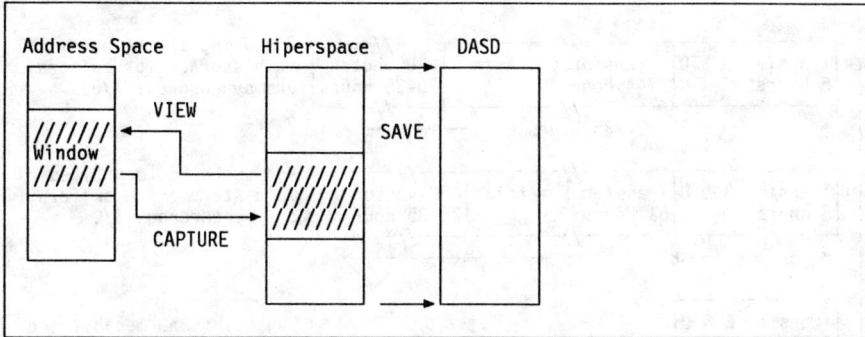

Figure 12. Hiperspace Model

View: Causes MVS to establish a view of the object in a virtual storage window. MVS sets up control information that relates pages in virtual storage with blocks in expanded storage. No data is transferred at this time. Data transfer occurs when the application program references the individual virtual pages in the window.

Capture: Causes MVS to capture the current view in the window for subsequent reviews or saves. MVS moves changed pages in the window to expanded storage.

Save: Causes MVS to move the information in expanded storage to DASD. Saves are made only for permanent data objects which are backed by DASD.

Sparse Matrix Example: Table C of Figure 4 indicates that a speedup of about 14 is obtained in the time required to read the E matrix when the main storage buffer size is increased from 17 to 250 megabytes. The amount of real storage made available to the program may vary from run to run however. For example, installation policy may restrict the amount of real storage that may be used during development runs, but allow the use of all available storage for production runs when only one job is active. Thus, a good strategy is to design the program so that it uses as much real main storage as is found to be available at run time.

After all available real storage has been used, one method of further reducing the execution time might be to use asynchronous I/O procedures in reading the columns of the E matrix into main storage. The Fortran READ statement - asynchronous [1] transmits unformatted data from DASD. The statement is asynchronous because other program statements are executed while data transfer is taking place. By using two input buffers, computations using the columns in one of them may proceed while the other is being filled. This overlap of I/O and processor operations eliminates some of the delays associated with I/O waits.

Another technique for reducing the I/O component of job time is to keep part of the E matrix in expanded storage. Here, a combination of asynchronous accessing of parts of the E matrix kept on DASD and synchronous accessing of parts in expanded storage then may be used to advantage.

A specific example is shown in Figure 13, using estimates taken from Figure 4 for a matrix dimension of 50,000. Six hours are needed to transfer the B matrix from DASD to main storage, and to transfer the D matrix from main storage to DASD, assuming

Transfer B & D / Transfer E matrix / CPU-vector — Run using 17 megabytes of
6 hours — 744 hours — 25-35 hours — main storage for buffers,
— and synchronous I/O
← 775-785 hours →

Transfer B & D / Transfer E matrix / CPU-vector — Run using 250 megabytes of
6 hours — 51 hours — 25-35 hours — main storage for buffers,
— and synchronous I/O
← 82-92 hours →

Transfer B & D — Run using 250 megabytes of
6 hours — main storage for buffers,
— and asynchronous I/O
CPU-vector
25-35 hours
Transfer E matrix
51 hours
← 51 hours →

Transfer B & D — Run using 250 megabytes of
6 hours — main storage for buffers,
— asynchronous I/O, and
CPU-vector — expanded storage
25-35 hours
Transfer E matrix
26-36 hours
← 26-36 hours →

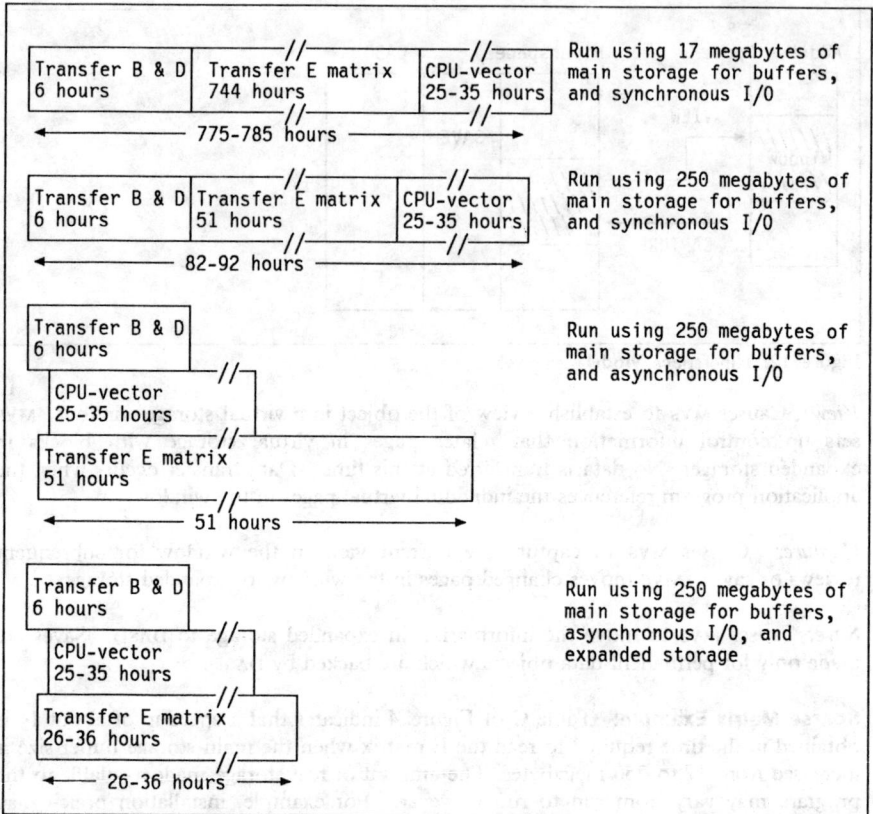

Figure 13. Hypothetical Runs Showing Reduction of I/O Delays

the transfers occur at the same time. 51-744 hours pass while reading the E matrix from DASD, depending on the size of the main storage buffers, and 25-35 hours are used for processor instructions, depending on the average MFLOPS rate achieved. A good design point might therefore be to keep half of the E matrix on DASD and the other half in expanded storage. This allows the overlapping with processor instructions of half of the 51 hours used to transfer the B matrix, while reducing the other half of the transfer time by a factor of 15 or more.

Summary

There are many ways of improving the performance of engineering and scientific programs. General speedup techniques are available with systems using ESA/370. The techniques include the following:

- I/O avoidance using large real storage and expanded storage
- I/O overlap with CPU and vector computations
- Parallel processing and vector processing
- Use of stride 1 algorithms
- Vector register exploitation
- Use of compound vector operations
- Locality of reference

Programs tuned for high performance using these techniques will obtain identical results when executed on newer, more powerful members of the ESA/370 line. Thus, the investment in improved application software is preserved.

References

1. *Designing and Writing Fortran Programs for Vector and Parallel Processing,* order no. SC23-0337, pp. 77-87, and *VS Fortran Version 2: Programming Guide, Release 3,* order no. SC26-4222-3, pp. 349-372, IBM Corp., San Jose, CA.

2. J. L. Gustafson, "Reevaluating Amdahl's Law," *Communications of the ACM,* Vol. 31, No. 5, pp. 532-533, May 1988.

3. B. L. Buzbee and D. H. Sharp, "Perspectives on Supercomputing," *Science,* vol. 227, pp. 591-597, Feb. 8, 1985.

4. *Programming Announcement, ESA/370 and MVS SP Version 3,* No. 288-059, IBM Corp., Rye Brook, NY, Feb. 15, 1988.

5. *IBM System/370 Vector Operations,* order no. SA22-7125, available through IBM branch offices.

6. *System Reference Library - IBM 3380 Direct Access Storage: Reference Summary,* order no. GX26-1678-4, Oct., 1987, IBM Corp., San Jose, CA. Seek times, rotational delays and transfer rates.

7. D. H. Gibson, D. W. Rain and H. F. Walsh, "Scientific and engineering processing on the IBM with 370/XA vector architecture," *IBM Systems Journal,* Vol. 25, No. 1, pp. 36-50, June, 1986.

8. *IBM 3090 Engineering/Scientific Performance,* IBM Washington Sys. Cent. Tech. Bull., order no. GG66-0245, available through IBM branch offices.

9. A. Padegs, B. B. Moore, R. M. Smith and W. Buchholz, "The IBM System/370 Vector Architecture: Design Considerations," *IEEE Transactions on Computers,* Vol. 37, No. 5, pp. 509-520, May, 1988.

10. B. B. Moore, A. Padegs, R. M. Smith and W. Buchholz, "Concepts of the System/370 Vector Architecture," *Proc. 14th International Symposium on Computer Architecture,* pp. 282-289, June, 1987.

11. W. Buchholz, "The IBM System/370 Vector Architecture," *IBM Systems Journal,* Vol. 25, No. 1, pp. 51-62, June, 1986.

12. B. Liu and N. Strother, "Programming in VS Fortran on the IBM 3090 for Maximum Vector Performance," *Computer,* Vol. 23, No. 6, pp. 65-76, June, 1988.

Fundamentals of the Boundary Element Method

C. A. Brebbia

Computational Mechanics Institute, Southampton, U.K.

1 Motivation for Boundary Elements

The origins of boundary elements started in the early 1970s as a result of the dissatisfaction of many researchers with some finite elements capabilities. In particular finite element codes were shown to be cumbersome to use in three dimensional problems and produced inaccurate results in many cases. During the early 1970s researchers at Southampton University worked on the development of boundary elements culminating in the first international conference on the technique [1] and the first book published on the method [2], both of which took place in 1978. Since then, another nine conferences have taken place on the analytical aspects of the method [2-11] and a series of three conferences have dealt with the applications in engineering [12-14].

The 1980s saw the consolidation of the method and the development of further engineering applications. Of particular interest were a series of state of the art books on boundary elements which followed the one published by Brebbia in 1978 [1]. As the field was rapidly advancing there was a need for a more definite book such as the one published in 1984 and written by Brebbia, Telles and Wrobel [15]. Other books on the topic of boundary elements are now available; for a complete list see reference [16]. State of the art applications started to be reported in a two volume series [17-18], but are now published in the Topics in Boundary Element Research books, six of which have been published so far [19-24].

It is now generally accepted that boundary elements are easier to implement in Computer Aided Engineering Systems and this constitutes a very important advantage in engineering practice. Finite Elements by contrast involve a still comparatively cumbersome and slow process, due to the need to define or redefine meshes in the domain under study, or the piece being designed. As the boundary element method requires only the discretization of the surface rather than the volume, BE codes are easy to use with existing solid modellers and mesh generators. This advantage is particularly important in designing, as the process usually involves a series of modifications which are more difficult to carry out using finite elements. Meshes can easily be generated and design changes do not require a complete remeshing.

Boundary element nodes, especially three dimensional ones, can easily be linked to CAE systems as the structure is defined using only the boundary. The discretization process is even simpler when using discontinuous elements of the type proposed by the

Computational Mechanics Institute group at Southampton [15, 25], and which are not admissible in finite elements. The mesh shown in Figure 1 represents the surface discretization of one eighth of a problem, i.e. a cylinder with a cylindrical perforation across. Notice that the use of elements which sometimes do not meet at corners and are consequently discontinuous in terms of their variables facilitate the meshing. In addition there is no need to use elements on the planes of symmetry.

Figure 1* is a rather academic case, but more complex three dimensional structures such as part of the piston shown in Figure 2*can be discretized relatively easily using a combination of continuous and discontinuous elements. Figure 3*describes the mesh used to analyse part of a crankshaft and Figure 4*the discretization on the surface of an aircraft component (part of the mechanism of the rollers used for the aircraft flaps). It is evident from these examples that boundary elements are an ideal tool for CAD mainly because it is easy to generate the data required to run a problem and to carry out the modifications needed to achieve an optimum design. While computing costs continue to decline savings in engineering time are becoming more important and hence the advantages of boundary elements over finite elements are more marked. Boundary elements can significantly shorten the "turn around" time taken by the analysis and design and can bring forward the completion date of a project.

Another important advantage of boundary elements over finite elements is their accuracy. This is particularly important when analysing problems with stress (or other) concentrations. Many studies in recent years have pointed out the poor accuracy of finite elements in such cases, the most notable of them have appeared in the reports published by NAFEMS (National Association for Finite Element Standards, UK, [26]. In addition many other studies have now been carried out and they tend to demonstrate the high accuracy of boundary elements for problems such as re-entry corners, stress concentration regions and even fracture mechanics applications. Studies of re-entry corners in particular [27-29] show that excellent results can be obtained with boundary elements using a very small number of elements, while the same degree of accuracy was only found using refined finite element meshes.

The development of more powerful hardware, especially supercomputers, favours the use of boundary elements. These computers are better suited to dealing with fully populated matrices and with the type of 'global' operations which are characteristic of boundary elements. Large problems can now be solved in a comparatively short time using supercomputers in a very efficient manner.

Problems other than stress or temperature analysis can be solved using boundary elements. Other typical applications include torsion, diffusion, seepage, fluid flow and electrical problems. Corrosion engineers for instance now use boundary elements to design better cathodic protection systems for offshore structures, ships and pipelines. Many of these, in common with other problems, are three dimensional and the region of interest extends to infinity. Early attempts to use finite difference or finite elements to solve these problems met with little success, as the region of interest was precisely at the interface between the structure and the medium - in this case the water surrounding the structure. Therefore use of finite elements to analyse the problem would require the subdivision of the medium surrounding the structure. The use of boundary elements instead represents the only practical solution for this type of problem.

The future of BEM hinges on its acceptance by practising engineers in particular as a design tool. Developers should aim to make the method more accessible to engineers by writing codes which are easy to use and take full advantage of the substantial recent

*See Figures 1 and 2 in colour section, page 160.

*See Figures 3 and 4 in colour section, page 161.

improvements in hardware, particularly with the advent of vector and parallel processors.

2 Potential Problems

Since 1978 many works have appeared in the literature, some dealing with potential problems such as seepage, electrostatics, electromagnetics and many other engineering applications. The importance of 1978 is that on that date the method started to become well established, although the name in the context of the now classical BEM appears to have been used for the first time in two papers by Brebbia and Dominguez and dated 1977 [30,31]. Up to that time boundary integral equations were almost exclusively the domain of mathematicians and physicists with very little work being done to apply them to realistic engineering problems.

Boundary elements are nowadays usually associated with the direct boundary integral formulation which, in the context of potential problems, can be traced to Jaswon [32] and Symm [33]. They proposed as early as 1963 to solve Fredholm boundary integral equations by discretizing the boundary into a series of small segments assuming a constant source density within each segment. They employed collocation to obtain the governing systems of equations and computed the influence coefficients using numerical techniques. They even proposed a more general formulation through the application of Green's identity with potentials and derivatives as boundary unknowns and results for this formulation were presented in [33] and [34]. All the elements of BEM were there, but somehow their work failed to attract the attention it deserved, probably due to the simultaneous emergence of the finite element method.

Since 1978 the boundary element method is seen as related to other numerical techniques such as finite elements and finite differences, mainly through the work of Brebbia and his collaborators. This relationship is sometimes highlighted by using weighted residuals of the type used in this chapter [30].

Basic Integral Equation

The starting boundary integral equation required by the method can be deduced in a simple way based on weighted residuals. Consider that one is trying to find the solution of a Laplace equation in a domain, such that:

$$\nabla^2 u = 0 \qquad \text{in } \Omega \tag{1}$$

where u is the potential. Equation (1) needs to be solved in conjunction with the appropriate conditions on the Γ boundary. These conditions will be assumed to be of the following two types for simplicity.

$$i) \quad \text{'Essential' conditions of the type } u = \overline{u} \quad \text{on } \Gamma_1 \tag{2}$$

$$ii) \quad \text{'Natural' conditions such as } q = \frac{\partial u}{\partial n} = \overline{q} \quad \text{on } \Gamma_2$$

where n is the normal to the boundary, $\Gamma = \Gamma_1 + \Gamma_2$, the total boundary, and the dashes indicate that the corresponding values are known.

Satisfaction of (1) and (2) in Ω and on Γ implies that the following equation is valid [30],

$$\int_\Omega (\nabla^2 u) u^* \, d\Omega = -\int_{\Gamma_1} (u - \overline{u}) q^* \, d\Gamma + \int_{\Gamma_2} (q - \overline{q}) u^* \, d\Gamma \tag{3}$$

where u^* is a weighting function and q^* is its derivative with respect to the normal, i.e. $q^* = \partial u^* / \partial n$.

Integrating equation (3) by parts twice produces the following relationship:

$$\int_\Omega (\nabla^2 u^*)\, u\, d\Omega = -\int_{\Gamma_2} \bar{q} u^*\, d\Gamma - \int_{\Gamma_1} q\, u^*\, d\Gamma + \int_{\Gamma_2} u\, q^*\, d\Gamma + \int_{\Gamma_1} \bar{u}\, q^*\, d\Gamma \quad (4)$$

Fundamental Solution

Our aim is to render equation (4) into a boundary integral equation. This is done by using a special type of weighting function u^* called the fundamental solution, which satisfies Laplace equations and represents the field generated by a concentrated unit charge acting at a point 'i'. The effect of this change is propagated from i to infinity without considering any boundary conditions. Because of this the solution can be written as:

$$\nabla^2 u^* + \Delta^i = 0 \quad (5)$$

where Δ^i represents a Dirac delta function which tends to infinity at the point 'i' and is equal to zero anywhere else. The integral of Δ^i is equal to one. The integral of a Dirac delta function multiplied by any other function is equal to the value of the latter at the point 'i'. Hence:

$$\int_\Omega u(\nabla^2 u^*)d\Omega = \int_\Omega u(-\Delta^i)d\Omega = -u^i \quad (6)$$

Equation (4) can now be written as:

$$u^i + \int_{\Gamma_2} u\, q^*\, d\Gamma + \int_{\Gamma_1} \bar{u}\, q^*\, d\Gamma = \int_{\Gamma_2} \bar{q}\, u^*\, d\Gamma + \int_{\Gamma_1} q\, u^*\, d\Gamma \quad (7)$$

It needs to be remembered that equation (7) applies for a concentrated charge at 'i' and consequently the values of u^* and q^* are those corresponding to that particular position of the charge. For any other position 'i' one will obtain a new integral equation similar to (7).

For an isotropic three dimensional medium the fundamental solution of (5) is:

$$u^* = \frac{1}{4\pi r} \quad (8)$$

and for a two dimensional isotropic domain, it is

$$u^* = \frac{1}{2\pi} \ln\left(\frac{1}{r}\right) \quad (9)$$

where r is the distance from the point 'i' of application of the Delta function to any other point under consideration.

Equation (7) is valid for any point inside the Ω domain. When the point is taken to the boundary - as it is usually done in boundary elements to obtain the algebraic system of equations - one needs to consider the behaviour of the singularity for this limiting case. This procedure has been discussed fully in [2, 15] and [35] amongst others. Here we will only show the final result. Consider for simplicity equation (7) before any boundary conditions have been applied, i.e.

$$u^i + \int_\Gamma u\, q^* \, d\Gamma = \int_\Gamma q\, u^* \, d\Gamma \tag{10}$$

where now $\Gamma = \Gamma_1 + \Gamma_2$. If the point '$i$' is on the Γ boundary one needs to find the limits of integrals containing the q^* function. This results in the following expression

$$c^i u^i + \int_\Gamma u\, q^* \, d\Gamma = \int_\Gamma q\, u^* \, d\Gamma \tag{11}$$

where the integrals are now interpreted in the sense of Cauchy Principal Value. The c^i coefficient is $\frac{1}{2}$ for a smooth body, or related to the solid angle for any other cases. For an internal point 'i' the value of c^i is equal to one as in equation (10) and for a point 'i' external to the Ω domain, it becomes zero.

The Boundary Element Method

Next one needs to consider how equation (11) can be discretized to to find the algebraic system of equations from which the boundary values can be found. Assume for simplicity that the body is two dimensional and its boundary is divided into N segments or elements as shown in Figure 5. The points where the values of u or q are considered are called nodes and taken to be at the centre of the element for the 'constant' type element (Figure 5a). These are going to be the elements considered in this chapter for simplicity, the reader interested in linear, quadratic or higher order elements is referred to [35].

For the constant elements considered here the boundary is divided into N elements over each of which the values of u and q are taken to be constant. Equation (11) can be discretized for a given point 'i' before applying any boundary conditions as follows,

$$c^i u^i + \sum_{i=1}^{N} \int_{\Gamma_j} u\, q^* \, d\Gamma = \sum_{j=1}^{N} \int_{\Gamma_j} u^*\, q \, d\Gamma \tag{12}$$

(where $c^i = \frac{1}{2}$ in this case as the boundary is always smooth for the node at the centre of the element) Γ_j is the boundary of the 'j' element.

The u and q values can be taken out of the integrals in this case as they are constant over each element. They can be called u^j and q^j for element 'j'. This gives

$$\frac{1}{2}u^i + \sum_{j=1}^{N} \left(\int_{\Gamma_j} q^* \, d\Gamma \right) u^j = \sum_{j=1}^{N} \left(\int_{\Gamma_j} u^* \, d\Gamma \right) q^j \tag{13}$$

Notice that there are now two types of integrals to be carried out over the elements, i.e. those of the following types

$$\int_{\Gamma_j} q^* \, d\Gamma \quad \text{and} \quad \int_{\Gamma_j} u^* \, d\Gamma \tag{14}$$

These integrals relate the 'i' node where the fundamental solution is acting to any other 'j' node. Because of this their resulting values are sometimes called 'influence' coefficients. They will be denoted as follows

$$\hat{H}^{ij} = \int_{\Gamma_j} q^* \, d\Gamma \quad ; \quad G^{ij} = \int_{\Gamma_j} u^* \, d\Gamma \tag{15}$$

Notice that we are assuming throughout that the fundamental solution has been applied at a particular 'i' node, although this is not explicitly indicated in u^* and q^* to avoid proliferation of indices. Hence for a particular 'i' one can write,

$$\frac{1}{2} u^i + \sum_{j=1}^{N} \hat{H}^{ij} \, u^j = \sum_{j=1}^{N} G^{ij} \, q^j \tag{16}$$

If one now assumes that the position of 'i' also varies from 1 to N, i.e. one assumes that the fundamental solution is applied at each node successively, one obtains a system of equations from applying (16) to each boundary node in turn.

Let us now define

$$H^{ij} = \begin{cases} \hat{H}^{ij} & \text{when } i \neq j \\ \hat{H}^{ij} + \frac{1}{2} & \text{when } i = j \end{cases} \tag{17}$$

Hence equation (16) can now be written as,

$$\sum_{j=1}^{N} H^{ij} \, u^j = \sum_{j=1}^{N} G^{ij} \, q^j \tag{18}$$

This set of equations can be expressed in matrix form as

$$\mathbf{H U} = \mathbf{G Q} \tag{19}$$

where $\mathbf{H}$ and $\mathbf{G}$ are two $N \times N$ matrices and $\mathbf{U}, \mathbf{Q}$ are vectors of known length N.

Notice that N_1 values of u and N_2 values of q are known on Γ_1 and Γ_2 respectively ($N = N_1 + N_2$), hence there are only N unknowns in the system of equations (19). To introduce these boundary conditions into (19) one has to rearrange the system by moving columns of **H** and **G** from one side to the other. Once all unknowns have been passed to the left-hand side one can write

$$\mathbf{AX} = \mathbf{F} \qquad (20)$$

where **X** is a vector of unknown u's and q's boundary values. **F** is found by multiplying the corresponding columns by the known values of u's and q's. It is interesting to point out that the unknowns are now a mixture of the potential and its derivative, rather than the potential only as in finite elements. This is a consequence of boundary elements being a 'mixed' formulation and gives an important advantage to the method over finite elements, as both unknowns are given with the same degree of accuracy.

Formula (20) represents a non-symmetric and fully populated system of equations which can be very conveniently solved using supercomputers. Notice that the function of the system of equations can also be done in parallel as the equation for each point 'i' can be computed independently of the others.

Once the system (20) is solved and all boundary values are known, one can find the values of potential and its derivatives at any internal point. This can also be done conveniently in supercomputers as the operation for any internal point is independent of the other internal points.

Higher Order Elements

Constant elements are seldom used in engineering practice and most commercial codes will have higher order elements, not only linear but also quadratic as the latter can follow more closely the geometry of the surface or volume to be represented (Figure 5c).

For three dimensional problems in can use surface elements which can be triangular or quadrilateral and also of different orders. Well written boundary element codes offer a selection of elements including discontinuous as well as continuous elements (Figure 6).

3 Elastostatics

In this section we will generate the boundary integrals for elastostatics using Kelvin's fundamental solution for an infinite space. Then we will discretize the boundary into elements and create the algebraic system of equations following the same idea as described earlier. Problems related to body forces, in-homogeneities and anisotropy will not be discussed here. For these problems the reader is referred to [35]. Non-linear and time dependent problems are studied in [15] as well as in the series of books from [17] to [24]. The Proceedings of the Boundary Element Conferences [1] to [14] regularly report further research work into new advances and applications of BEM.

Elastostatics problems are more complex than potential cases, because we are now working with vector quantities (i.e. displacements or tractions vectors) rather than a scalar quantity u or q as was the case in Section 2. Because of this we will use from now on the indicial notation which takes into consideration the components of the different functions.

The governing equilibrium equation for elastostatics can be written as

$$\sigma_{ij,j} + b_i = 0 \qquad \text{in } \Omega \tag{21}$$

where i and j vary from 1 to 3 for three dimensional cases and from 1 to 2 for two dimensional problems. The comma indicates derivatives with respect to x_j.

The stress components on the surface Γ can be projected to produce a set of surface forces or tractions such that

$$p_i = \sigma_{ij}\, n_j \qquad \text{on } \Gamma \tag{22}$$

where n_1, n_2 and n_3 are the direction cosines of the outward normal n with respect to the x_1 x_2 x_3 axis, i.e.

$$n_i = \cos(n, x_i) \tag{23}$$

The applied tractions denoted as $\bar{p}_i$ have to be in equilibrium with the traction components obtained from the internal stresses at the boundary, i.e.

$$p_i = \sigma_{ij}\, n_j = \bar{p} \qquad \text{on } \Gamma_2 \tag{24}$$

Γ_2: part of the boundary value where the $\bar{p}_i$ or 'natural' conditions are applied.

The strains are defined in terms of displacements as

$$\varepsilon_{ij} = \frac{1}{2}(u_{i,j} + u_{j,i}) \tag{25}$$

The corresponding boundary conditions are given in terms of the displacement components on the part Γ_1 of the boundary, i.e.

$$u_i = \bar{u}_i \qquad \text{on } \Gamma_1 \tag{26}$$

Finally the state of stresses and strains are related to each other through the constitutive equations for the linearly elastic material. For the isotropic case one can write,

$$\sigma_{ij} = \lambda\, \delta_{ij}\, \varepsilon_{kk} + 2\mu\, \varepsilon_{ij} \tag{27}$$

where λ and μ are the Lame constants for the volumetric and shear components of strains. ε_{kk} is the volumetric strain and $\triangle_{ij}$ is the Kronecker delta ($\equiv 1$ for $i = j; \equiv 0$ for $i \neq j$).

Fundamental Solution

The fundamental solution for this case corresponds to the solution of the elastic problem with the same material properties as the boundary under consideration but corresponding to an infinite domain with a concentrated unit point load. This is called the Kelvin solution and represented by an asterisk, such that

$$\sigma^*_{ij,j} = \Delta^i e_\ell = 0 \tag{28}$$

where Δ^i represents the Dirac delta function at the point 'i' of application of the load and e_ℓ is a unit vector in the direction 'ℓ' in which the unit load is acting. The components of traction and displacements for the fundamental solution are illustrated in graphical form in Figure 7.

Boundary Integral Equation

Consider now the weighted residual statement corresponding to equations (21) with boundary conditions (24) and (26), using the fundamental solution as the distribution function. This gives [35],

$$\int_\Omega (\sigma_{kj,j} + b_k) u^*_k \, d\Omega = \int_{\Gamma_2} (p_k - \overline{p}_k) u^*_k d\Gamma + \int_{\Gamma_1} (\overline{u}_k - u_k) p^*_k d\Gamma \tag{29}$$

One can integrate by parts twice the left hand side of the above equation to give

$$\int_\Omega \sigma^*_{kj,j} \, u_k \, d\Omega + \int_\Gamma p_k \, u^*_k \, d\Gamma = \int_\Gamma p^*_k \, u_k \, d\Gamma \tag{30}$$

(Notice that now $\Gamma = \Gamma_1 + \Gamma_2$, which implies that the boundary conditions in $\overline{u}_k$ and $\overline{p}_k$ are included in these integrals for simplicity).

Taking into consideration (28) the first integral in equation (30) can now be written as

$$\int_\Omega \sigma^*_{kj,j} \, u_k \, d\Omega = - \int_\Omega (\Delta^i e_\ell) u_k \, d\Omega = -u^i_\ell \, e_\ell \tag{31}$$

where u^i represents the ℓ component of the displacement at the point of application of the load.

Equation (30) can now be interpreted as representing the three separate components of the displacement at 'i' by taking the three divisions of the point load at 'i' independently, i.e.

$$u^i_\ell + \int_\Gamma p^*_{\ell k} \, u_k \, d\Gamma = \int_\Gamma u^*_{\ell k} \, d\Gamma \tag{32}$$

Notice that when one applies a unit point load along a particular direction 'ℓ' the tractions and displacements at any point in the domain have components along the three (or two) directions and terms of the type $\sigma_{\ell j,j}$ only are different from zero along the directions of the load.

Equation (32) is sometimes known as Somigliana's identity and gives the values of the displacements at any internal point 'i' in terms of the boundary values u_k and p_k.

The integral expression corresponding to the case for which the point 'i' has been taken to the boundary results from taking the limit of equation (32). When 'i' is taken to the boundary, however the integrals have a singularity and we need to analyse this behaviour in the same way as was done for the potential problem. The deduction is beyond the scope of this chapter and the interested reader is referred to [3] for a full explanation. The result is

$$c_{\ell k}^i \, u_k^i + \int_\Gamma p_{\ell k}^* \, u_k \, d\Gamma = \int_\Gamma u_{\ell k}^* \, p_k \, d\Gamma \tag{33}$$

where the integrals are in the sense of Cauchy principal value. If Γ is smooth at 'i' the value of the constant is simply $c_{\ell k}^i = \frac{1}{2}\delta_{\ell k}$. When '$i$' is at a point where the boundary is not smooth, the value of the constant is generally more difficult to obtain. Fortunately explicit calculations of these values are not usually necessary as they can be obtained using rigid body considerations once the off diagonal coefficients of the system matrix have been calculated [35].

Boundary Element Formulation

In order to solve the integral equations numerically, the boundary will be discretized into a series of elements over which the displacements and tractions are written in terms of their values at a series of nodal points. Writing the discretized form of (33) a system of linear algebraic equations is obtained. Once the boundary conditions are applied the system can be solved to obtain all the unknown values and consequently an approximate solution to the boundary problem is found.

Following the notation in Section 2, the system of equations can be written as

$$\mathbf{HU} = \mathbf{GP} \tag{34}$$

where $\mathbf{H}$ and $\mathbf{G}$ are $3N$ x $3N$ matrices (or $2N$ x $2N$ in two dimensional cases), N being the number of boundary nodes in the system. $\mathbf{U}$ is a vector of surface displacements and $\mathbf{P}$ of tractions.

Applying the displacements and tractions boundary conditions the system becomes

$$\mathbf{AX} = \mathbf{F} \tag{35}$$

Notice that $\mathbf{H}$ will have a series of c coefficients on the 3 x 3 submatrices on the diagonal. These elements - except for the smooth boundary case - are not simply given by the solid angle but can be very cumbersome to compute analytically. Fortunately this is not required as they can usually be computed by consideration of rigid body movement [35].

Once all the displacement and traction unknowns are found (i.e. vector $\mathbf{X}$) one can use these values to calculate the values of displacements and stresses at internal points as explained in detail in [35].

Subdivision of the domain into subregions or zones, as is usually done in large elasticity problems, will provide a block of sparse matrices for **A** but the blocks are still fully populated sparse type matrices which are very suitable for parallel processing.

4 Applications of BEM in Supercomputers

Supercomputers are becoming a major tool for engineering analysis and design, because of the increase in capacity and speed which they offer. While both finite and boundary elements techniques benefit from supercomputers, the unique characteristics of boundary elements make the method specially suited to vector and parallel processors. In this regard the boundary element code BEASY [25] has been found to be several times more efficient in its use of supercomputer resources than finite element programs. There are some major reasons for this:

Full matrix - The boundary element is characterized by an unbanded final matrix equation. This is normally one of the reasons why, although a boundary element node has far fewer degrees of freedom than a corresponding finite element node, the CPU times are comparable in sequential machines. But while this could be considered a disadvantage on smaller machines, the vector facility turns this feature into a positive advantage. The inner DO-loop considered in the matrix solution is always the bandwidth of the matrix. As the vector operations reduce the computation time inside the inner loop to almost nothing, it does not matter what the bandwidth is - the CPU time will be almost identical for banded or unbanded systems. Furthermore, because it is known that the matrix will be unbanded, no extra computation is necessary in deciding which parts of the matrix are already zero and need not be considered. No 'skylines' need to be stored and analysed and the reduction of this type of overhead calculation makes the boundary element solution considerably more efficient.

Independence of Matrix Terms - The boundary element method is formulated such that every term in the matrix equation can be computed completely independently of every other. There is therefore considerable potential for the use of multiple processors in generation of these equations, as well as in the solution of the system. Similar considerations apply when computing the equations corresponding to the internal points results, each of which can be computed independently of the other.

In conclusion when combined with supercomputers, the special features of the boundary element method are further emphasized. The run-time is reduced and models of increased complexity can be handled. This produces an ideal environment for the engineering designer. The modelling time and the run-time are both small. Changes can be made quickly and the analysis run again to assess the effect of design changes.

5 Conclusions

The development of powerful supercomputers favours the use of boundary elements. These machines are better suited to dealing with fully populated matrices and with the type of 'global' integrations which are characteristics of boundary elements.

The reduction in run-time and the possibility supercomputers offer of designing interactively, using boundary element models, provides an ideal environment for engineering design. Because of this it is envisaged that engineering work stations connected to large supercomputers will be increasingly used in practice. While the work stations can be used to run simple problems or generate the data required to run a complex boundary element model, the supercomputers will provide the necessary power to carry out the analysis in a comparatively short time.

References

1. Brebbia, C.A.(Ed.); Recent Advances in Boundary Element Methods, Proc. 1st Conf., Pentech Press, London, Computational Mechanics Publications, Southampton, 1978.

2. Brebbia, C.A. The Boundary Element Method for Engineers, Pentech Press, London, 1978.

3. Brebbia, C.A. (Ed.); New Developments in Boundary Element Methods, Proc. of 2nd Conf., Computational Mechanics Publications, Southampton, 1980.

4. Brebbia, C.A. (Ed.); Boundary Element Methods, Proc. 3rd Conf.; Springer-Verlag, Berlin and New York, Computational Mechanics Publications, Southampton, 1981.

5. Brebbia, C.A. (Ed.); Boundary Element Methods in Engineering, Proc. 4th Conf.; Springer-Verlag, Berlin and New York, Computational Mechanics Publications, Southampton, 1982.

6. Brebbia, C.A., Futagami, T. and Tanaka, M. (Eds.); Boundary Elements, Proc. of 5th Conf.; Springer-Verlag, Berlin and New York, Computational Mechanics Publications, Southampton, 1983.

7. Brebbia, C.A. (Ed.); Boundary Elements VI, Proc. 6th Conf.; Springer-Verlag, Berlin and New York, Computational Mechanics Publications, Southampton, 1984.

8. Brebbia, C.A. and Maier, G. (Eds.); Boundary Elements VII, Proc. 7th Conf.; Springer-Verlag, Berlin and New York, Computational Mechanics Publications, Southampton, 1985 .

9. Tanaka, M. and Brebbia, C.A. (Eds.); Boundary Elements VIII, Proc. 8th Conf.; Springer-Verlag, Berlin and New York, Computational Mechanics Publications, Southampton, 1986.

10. Brebbia, C.A. and Wendland W. (Eds.); Boundary Elements IX, Proc. 9th Conf.; Springer-Verlag, Berlin and New York, Computational Mechanics Publications, Southampton, 1987.

11. Brebbia, C.A. (Ed.); Boundary Elements X, Proc. 10th Conf.; Springer-Verlag, Berlin and New York, Computational Mechanics Publications, Southampton, 1988.

12. Brebbia, C.A. and Noye, B.J.; BETECH/85, Proc. 1st Conf. on BEM Technology; Springer-Verlag, Berlin and New York, Computational Mechanics Publications, Southampton, 1985.

13. Brebbia, C.A. and Connor, J.J.; BETECH/86, Proc. 2nd Int. Conf. on BEM Technology; Computational Mechanics Publications, Southampton, 1986.

14. Brebbia, C.A. and Venturini, W.; BETECH/87, Proc. 3rd Int. Conf. on BEM Technology; Computational Mechanics Publications, Southampton, 1987.

15. Brebbia, C.A., Telles, SJ. and Wrobel, L.C.; Boundary Element Techniques - Theory and Applications in Engineering, Springer-Verlag, Berlin and New York, 1984.

16. Brebbia, C.A. (Ed.); The Boundary Element Reference Book, Springer-Verlag, Berlin and New York, Computational Mechanics Publications, Southampton, 1988.

17. Brebbia, C.A. (Ed.); Progress in Boundary Elements, Vol.1, Pentech Press, London, 1978.

18. Brebbia, C.A. (Ed.); Progress in Boundary Elements, Vol.2, Pentech Press, London, Springer-Verlag, New York, 1981.

19. Brebbia, C.A. (Ed.); Topics in Boundary Element Research, Vol. 1 - Basic Priciples and Applications, Springer-Verlag, Berlin and New York, 1983.

20. Brebbia, C.A. (Ed.); Topics in Boundary Element Research, Vol. 2 - Time Dependent and Elastodynamics, Springer-Verlag, Berlin and New York, 1985.

21. Brebbia, C.A. (Ed.); Topics in Boundary Element Research, Vol. 3 - Artificial Aspects, Springer-Verlag, Berlin and New York, 1986.

22. Brebbia, C.A. (Ed.); Topics in Boundary Element Research, Vol. 4 - Geomechanics Applications, Springer-Verlag, Berlin and New York, 1987.

23. Brebbia, C.A. (Ed.); Topics in Boundary Element Research, Vol. 5 - Electrostatics, Springer-Verlag, Berlin and New York, 1988.

24. Brebbia, C.A. (Ed.); Topics in Boundary Element Research, Vol. 6 - Electromagnetics, Springer-Verlag, Berlin and New York, 1989.

25. Brebbia, C.A., R. Adey and H. Mizoguchi; BEASY: An Advanced Boundary Element Alalysis System, Proc. of the US/JAPAN Seminar on Boundary Element Methods, Tokyo, October 1988. Pergamon Press, Oxford, 1988.

26. NAFEMS; National Agency for Finite Elements Methods and Standards. Benchmark Reports, UK., 1987.

27. Floyd, C.G.; The Determination of Stress using a Combined Theortical and Experimental Analysis Approach. Computational Methods and Experimental Measurements, Proc. of 2nd Int. Conf. June/July 1984, (Ed. Brebbia, C.A.) Springer-Verlag, Berlin and New York, Computational Mechanics Publications, Southampton, 1984.

28. Sussman, T. and Bathe, K.J.; Studies in Finite Element Procedures - On Mesh Selection, Computers and Structures, 21, pp. 257-264, 1985.

29. Brebbia, C.A. and Trevelyan, J.; On the Accuracy and Convergence of Boundary Element Results for the Floyd Pressure Vessel Problem, Technical Note, Computers and Structures, 24, pp. 513-516, 1986.

30. Brebbia, C.A. and Dominguez, J.; Boundary Element Methods versus Finite Elements, Proc. Int. Conf. on Applied Numerical Modelling, Southampton 1977, (Ed. Brebbia, C.A.), Pentech Press, London, 1977.

31. Brebbia, C.A. and Dominguez, J.; Boundary Element Methods for Potential Problems, Applied Mathematical Modelling, 1, 7, December 1977.

32. Jaswon, M.; Integral Equation Methods in Potential Theory I, Proc. Roy. Soc. Ser. A, 275, pp. 23-32, 1963.

33. Symm, G.T.; Integral Equation Methods in Potential Theory II, Proc. Roy. Soc. Ser. A, 275, pp. 33-46, 1963.

34. Jaswon, M. and Ponter, A.R.; An Integral Equation Solution of the Torsion Problem, Proc. Roy. Soc. Ser. A, 273, pp. 237-246, 1963.

35. Brebbia, C.A. and Dominguez, J; Boundary Elements - An Introductory Course, McGraw - Hill, New York, Computational Mechanics Publications, Southampton, 1989.

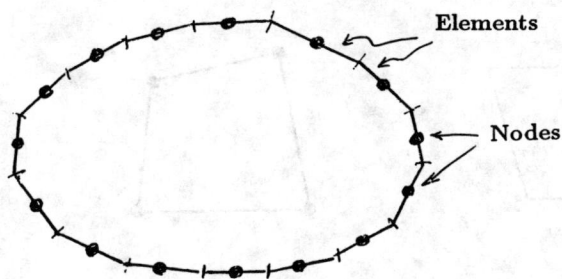

a) Constant Elements

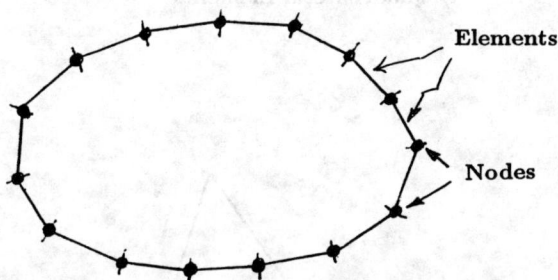

b) Linear Elements

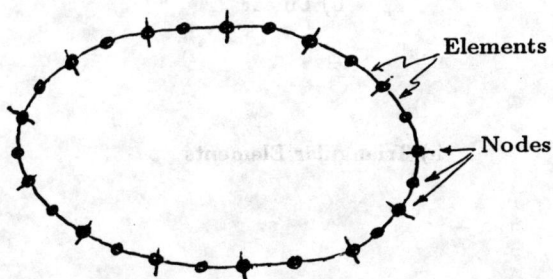

c) Quadratic Elements

Figure 5   Different Types of Elements

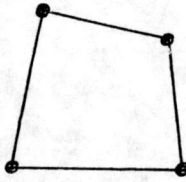
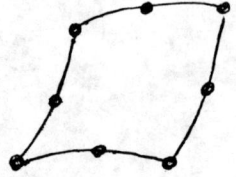

a) Constant   b) Linear   c) Quadratic

i) Quadrilateral Elements

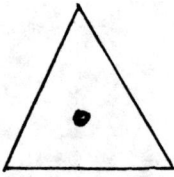
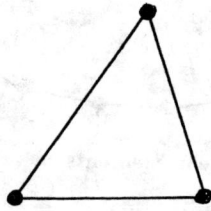
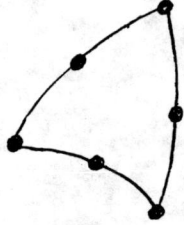

a) Constant   b) Linear   c) Quadratic

ii) Triangular Elements

Figure 6   Three Dimensional Elements

a) Geometric definition

b) Displacement components of the
fundamental solution on the
surface
(unit load acting in x_1 direction)

c) Traction components of the
fundamental solution on the
surface
(unit load acting in x_2 direction)

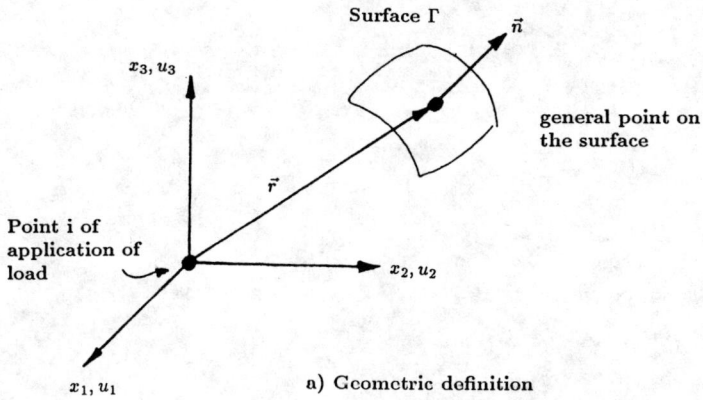

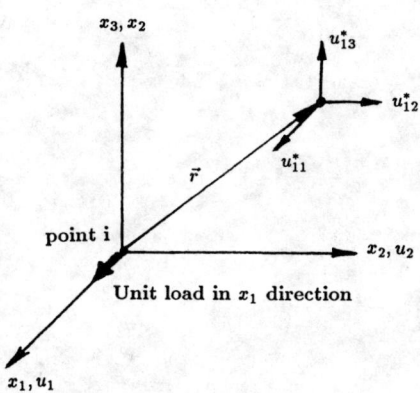

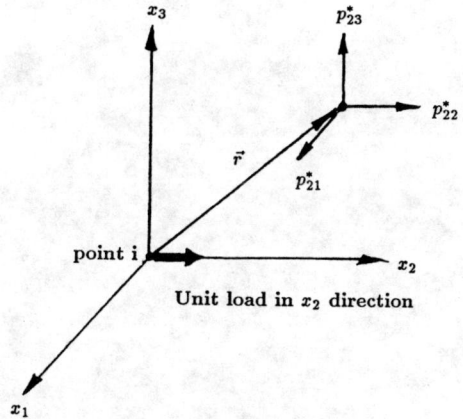

Figure 7   **Geometrical interpretation of the components of the
fundamental solution**

Computational Aspects and Applications of Boundary Elements on Supercomputers

R A Adey
*Computational Mechanics Institute, Ashurst Lodge,
Ashurst, Southampton, England*

The boundary element method has developed into a powerful technique for solving engineering analysis problems. The main feature of the method is that only the boundary or surface of the problem to be analysed has to be described with elements. Therefore, for example, sophisticated stress analyses can be performed on geometrically complex mechanical components quite easily because the engineer has only to describe the components surfaces.

It has been found in an industrial environment that engineers can exploit this reduced modelling requirement to make dramatic savings in the time required to build a model. Trials have shown that engineers can develop the BEM model up to twenty six times faster than the equivalent finite element model although factors of between three and ten are more common.

The advent of supercomputers has had a similar impact on the computational aspect of engineering analysis in providing orders of magnitude reduction in the computing time to perform a calculation. However, to achieve higher level of performance particularly with vector and parallel computing, algorithm and programming style have had to be changed in order to achieve anything near the theoretical performance of these new computing systems.

In this paper we will review the particular features of the boundary element method and its application on supercomputer systems.

The Boundary Element Method

If a typical large three dimensional stress analysis calculation using the boundary element method is examined in detail the key computational stress can be clearly defined. It is the evaluation of the element integrals to form the system of equations and the solution of the equations. In general the smaller the problem, the more floating point operations are required for the evaluation of the integrals whereas for larger problems more floating point operations are required for solving the equations.

Evaluation of Element Integrals

The basic equation of the boundary element method can be written as in [1] and [2], ie

$$C_i U_i + \int_S T^* U \, dS = \int_S U^* T \, dS \tag{1}$$

where

$$C_i = \text{constant} \tag{2}$$

$$U_i = \text{displacement}$$
$$T^*U^* = \text{influence functions}$$
$$T = \text{tractions}$$
$$S = \text{boundary surface}$$

Converting to elements

$$C_i U_i + \sum_n \left\{ \int_{S_j} T^*U \ dS = \int_{S_j} U^*T \ dS \right\} \qquad (3)$$

where

$$n = \text{number of elements} \qquad (4)$$
$$S_j = \text{surface area of jth element}$$

The main computational task is therefore for each element to perform the integration from each node (source point). The complete details of the boundary element method are described in [1] and [2].

The normal technique used to evaluate the integrals is gaussian quadrature as shown in (5).

$$\int_{S_j} T^*U \ dS = \left\{ \sum_m^k T^*(\zeta)\Phi(\zeta)|G(\zeta)| \ W_k \right\} U^n \qquad (5)$$

where

$$m = \text{number of gauss points} \qquad (6)$$
$$\zeta = \text{local coordinate on the element}$$
$$\Phi(\zeta) = \text{interpolation function}$$
$$|G(\zeta)| = \text{surface area (from Jacobian)}$$
$$W_k = \text{weighting function}$$
$$U^n = \text{vector of nodal displacements}$$

Therefore for a typical problem the above integral has to be evaluated for each element for each node, (eg number of elements x number of nodes).

The first observation that can be made is that all of the integrals are independent of each other and could therefore be carried out in parallel.

For vector processing the position is not so clear as $T^*(\zeta)$, $|G(\zeta)|$ and W_k are scalars and only $\Phi(\zeta)$ is a vector. One obvious solution would therefore be to evaluate a number of elements simultaneously in order to increase the vector length. However, this is difficult as normally the number of gauss points used per element varies.

The major computational cost in this step is however the evaluation of $|G(\zeta)|$ which is derived from the Jacobian. These operations can be expressed in terms of vector products but the vector length is small. For example, in three dimensional analysis the vector length is equal to the number of nodes per element which in the case of a quadratic quadrilateral element is only 9.

To fully exploit vector processing, it is therefore necessary to make major changes to the algorithms as the minimum vector length required to achieve an increase in performance on current computers is in the order of twenty.

Solution of Equations

The solution of equations is a subject area in supercomputing where much work has been carried out. Many numerical techniques like finite elements and finite difference require the solution of a large set of equations. The boundary element equations have some major differences which make them particularly suited to supercomputers.

Finite element models systems have, in general, a large number of equations, but they are very sparse and symmetric. In contrast, boundary elements have a small number of equations but the equations are unsymmetric and have a blocked sparse structure. This last characteristic is very advantageous on supercomputers because this very regular structure leads to long vector lengths and a uniform data structure.

Figures 1 and 2 show a diagramatic representation of a three zone boundary element model and the resulting set of equations. The very regular sparseness introduced by the zoning and the long vector lengths and denseness of the data can be clearly seen.

A scheme like this has been implemented in the BEASY system [3] and run on a vector processing computer and achieved the results shown in the next section.

Finally the equation solution is also very suited to parallel computing.

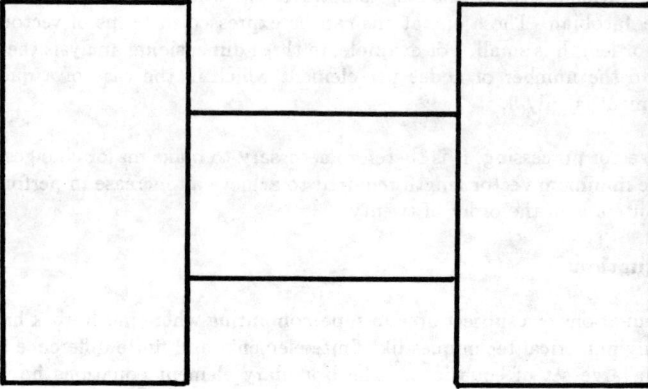

Figure 1 - Boundary Element 3 Zone Model

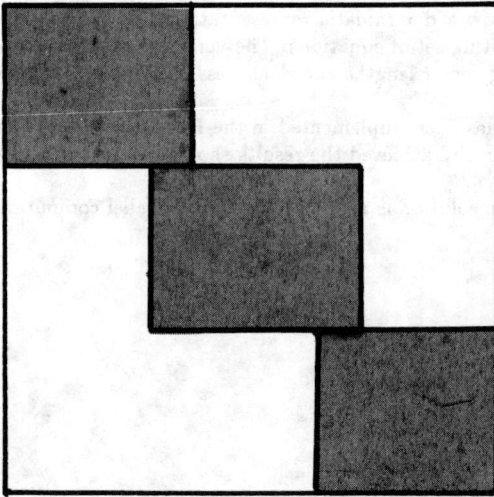

Figure 2 - BEM Equations for 3 Zone Model

Example Calculation on Supercomputer

A large three dimensional stress calculation was performed on a supercomputer with a single vector processor. The problem had 12000 degrees of freedom and eight zones. The computer times for the various operations were as follows:-

- SUPERCOMPUTER

  Element matrix formation 74%

  Solution of equations 26%

- SCALAR COMPUTER

  The same example was run on a scalar computer (VAX) and the following distribution of times achieved:-

  Element matrix formation 15%

  Solution of equations 85%

As was stated in the previous section the ony part of the code which was vectorised was that associated with the equation solution. This change achieved a factor of eighteen reduction in the proportion of time required to solve the equations. On the scalar computer the solution of equations took nearly six times longer than the formation of equations. However, on the supercomputer the solution of equations took one third of the time required by the formation of equations. (Note some of the improvements may be attributable to the improved input/output performance of the supercomputer.)

Applications of Boundary Elements

Supercomputers, coupled with boundary element programs like BEASY, enable large complex three dimensional problems to be solved at a fraction of the cost previously associated with large analyses. BEASY, for example, can solve stress analysis, thermal analysis and field problems (eg electrostatics, electromagnetics) for two dimensional, three dimensional and axisymmetric models. The key benefits introduced by boundary elements are the rapid reduction in the engineers time to build a model and the improved quality of the solution.

A typical example of the type of problem that can be solved using BEASY is shown in Figure 3.* This analysis was performed as part of a checking stress calculation of part of a modern airliner, [4] (courtesy of British Aerospace and Cray Research). The engineer preparing the model simply describes the surface of the component and applies the loads. In this case 18 load cases and three cases of symmetry and antisymmetry were considered and the Von Mises stresses generated by one of the load cases are shown in Figure 4.* The results were displayed using the PATRAN program from PDA Engineering Inc.

An interesting thermal problem solved using BEASY is shown in Figures 5*and 6.* This model uses symmetry to model half the mould and shows the circular shape of the casting. The lower half of the mould shown in Figure 6 shows clearly how only the surface of the casting is modelled.

The loading in this example is the heat from the casting and the cooling holes which run through the casting. These cooling holes would be very difficult to model with other techniques, but are very easy to model with boundary elements.

*See Figures 3 and 4 in colour section, page 162.

*See Figures 5 and 6 in colour section, page 163.

A final example shown in Figure 7 'and 8'shows an analysis of a lifting lug. The boundary element method is very effective at modelling stress concentrations such as this case. It can also solve other problems involving high potential gradients, in particular those found in damage tolerance calculations and thermal and electromagnetic problems.

Conclusions

The boundary element technique has great potential on supercomputers, as it is well suited to vector and parallel computing. The initial studies in this paper show major reductions in computer costs that have been achieved by vectorising the equation solution.

Supercomputers and boundary elements are complementary in that these machines reduce the computer time and the boundary element method minimizes the engineers time required to perform an analysis.

REFERENCES

1. BREBBIA, C A "The Boundary Element Method for Engineers", Pentech Press and Computational Mechanics Publications, Southampton, 1978.

2. BREBBIA, C A "Boundary Elements - A Course for Engineers", Computational Mechanics Publications, Southampton, McGraw-Hill (in USA), 1988.

3. BEASY USERS GUIDE, Computational Mechanics, Southampton, England, 1988.

4. "Stress Analysis of an Aircraft Fitting Using BEASY" M Z Mannan. Proceedings of Boundary Elements 10, Vol. 3, pp 601-613, Computational Mechanics Publications, Southampton, 1988.

*See Figures 7 and 8 in colour section, page 164.

Modeling and Solution Strategies for Nonlinear Braced Frames

G.H. Powell

University of California, Berkeley, CA 94720, USA

INTRODUCTION

Although linear structural analysis is still the basis of most engineering design, there is an increasing need for practical nonlinear analysis. Unfortunately, nonlinear analysis is vastly more complex, both theoretically and computationally, than linear analysis, and is presently as much an art as a science. Major efforts in education and software development will be needed before nonlinear analysis can be used routinely by practicing engineers. The education effort will, incidentally, need to be directed as much to theoreticians and software developers as to engineering users, to ensure that the true practical needs are being met.

This paper is concerned with both theory and practice, and in particular with the analysis of large steel frame structures of the type used for offshore oil production platforms (Figure 1). Large numbers of these structures exist around the world and many of them are becoming quite old. There is a need to estimate their strengths and to devise ways of upgrading those which may be substandard.

For the nonlinear analysis of two- and three-dimensional finite element systems, the current state of the art is quite advanced (although much improvement is needed in the modeling of material behavior). This is not so true, however, for frame structures. Paradoxically, although frame structures may seem to be relatively simple to model and analyze, they can pose greater challenges than structures which are modeled with solid finite elements.

For frames of the type shown in Figure 1, an important property is the behavior under horizontal static loads (representing wave loads). This behavior is controlled largely by the yielding, buckling and post-buckling behavior of the diagonal brace members, and possibly by yielding, etc. of the connections between members. The modeling is complicated by the need to consider such effects as denting and corrosion damage to the members. The numerical analysis is complicated by the severe local non-linearities which develop as members buckle. Special care must be taken in modeling, to ensure that the important modes of nonlinear behavior are captured in the model, and special analysis strategies are needed to account for the local nonlinearities. This paper reviews some basic modeling concepts, indicates how these concepts apply for the modeling of tubular steel brace members, and outlines an analysis strategy which accounts effectively for both local and global nonlinearities.

REVIEW OF MODELING PRINCIPLES

Since it is easy to get absorbed in the details and forget the essentials, it is useful to review the principles on which nonlinear element modeling is based. Frame elements can be based on force (flexibility) or displacement (stiffness) formulations. Only displacement formulations are considered here. The essential steps are as follows.

(1) Define the lowest level element deformations, ε, and corresponding element actions, σ, which are to be considered, and the locations in the element at which they are to be monitored. The deformations and actions may be strains and stresses, or strain and stress resultants and must be energy conjugates.

(2) Define the highest level element degrees of freedom (dofs), q. These will usually be nodal translations and rotations which are a direct subset of the displacement dofs of the complete structure, r. However, they may be related to r by rotation and/or slaving transformations. The corresponding forces, Q, will usually be nodal forces and moments, acting on the element.

(3) Define the kinematic relationship between ε and q. This relationship should ideally satisfy geometric compatibility (material continuity) within the element. It may be defined in a number of stages, for example, in a beam type element, from nodal displacement to curvature to strain. The relationship in each stage may be linear or nonlinear, depending on whether or not small displacements can be assumed. A key point is that given any state of the element and a finite increment Δq, it must be possible to calculate the corresponding $\Delta\varepsilon$ at all monitored points. It also follows that given any state of the element, a transformation a can be computed, where

$$d\varepsilon = a \, dq \qquad (1)$$

If the $\varepsilon - q$ relationship is linear, then a will be constant. If not, it will be displacement dependent.

(4) Define the constitutive relationship between σ and ε. This may be linear or nonlinear, and for the nonlinear case may be defined as a state-dependent relationship between $d\sigma$ and $d\varepsilon$. Given any state at a monitored point and a finite increment $\Delta\varepsilon$, it must be possible to calculate the corresponding $\Delta\sigma$. This may involve integration along an assumed strain path between ε and $\varepsilon+\Delta\varepsilon$.

(5) It follows from the virtual displacements principle that for any given state, the element "resisting" forces, Q_r, required to satisfy element equilibrium, are given by

$$\underline{Q}_r = \int (\underline{a}'\underline{\sigma})dV \tag{2}$$

where V = element volume if $\underline{\sigma}$ is stress, and area or length if $\underline{\sigma}$ is a stress resultant.

These relationships define the behavior of the element model. Given an element state and an increment Δq, the evaluation of $\Delta \underline{\varepsilon}$, $\Delta \underline{\sigma}$ and $\underline{Q}_r$ constitutes the "state determination" calculation. The $\underline{Q}_r$ vectors for all elements can be assembled to give the structure resisting force, $\underline{R}_r$. The difference between the external nodal loads on the structure, $\underline{R}_e$, and $\underline{R}_r$ is then the unbalanced load vector, $\underline{R}_u$, which defines the equilibrium error. The essential goal of nonlinear analysis is to determine a displacement increment, $\Delta \underline{r}$, for which $\underline{R}_u$ is acceptably small.

MODELING STRATEGY : GENERAL

An analysis model will inevitably include certain features of the real structure and ignore others. As a general rule, the important modes of structural behavior must be modeled accurately, and the less important modes need to be modeled only approximately. This presupposes that the important modes can be identified, that they can be modeled using available computer programs, and that the resulting model can be analyzed.

Because of the complexity of the problem, budgetary restrictions, and the limited modeling and analysis capabilities of computer programs, there is a tendency for one of two extremes to be followed. At one extreme, the important nonlinear modes are identified, but an over-simplified model is used because of limitations in the available modeling capabilities. An example is the use of truss-bar type elements to model braced frame members, as described below. This approach is understandable, but it may not produce meaningful results. At the other extreme, the important modes are not identified. Instead, an attempt is made to model all possible modes, and to have the analysis identify those which are important. At best this approach is unfortunate, since it means abdicating responsibility to a computer program. At worst, it can be dangerous. The reason is that in order to model all modes of behavior, an extremely large and complex model will usually be needed, with huge numbers of solid elements and very sophisticated material models. Since this is usually impractical, there is a tendency to choose the most elaborate model (e.g., the finest mesh) which is permitted by the budget and is within the computer program capabilities, and to hope that the important modes of behavior will still be captured. All too often this can be a forlorn hope. It is also important to remember that even if it were possible to model all modes of behavior, there would always be uncertainties, in the geometry of the structure, the material properties and the loading. Hence, no matter how elaborate the model, true accuracy is almost never possible. (This is, incidentally, an argument in favor of designing a structure to have well defined "weak links" which control its nonlinear behavior. The modelling effort can then emphasize these weak links, paying less attention to the strong ones. The result is not only a more accurate analysis, but a more resilient design.)

It is, of course, easy to point out problems, and more difficult to provide solutions. Nevertheless, there is a real danger that nonlinear modelling and analysis may be little more than an academic exercise. As nonlinear analysis is used increasingly in practice, increasing attention will have to be paid to the selection of appropriate models. In constructing a nonlinear model it is important (a) to identify the important modes of behavior, (b) to ensure that these modes are modeled adequately, and (c) to assess the effects of uncertainties. These are difficult requirements to satisfy, especially if there are no experimental results to use in calibrating the model. A great deal of engineering judgement is generally needed, to determine what level of accuracy is required, and what level is realistically achievable.

TUBULAR BRACE MODELING : GENERAL

If a tubular brace member is isolated, clamped against rotation at both ends, and subjected to axial force only, it will behave as indicated in Figure 2. At point A the member is essentially straight and elastic. At point B it buckles, elastically or inelastically, depending on its slenderness. In the region containing point C, plastic hinges form under a combination of axial force (P) and bending moment (M), where the moment is due to the axial force acting through the lateral displacement. As the lateral displacement increases, the axial force must decrease, since the moment capacity is limited by the cross section strength. The member thus loses strength in the post-buckling range. On reversal of the load, the member returns to an elastic state (point D), but because it is bent, its axial stiffness is less than the stiffness at Point A. Hinges then form under tension load (point E). This causes an initial softening, but as the lateral displacement decreases, the axial strength and stiffness increase. Ultimately the member becomes nearly straight, and its strength is governed by the material yield strength (point F). At the second load reversal, the member again becomes elastic, regaining essentially its original elastic stiffness. It does not, however, regain its original buckling strength (compare points G and B). This is mainly because of a Bauschinger effect in the steel, which causes earlier yielding under cyclic loading than under the initial loading.

Figure 2 shows the behavior for large amplitude cycling. For arbitrary loading on a frame, the load cycles on a member can also be arbitrary (large, small, partial, etc., in any sequence), and this must be accounted for in the model. For static push-over of a frame, the external load consists of constant gravity forces and monotonically increasing horizontal forces. However, it does not follow that the member forces will increase monotonically, because of force redistribution as members yield and buckle. It is thus necessary to account for unloading, if not for cyclic loading.

In a braced frame of the type shown in Figure 1, the main legs are typically of much larger section than the braces. Hence, two possible modeling assumptions are (a) that all brace members are restrained at their ends against rotation, and (b) that the only significant load they carry is axial force. Brace members can then be modeled with only axial strength and stiffness (i.e. as truss bar elements), exhibiting the behavior shown in Figure 2. An empirical model (the "Marshall-Maison strut", Marshall [1], Maison-Popov [2]) approximates the behavior in Figure 2 by a series of straight lines, with empirical rules defining the slopes and limits of each line. A semi-rational model (the "SSD strut", Powell-Row-Hollings [3]) models the member as rigid links connecting deformable hinges which yield under combined axial force and moment. This model is only semi-rational since, among other things, certain properties of the hinges must be assigned empirically.

Models of truss bar type have been used for estimating the static strength and seismic response of offshore platforms. They are fundamentally flawed, however, because it is not sufficiently accurate to assume that brace members carry only axial force. One reason is that brace members are not connected only to the main legs of the frame, but may also be connected to other brace members. As a result, at many joints there is no member which obviously dominates the behavior. Another reason is that brace members may be subjected to wave forces applied within their length, which can not be included in a truss bar model. Truss bar models thus err in the direction of over-simplification.

An alternative is to regard each tubular brace member as a steel shell, and to model it with a number of shell elements. This would test the limits of currently available analysis capabilities, in particular because of the local-global nonlinearity problem discussed below. With a sufficiently fine mesh it would also test the limits of any reasonable budget, and with anything less than a fine mesh would probably not capture the important nonlinear modes of behavior. Some potentially important effects include (a) local weakening due to denting or corrosion, and (b) the fact that where the end of a brace member is welded to the face of a main leg, local deformation of the leg in the joint region may have a significant effect on brace behavior. Effects such as these can be captured only with sophisticated finite element meshes. Even with such meshes, the accuracy might be more illusory than real, because of uncertainties in the structure properties and approximations in the loading.

It follows that neither a simplified truss-bar model nor a sophisticated finite element model is appropriate for the analysis of brace behavior. Instead, an intermediate type of model is needed, based on an engineering assessment of the important behavior modes. From such an assessment, it appears that a brace member model must account for (a) material yield, (b) bending and torsional, as well as axial, resistance, (c) buckling, postbuckling and unloading, (d) local weakness due to dents or corrosion, (e) local deformation in joint regions, and (f) loads applied within the member length. In addition, buckling deformations within a member and overall deformations of the structure may both be significant, so that finite displacement (or at least second order) effects must be considered. It appears to be reasonable to assume that longitudinal fiber stresses are dominant in causing inelastic material behavior, and not torsional, hoop or radial stresses. This is an important simplification because it allows a uniaxial stress-strain relationship to be used. With a uniaxial relationship, rules can be devised to follow virtually any stress-strain path. With a multiaxial relationship, flow rules or similar constraints must be observed, and devising rules for complex behavior is far more difficult. Local bending of the tube wall at dents must be taken into account. It is reasonable to do this by semi-empirical modification of the uniaxial material law (Taby-Moan-Rashed [4]), since there are so many uncertainties that even the use of a sophisticated finite element mesh in the region of a dent would be substantially empirical. For the same reason, it also appears to be reasonable to model local joint deformations by simple means, as outlined below.

The need to consider bending and torsion, and the assumption that axial stresses dominate, indicates that a three-dimensional beam-column type of element is appropriate. The fact that finite displacements must be considered within a member then dictates that each member be subdivided into several elements, each undergoing only small deformations. One reason for this is that it is difficult to devise a shape function which accounts for finite internal displacements in a single element. A second reason (which also applies for members with small internal displacements) is that the internal shape of a yielding frame member changes progressively, so that a shape function covering the entire member would need to be *variable* (Mahasuverachai-Powell [5]). Subdivision into several elements also makes it much simpler to apply loads within the member length.

One type of inelastic beam-column element is based on *lumped plasticity* assumptions, in which the plastic deformation is concentrated in "generalized" plastic hinges, with elastic elements in between (e.g. Powell-Chen [6]). With this type of element, it is typical to require that the hinge (i.e. cross section) properties be defined in terms of stress and strain resultants (e.g. moment-rotation relationships), and that yield interaction relationships (e.g. between axial force and moment) be defined. A more accurate and convenient approach is to monitor stresses and strains at several points in the cross section, and to obtain the section properties by integrating over these points. This approach can be combined with the plastic hinge assumption, by assigning some "tributary" element length to each hinge (Powell-Hollings-Row [3]). However, it is more consistent to use a "distributed plasticity" approach, using shape functions. This is the basis of the brace model outlined in the next section.

SUBSTRUCTURED BRACE MODEL

The main ingredients of a brace model are illustrated in Figure 3. The most basic element is a beam "fiber", for which longitudinal strains and stresses are monitored following a multi-linear stress-strain relationship. The fibers in a cross section are combined to define a "slice". Each slice has an axial force, bending moments about two axes, and a torsional moment, with corresponding axial strain, curvature and rate of twist deformations. The fiber strains are related to these deformations with the assumption that cross sections remain plane and undeformed. The fiber properties are integrated (using simple summation) to define the axial and flexural properties of each slice. The torsional properties are defined separately, assuming elastic, uncoupled behavior in shear.

A series of nodes is located along the brace member axis (which is not necessarily straight). A straight, constant-section beam-column element is defined between each pair of nodes, and slices are located at the quadrature points for 2-point Gauss integration in each element (other slice locations could be chosen). Each node has 6 displacement dofs, and the slice deformations are defined in terms of these displacements, using conventional beam shape functions (linear curvature; constant axial strain and rate of twist). The 6 displacements at each node are dofs of the element. They are not sufficient, however, to define element behavior adequately, because they allow only constant strain variation along the axis of each element. This is sufficient for an elastic element under constant axial force, since the strain along the centroidal axis will indeed be constant. It is not sufficient, however, for an inelastic element, because the effective centroidal axis may shift. It is necessary, therefore, to add one additional dof for each element, to allow a linear axial strain variation along any fiber. This is conveniently chosen as an axial displacement at the element midpoint, defined relative to the axial displacements at the nodes. These displacements also are a part of q. Flexural shear deformations of a slice may be significant, and can be added at this stage, assuming elastic behavior as for torsion.

A series of beam-column elements makes up the clear length of the brace member between joint faces. A rigid link then connects the member end to a 6-dof node at the joint center. Finally, an extra node and a zero-length element can be added at each joint face, to model joint deformations. This joint element is made up of several subelements, which are much like the fibers in a slice. However, they are assigned multi-linear stress-extension relationships rather than stress-strain relationships. These stress-extension relationships must be chosen to model the local stiffness and strength characteristics of the joint region, and hence must be largely empirical, based on experiments or separate analyses of joint behavior. Fiber stress-strain relationships for slices at dent locations must similarly be based on semi-empirical data (Taby-Moan-Rashed [4]). These are the main weaknesses of the model.

A brace member defined in this way forms a natural substructure, with external nodes and dofs at the joint centers, and internal nodes and dofs along the member length. Whether or not to treat a brace as a substructure for analysis is, however, more a matter of solution strategy (and computer program capability) than of modelling. Substructuring can have great advantages for *linear* analysis, simplifying the task of defining the structure geometry, reducing computational effort (e.g. Row-Powell [7], Dodds-Lopez [8]), and supporting parallel computation (e.g. Farhat-Wilson-Powell [9]). There can be similar advantages for *nonlinear* analysis, but much less work has been done on computational techniques and strategies. For example, the FACTS program (SSD [9]), for which the brace model described here has been developed, uses substructuring for those parts of the analysis which are common to linear and nonlinear problems, but not for those parts which are unique to nonlinear problems. Hence, the substructuring features described in the following sections have been "hard coded" for the brace model.

SOLUTION STRATEGY

The basis of essentially all nonlinear analysis strategies is as follows.

(1) Given a structure state (r, q, ε, etc.), *advance* the solution by selecting a displacement increment Δr_a, determining the new state, and calculating the unbalance R_u. The selection of Δr_a is based on some *advancing strategy*.

(2) If R_u exceeds the permissible tolerance, *correct* the solution, usually iteratively, by selecting one or more displacements Δr_c, and repeating the state calculations, until R_u is sufficiently small. The selection of Δr_c is based on some *correcting strategy*.

The advancing and correcting strategies are usually based on linearization of the problem, using a tangent, secant or some other stiffness matrix for the structure and solution of a set of equilibrium equations. If the structure model is known to have a *multi-linear* relationship between R and r (i.e., if the elements have multi-linear behavior and large displacement effects are negligible), Δr for each advancing step can be chosen to reach exactly the next stiffness change (e.g., a yield "event"). There is thus no equilibrium error, and no need for correcting steps. If it can be used, this "event-to-event" strategy is very reliable, although it can be computationally expensive for large structures with large numbers of events. For a *curvilinear* $R-r$ relationship, the number of events is infinite, and unbalances can not be avoided. Nevertheless, the use of events to control the size of advancing (and correcting) steps can be a valuable means of avoiding large unbalances, and hence of improving solution reliability.

If R_u exceeds the tolerance after the advancing step, iterative correcting steps are needed. For simple "load-controlled" Newton-Raphson iteration, the equations for the advancing step and each correcting step are

$$\underline{K}_t \, \Delta \underline{r}_a = \Delta \underline{R}_e \tag{3a}$$

$$\underline{K}_t \, \Delta \underline{r}_c = \underline{R}_u \tag{3b}$$

where $\underline{K}_t$ is the current tangent stiffness and $\Delta \underline{R}_e$ is a specified increment of external load. This strategy tends to be unreliable, and "displacement controlled" strategies are more commonly used (e.g., Riks [11], Ramm [12], Crisfield [13]). The equation for the advancing step in a typical strategy is

$$\underline{K}_t \, \Delta \underline{r}_a = \alpha_a \Delta \underline{R}_e \tag{4}$$

where α_a, the *advancing event factor*, is calculated to reach some event such as (a) exceeding a specified displacement increment, (b) significant stiffness change due to yield, or even (c) exceeding a specified unbalanced load. The equations for each correcting step is typically of the form

$$\underline{K}_t \, \Delta \underline{r}_{cu} = \underline{R}_u \tag{5a}$$

$$\underline{K}_t \, \Delta \underline{r}_{ce} = \Delta \underline{R}_e \tag{5b}$$

$$\Delta \underline{r}_c = \alpha_u \Delta \underline{r}_{cu} + \alpha_e \Delta \underline{r}_{ce} \tag{5c}$$

where α_u and α_e (the *correcting event factors*) are chosen to satisfy a displacement constraint. A common strategy (the arc length method) is to apply the entire unbalance in each correcting step (i.e., $\alpha_u = 1$), and to maintain a constant length for Δr, by choosing α_e such that the length of $\Delta \underline{r}_a + \Sigma \Delta \underline{r}_c$ remains equal to the length of $\Delta \underline{r}_a$. Note that the above equations contain the tangent stiffness, $\underline{K}_t$. However, other stiffnesses can be used.

A number of nonlinear analysis programs allow for displacement controls, imposing them on the complete structure displacement, Δr. Experience with the FACTS program showed, however, that such "global" controls are not very effective in the analysis of braced frames, especially when single brace members were divided into several beam-column elements. The reason is that as brace members buckle, become unstable and lose strength, substantial load redistributions occur. There can thus be substantial shape changes inside a brace, yet only small changes in shape of the structure as a whole. Displacement controls applied at the complete structure level do not prevent substantial unbalanced loads from developing locally, and do not significantly constrain the local displacements due to these unbalances. As a result, convergence in the correcting phase can be difficult or impossible to obtain. It is possible to introduce multiple displacement controls and to let the most critical one govern in each iteration (Simons-Powell [14]). In principle, therefore, displacement controls could be defined separately for each brace member, preventing large increments of local deformation. In practice, however. this is not feasible, because there are too many brace members. It was apparent, therefore, that a method was needed to control the local unbalances directly, treating each brace member as a substructure. The approach adopted was to treat each brace member not only as a substructure but as a single element, and to devise an algorithm for the state determination calculations which avoids unbalances on the substructure internal dofs. The complete structure then sees only unbalances on the external nodes (those at the joint centers in Figure 3). These can be managed effectively by global displacement controls.

The state determination calculation thus involves a nonlinear analysis strategy. However, this is applied at the element level, and is specialized for the problem of brace behavior. This specialization also involves a new iteration concept. The essential steps are as follows.

(1) The state of the brace is known. This includes the displacements at all internal nodes and the states of all beam-column elements. From the global analysis, displacement increments are provided at the external nodes. It is necessary to update the state of the brace, ensuring negligible unbalanced loads on the internal dofs. The resisting forces, Q_r, returned to the global analysis are then the forces at the external nodes which are in equilibrium with the internal element forces.

(2) The substructure tangent stiffness matrix is formed. Important ingredients of this stiffness are geometric stiffnesses for the beam-column elements, which depend on the current axial force and can dominate the behavior for a buckled brace.

(3) The internal displacement increments are calculated, and scaled to satisfy an event criterion. Events include yield and reaching a displacement limit. A new state is determined, based on the scaled increment. In particular, a new axial force is calculated.

(4) Because of the displacement control, it is reasonable to assume a linear relationship between $\Delta \varepsilon$ and Δq, and because of the yield control there are no changes in the elastic-plastic part of the stiffness. However, if the axial force changes there is a change in the geometric stiffness. Hence, there can be an equilibrium unbalance. However, instead of attempting to correct for this unbalance (which, incidentally, is difficult to do), the strategy estimates a new value of the axial force and repeats the analysis from step 2. When the axial force at the beginning and end of the step are essentially equal, the

step is a linear one and there is no unbalance. Iteration on the axial force is done by secant interpolation. The strategy is a form of secant stiffness iteration.

(5) That part of the external displacement which has been accounted for is subtracted from the remaining external displacement. The solution continues from step 2 until all of the displacement has been considered.

COMPUTATIONAL ASPECTS

The element has been implemented in the FACTS program for the CRAY computer, with no attempt to vectorize either the element or the solution strategy. Over the short term, significant gains in execution speed could undoubtedly be obtained by vectorizing the code, although this is not necessarily a simple task. Over the long term, the most promising approach appears to be concurrent processing. Effective computational strategies have been devised for linear substructured systems (e.g., Farhat-Wilson-Powell [9]). Substantial additional research is needed, however, for nonlinear systems of the type considered in this paper. In particular, the strategy in [9] relies on balancing the load among processors. This is relatively easy for linear systems, but less so for nonlinear systems, because different substructures are likely to require different numbers of iterations to converge, and the advantages of synchronization are lost.

CONCLUSION

This paper has summarized the displacement formulation for modeling nonlinear elements, pontificated on modelling strategies, and reviewed a number of solution strategies. It has been emphasized that the goal of the modeling strategy is to define appropriate kinematic and constitutive relationships, and that the goal of the solution strategy is to satisfy equilibrium. In the modeling phase, the emphasis is on selecting levels of accuracy and complexity which are appropriate to the problem at hand. In the solution phase, the emphasis is on efficiency and reliability. An application to the modeling of brace members for steel offshore platforms has been presented, (a) to illustrate what the author believes to be appropriate modeling practice, and (b) to show that more reliable and efficient strategies are needed if nonlinear analysis is to be used routinely.

Space does not permit an example to be presented (there are also concerns about the proprietary information of clients). It may be stated, however, that the element which has been described appears to model brace behavior very well, and that it has significantly improved solution reliability and efficiency. It has not, however, solved all of the problems, and work is continuing. Finally, it may be noted that the substructured element is suitable for modeling members that do not buckle, including the main legs of a frame. With modifications to account for cross sections other than tubular ones, and for materials other than steel, the element could also be used for building frames.

REFERENCES

1. Marshall P.W. (1978), Design Considerations for Offshore Structures Having Nonlinear Response to Earthquakes, Proceedings, ASCE

Annual Convention and Exposition, Chicago.

2.   Maison B. and Popov E.P. (1980), Cyclic Response Prediction for Braced Steel Frames, Journal of the Structural Division, ASCE, Vol.106, No.ST7, pp.1401-1446.

3.   Powell G.H., Row D.G. and Hollings J.P. (1984), Improved Modelling of Brace Elements Under Severe Cyclic Loading, Journal of Energy Resources Technology, ASME, Vol.106, pp.240-245.

4.   Taby J., Moan T. and Rashed S.M.H. (1981), Theoretical and Experimental Study of the Behaviour of Damaged Tubular Members in Offshore Structures, Norwegian Maritime Research, No.2, pp.26-33.

5.   Mahasuverachai M. and Powell G.H. (1982), Inelastic Analysis of Piping and Tubular Structures, Report No. UCB/EERC-82/27, Earthquake Engineering Research Center, Univ. of California, Berkeley.

6.   Powell G.H. and Chen P.F-S. (1986), 3D Beam-Column Element with Generalized Plastic Hinges, Journal of Engineering Mechanics, ASCE, Vol.112, No.7, pp.627-641.

7.   Row D.G. and Powell G.H. (1978), A Substructure Technique for Nonlinear Static and Dynamic Analysis, Report No. UCB/EERC-78/15, Earthquake Engineering Research Center, Univ. of California, Berkeley.

8.   Dodds R.H. and Lopez L.A. (1980), Substructuring in Linear and Nonlinear Analysis, International Journal of Numerical Methods in Engineering, Vol.15, pp.583-597.

9.   Farhat C., Wilson E.L. and Powell G.H. (1987), Solution of Finite Element Systems on Concurrent Processing Computers, Engineering with Computers, Vol.2, pp.157-165.

10. FACTS: Finite Element Analysis of Complex Three-Dimensional Systems: User Guide, SSD Inc., Berkeley, CA.

11. Riks E. (1978), An Incremental Approach to the Solution of Snapping and Buckling Problems, International Journal of Solids and Structures, Vol.15, pp.529-551.

12. Ramm E. (1980), Strategies for Tracing Nonlinear Response Near Limit Points, Proceedings, U.S.-Europe Workshop on Nonlinear Finite Element Analysis in Structural Mechanics, Bochum, Germany, pp.63-89.

13. Crisfield M.A. (1981), A Fast Incremental/Iterative Solution Procedure that Handles "Snap-Through", Computer and Structures, Vol.13, pp.55-62.

14. Simons J.W and Powell G.H. (1982), Solution Strategies for Statically Loaded Nonlinear Structures, Report No. UCB/EERC-82/22, Earthquake Engineering Research Center, Univ. of California, Berkeley.

FIGURE 1. BRACED FRAME STRUCTURE

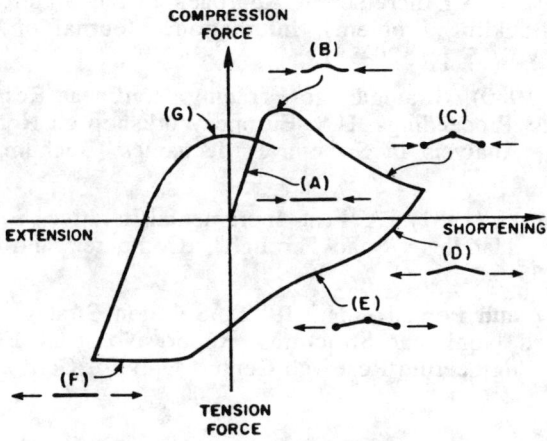

FIGURE 2. TYPICAL BRACE BEHAVIOR

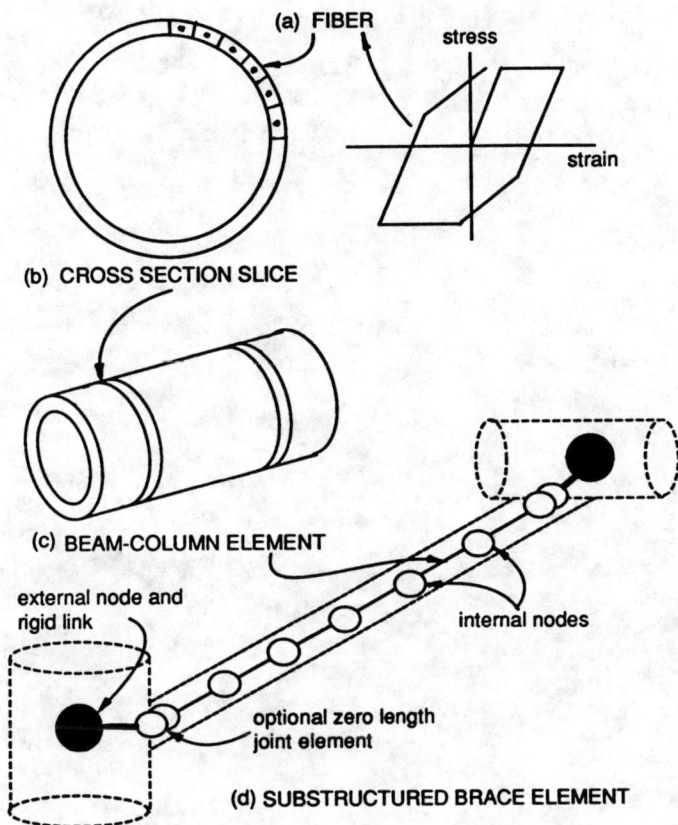

(a) FIBER

stress

strain

(b) CROSS SECTION SLICE

(c) BEAM-COLUMN ELEMENT

external node and rigid link

internal nodes

optional zero length joint element

(d) SUBSTRUCTURED BRACE ELEMENT

FIGURE 3. BRACE ELEMENT INGREDIENTS

Parallel Computational Strategies for Large Space and Aerospace Flexible Structures: Algorithms, Implementations and Performance

Charbel Farhat

Centre for Space Structures and Controls, University of Colorado at Boulder, Campus Box 429, Boulder, CO 80309, USA

ABSTRACT

New computer hardware with parallel processing capabilities is now commercially available and has the potential of revolutionizing scientific computing. The process of moving finite element applications to concurrent processors faces significant obstacles which center on methods, algorithms, languages and implementations. In this paper, we present a general approach to the design of next generation finite element software. We present several parallel computational strategies for both implicit and explicit computations. We describe their implementation on a large set of parallel environments. A portable parallel finite element prototype code has been designed at the Center for Space Structures and Controls. We report on its performance on a wide variety of multiprocessors including iPSC-32, ALLIANT FX/8, CRAY-2 and the CONNECTION MACHINE.

INTRODUCTION

The design and analysis of advanced space and aerospace structural systems will demand computational resources that go far beyond those presently provided by ordinary mainframes. Examples of such systems are next-generation composite airplane structures and space stations. Parallel computers have recently undergone rapid development spurred by architectural advances. As a result, the use of such machines for engineering applications has attracted increased attention. However, moving engineering applications to concurrent processors faces significant obstacles that will have to be resolved. The obstacles center on methods, algorithms, data structures and languages. In this paper, we begin with an overview of the present status of parallel computers that is pertinent to finite element computations. Through the examples of SIMD, MIMD, local memory and shared memory multiprocessors, we address the impact of hardware architecture on the design and implementation of parallel algorithms and parallel data structures. We present efficient parallel computational strategies for both implicit and explicit finite element computations on the new multiprocessors. Next we focus on the additional impact of difficulties inherent in large space and aerospace structures. These are typically highly geometrically nonlinear and lead to matrices that are extremely ill-conditioned. Also, we address several issues related to parallel disk I/O. Finally, we conclude with the importance of "code portability" and we report on our *hands on* experience with parallel nonlinear FE simulations of large space structures on a broad range of shared memory and local memory multiprocessors, including iPSC-32, ALLIANT FX/8, CRAY-2 and the CONNECTION MACHINE.

See Figure 1b in colour section, page 165.

CHARACTERISTICS OF NEW PARALLEL COMPUTING ENVIRONMENTS

Several parallel computers have already been marketed commercially. Here, we do not discuss these individually. We rather focus on presenting an overview of their architecture and emphasize the impact of their hardware features on the design and implementation of parallel computational strategies for finite element computations. A review of some of the commercially available parallel systems can be found in Babb [1], where programming examples are also provided.

Multiprocessors can be generally described by three essential elements, namely, granularity, topology and control.

• Granularity relates to the number of processors and involves the size of these processors. A fine-grain multiprocessor features a large number of usually very small and simple processors. The CONNECTION MACHINE (65,536 processors) is such a massively parallel supercomputer. NCUBE's 1024-node and iPSC's 128-node models are comparatively medium-grain machines. On the other hand, a coarse-grain multiprocessor is typically built by interconnecting a small number of large, powerful processors - usually vector processors. ALLIANT FX/8 (8 processors), IBM 3090 (6 processors), CRAY X-MP (4 processors), CRAY-2 (4 processors) and the ETA-10 (8 processors) are examples of such multiprocessors and supermultiprocessors. Granularity directly affects the parallel strategy. On a coarse-grain multiprocessor, finite element computations can be parallelized at the subdomain level. On a fine-grain machine, they are best parallelized at the element and some times at the degree of freedom level. When designing parallel algorithms for finite element computations on coarse grained vector supermultiprocessors, one should preserve vectorization. This is because the potential speed-up due to interconnecting a few processors cannot compete with the speed-up due to the vector capabilities of a single processor. A good strategy for maximizing speed-up consists of organizing the computations around two nested loops, where the outer one is multiprocessed and the inner one is vectorized.

• Topology refers to the pattern in which the processors are connected and reflects how data will flow. Currently available designs include hypercube arrangement, network of busses, and banyan networks. Usually, the interconnection topology is related to the memory organization. For example, iPSC, NCUBE and the CONNECTION MACHINE are local memory multiprocessors with a hypercube topology. On these systems, a processor is assigned its own (local) memory and can only access this memory. Independent processors communicate by sending each other messages. Efficient solution of finite element simulations on these machines requires minimizing the interprocessor communication bandwidth. This requires the mapping of adjacent elements as far as possible onto directly connected processors, which can be a non trivial problem. On the other hand, the processors on a shared memory system such as ALLIANT FX/8 are connected through a common memory bus and can access the same (global) large memory system. Adequate finite element parallel data structures are crucial for efficient computations on both shared and local memory multiprocessors. On a local memory machine, one has to introduce the concept of distributed data base and data structure. Each local memory is loaded only with the data relevant to the computational task assigned to its attached processor. For a system with thousands of processors, the total amount of available memory can be very large. Yet, it is the storage capacity of each local memory which really matters. Different finite elements require different amounts of data to be stored. For each finite element in the mesh, a material and geometrical nonlinear high order shell element may require an amount of data storage two orders of magnitude higher than a simple linear truss element. Hence, one may be able to assign one or several finite elements of a certain type to one processor but may fail in the attempt to assign one or several elements of another type to a similar processor. Also, in the case of MIMD (multiple instruction multiple data) machines such as iPSC and NCUBE, one has to ensure that the compiled subroutines can be accommodated on the local memory. Consider the case where a processor is mapped onto a submesh containing different types of elements. In this situation, one has to load into the processor's local memory all the element libraries for the types encountered in the assigned submesh. Generally, one can overcome these problems by devising an intelligent partitioning scheme and a compact data structure. Careful data structures must also be designed for shared memory multiprocessors to avoid potential serializations due to memory conflicts.

• Finally, control describes the way the work is divided up and synchronized. Of particular interest are the SIMD (single instruction multiple data) and MIMD (multiple instruction multiple data) machines. The CRAY-2 (4 processors) and iPSC (128 processors) are respectively a shared memory MIMD supermultiprocessor and a local memory MIMD hypercube. They can simultaneously execute multiple instructions which can operate on multiple data. The CONNECTION MACHINE is an SIMD system where a single instruction is executed at a time - an instruction which can operate on multiple data. Typically, on an SIMD machine a single program executes on the front end and its parallel instructions are submitted to the processors. On an MIMD parallel processor separate program copies execute on separate processors.

Practically, local memory parallel processors are more difficult to program than shared memory multiprocessors. However, it is believed that local memory systems are easier to scale to a large number of processors. Shared memory multiprocessors are usually coarse grained because the bus to memory saturates and/or becomes prohibitively expensive above a few processors. Note that on SIMD machines, one has to devise special tricks to be able to process in parallel finite elements of different types, since these do not involve the same instructions and only one instruction can be executed at a time.

Recommending one hardware architecture over another for parallel finite element computations is beyond the scope of this paper. Our goal is to fully exploit the computational power of the fastest currently available multiprocessors of any architecture for solving large nonlinear finite element dynamic simulations.

ARCHITECTURE OF A PARALLEL FINITE ELEMENT SOFTWARE

In this section we summarize a new concurrent computer program architecture capable of handling linear and nonlinear, static and dynamic analyses. This software architecture is flexible enough to be implemented on all local memory and shared memory SIMD and MIMD multiprocessors. Based on it, a prototype program has been designed at the Center for Space Structures and Controls. It includes several special purpose modules which constitute optimal algorithms for solving specific subproblems, such as domain decomposition, direct and iterative solution of systems of equations, solution of transient dynamics, and mesh refinement. These modules are designed to be consistent with the topology and data structure of the overall program architecture, so that a maximum overall efficiency can be achieved during a complete finite element analysis.

DIVIDE AND CONQUER

In structural analysis the technique of dividing a large system into a system of substructures is very old and is still used extensively. For large aerospace structures, its use is often motivated by the fact that different components are designed *in parallel* by different groups or companies. Therefore only the basic static and dynamic properties of the substructure need to be *communicated* between groups. This approach has also resulted in computational saving. If the design of one component is changed, only that substructure needs to be re-analyzed and the global system of substructures re-solved. In the case of limited nonlinear systems, only the substructures which are nonlinear need be studied incrementally with time.

It is clear that this traditional substructuring approach can be used with parallel processors if the complete finite element system is subdivided so that each group of elements within a small domain is assigned to one processor. The data structure for such an approach is very simple. On local memory multiprocessors, only the storage for the node geometry and element properties within the substructure need be stored within the RAM of the processor assigned to that substructure. In addition, concurrent formation and reduction of the stiffness matrix for that region require no interprocessor communication. On shared memory multiprocessors, the substructure data is accessed only by the processor assigned to that substructure, so that no memory conflict occurs. After the displacements are found, the postprocessing of substructure stresses can be done in parallel.

AUTOMATIC DOMAIN DECOMPOSER

Here, it becomes essential to distinguish between a substructure and a subdomain. By subdomain, we mean a collection of elements of the global domain (mesh) that is assigned to a corresponding processor. A subdomain may hence represent anything from one element of a physical substructure to several substructures.

Domain decomposition is attractive in finite element computations on parallel architectures because it allows individual subdomain operations to be performed concurrently on separate processors. Given a number of available processors, N_p, an arbitrary finite element domain is decomposed into N_p subdomains, where each of the following computations can be carried out independently of similar computations for the other subdomains, and hence performed in parallel:

- formation of element matrices;

- assembly of global matrices;

- partial factorization of the stiffness matrix (either in view of the solution of the equations of equilibrium with a direct method, or for preconditioning these before the use of an iterative numerical scheme for that purpose)

- state determination or evaluation of the generalized stresses.

All of the operations above can be executed concurrently without any synchronization or message passing between the processors. Since the time to complete a task will be the time to complete the longest parallel subtask (which is the restriction of a task to a subdomain), an algorithm for domain decomposition will be efficient only if it yields subdomains that require an equal amount of execution time for each of the operations discussed above. In other words, the algorithm has to achieve a load balance among the processors.

Whatever numerical scheme or computational strategy is selected, the parallel solution for the generalized displacements u usually requires explicit synchronization on shared memory multiprocessors and message passing on local memory ones. The reason is that the physical subdomains share physical information along their common boundaries that cannot be treated independently or asynchronously in each subdomain. This results in an overhead that, if not minimized, may ruin the sought after speed-up. Hence, when partitioning an arbitrary finite element mesh, one has to minimize the number of interface nodes.

The above observations suggest that an automatic finite element domain decomposer must meet three basic requirements in order to be successful: (1) it must be able to handle irregular geometry and arbitrary discretization in order to be general purpose; (2) it must yield a set of balanced subdomains in order to ensure that the overall computational load will be as evenly distributed as possible among the processors; (3) it must minimize the amount of interface nodes in order to reduce the cost of synchronization and/or message passing between the processors. A simple and automatic finite element decomposer that meets these requirements and that is completely transparent to the user is described in [2]. It is applied to the discretized flexible aircraft shown in figure 1.a. Color figure 1.b displays the resulting subdomains for an 8-processor machine (for the sake of clarity, only half of the domain decomposition is shown).

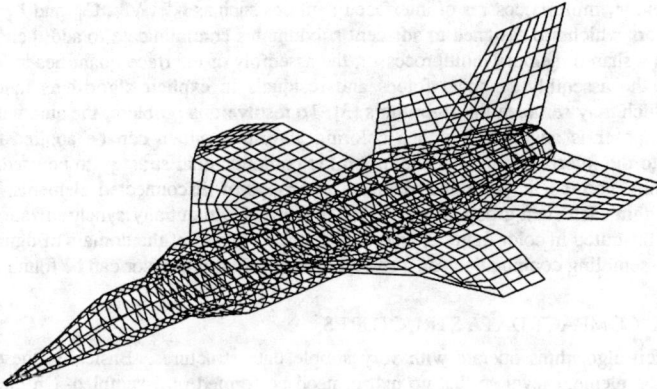

FIG.1.a. FE discretization of a flexible aircraft

PARALLEL SET UP OF THE FINITE ELEMENT SIMULATION

Consider the automatic decomposition of the mesh shown in figure 1.a into a system of subdomains D_j. The mesh nodes which are common to the subdomain interfaces define a unique global interface noted D_i. They are automatically identified by the program. The nodal point unknowns within the subdomains are numbered first, and the interface nodes are numbered last. The resulting stiffness matrix K has the following "arrow" pattern.

$$
K = \begin{bmatrix}
K_{11} & & & & & & K_{si}(1) \\
 & K_{22} & & & & & K_{si}(2) \\
 & & \cdots & & & & \cdots \\
 & & & K_{jj} & & & K_{si}(j) \\
 & & & & \cdots & & \cdots \\
K_{1i}^{T} & K_{2i}^{T} & \cdots & K_{ji}^{T} & \cdots & & K_{ii}
\end{bmatrix} \qquad (1)
$$

Each diagonal submatrix K_{jj} represents the local stiffness of a subdomain D_j. An off-diagonal submatrix K_{ji} denotes the coupling stiffness between D_j and the interface D_i. Block K_{ii} is the stiffness associated with D_i. Consequently, the vector of unknown responses u is partitioned into subvectors u_j which correspond to the degrees of freedom lying in region D_j; similarly, f_j denotes the part of the loading vector f associated with D_j. If a consistent mass matrix, M, is used for the dynamic analysis, M has the same pattern as K. If a lumped mass matrix is used, M has the same pattern as f. Each processor, p_j, is assigned to a subdomain D_j. It forms and stores in parallel with the other processors the local stiffness K_{jj}, its corresponding coupling term K_{ji}, the mass M_j, and the prescribed loading f_j. An eventual damping term, C, is treated in a similar fashion to the mass and stiffness terms. As noted earlier, the set up of the problem in each subdomain D_j does not require interprocessor communication or processor synchronization.

Implicit algorithms usually require the assembly of the stiffness, mass and damping matrices. Explicit algorithms require only the assembly of forces and residuals at the finite element nodes. Hence, both computational strategies do involve some kind and some amount of assembly operations. Inside each subdomain D_j, the assembly of all of K_{jj}, K_{ji}, M_j, C_{jj}, C_{ji} and f_j requires no interprocessor communication on a local memory multiprocessor and does not induce any memory conflict on a shared memory multiprocessor. On the other hand, the assembly on local memory multiprocessors of interface quantities such as K_{ii}, M_i, C_{ii} and f_i requires that the processors which are assigned to adjacent subdomains communicate to add their local contributions. On a shared memory multiprocessor, the assembly of interface quantities in implicit algorithms and the assembly of nodal forces and residuals in explicit algorithms induce memory conflicts which may serialize the algorithms [3]. To resolve this problem, the automatic finite element decomposer is augmented with a coloring algorithm which can be applied to the entire domain or to the interface only, depending on the computational strategy to be used. In essence, this algorithm reorders the specified elements in groups of disconnected elements. The disjoint elements within each such group are processed in parallel without any synchronization. This simple idea is illustrated in color figure 2,* where it is applied to half of the domain of figure 1.a. More details on assembling computations on a shared memory multiprocessor can be found in [3, 4, 13].

PARALLEL COMPACT DATA STRUCTURES

Explicit algorithms operate with very simple data structures. Basically, these algorithms operate at the element level so that no matrix need be formed or assembled. On the other hand, implicit computations often require the factorization of some matrix, for example, the stiffness matrix K.

Since all the subdomains D_j are separated by the interface, they constitute a set of disconnected meshes that can be numbered independently. Each diagonal block K_{jj} is stored in skyline/profile form. The degrees of freedom in each subdomain are renumbered in parallel in order to minimize the local storage requirements [5]. Note that the complexity for renumbering N_p disconnected subdomains D_j is much lower than for renumbering the entire domain D. Moreover, this phase is fully speeded up by the multiprocessing capabilities.

The coupling blocks K_{ji} are usually very sparse and do not introduce fill-in if a suitable direct method is selected for the factorization of K. Consequently, only their non zero values are stored column-wise in packed lists. In addition to K_{jj}, K_{ji} and F_j, each processor holds or is assigned to a set of columns of K_{ii} which is the stiffness associated with the subdomain D_i. These data structures and the processor assignments are summarized in figure 3.

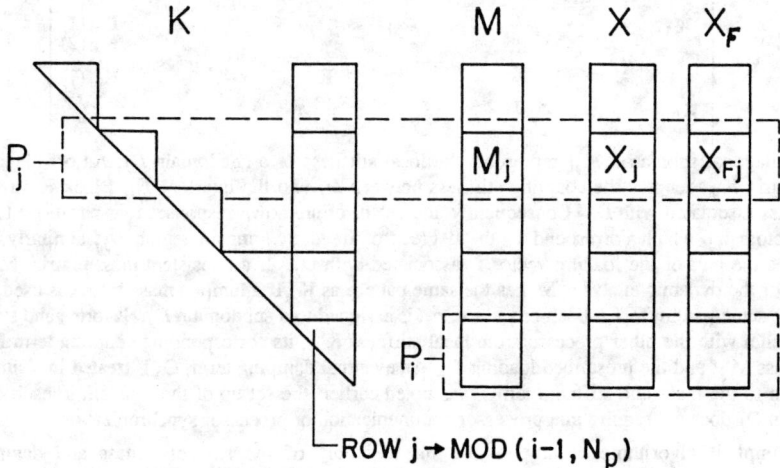

FIG.3. FE parallel data structures

*See Figure 2 in colour section, page 166.

POST PROCESSING - STATE DETERMINATION

The post processing phase - that is, the evaluation of field derivatives such as stresses and fluxes - is parallelized subdomain-by-subdomain or element-by-element on both local memory and shared memory architectures, without any communication or synchronization overhead. However, on local memory multiprocessors, state determination in nonlinear analysis can face load balancing problems. For example, the number of elements which may yield vary from subdomain to subdomain. Since the complexity of the computation of the internal forces depends on the constitutive equation that applies to the particular element in a specific subdomain, this implies that some processors would have to work less than others and hence would have to wait for each other. A load balancing strategy for problems where the nonlinearity is localized is described in [6]. It can be extended to the case where the analyst can predict the regions in the finite element mesh that will show only a linear behavior. The computational strategy is summarized below.

Let D be a given arbitrary finite element domain, and let $D^{(n)}$ be the known subset of D where nonlinearities are localized and/or predicted. With these definitions, $D^{(n)}$ may be a set of regions separated by linear substructures, and hence does not have to be a single physical substructure. We shall assume that $D^{(n)}$ has been somehow marked during the mesh generation phase, so that all its elements can be identified at any time by the software. In our program, these elements are marked with a specific element attribute value. Also, we define $D^{(l)}$ as being the collection of elements of D which are known (or predicted) to undergo only linear deformations. Hence, we can write:

$$D = D^{(l)} \bigcup D^{(n)}$$

$$D^{(l)} \bigcap D^{(n)} = 0$$

The automatic subdivision algorithm is extended to perform two passes on D. In the first pass, the marked elements are skipped and $D^{(l)}$ is split into N_p subdomains. The second pass decomposes $D^{(n)}$. As a result, D is subdivided into two sets of N_p subdomains each, namely:

- linear ones (or predicted so): $D^{(l)}_j$, $j = 1, 2, ..., N_p$; and
- nonlinear ones (or predicted so): $D^{(n)}_k$, $k = N_p+1, N_p+2, ..., 2N_p$.

Renumbering is done as for the linear case, the linear subdomains being treated first. Each processor p_j is assigned the pair of subdomains $D^{(l)}_j$, $D^{(n)}_{j+N}$, so that load balancing is achieved. Further details on the implementation of this strategy on local memory MIMD multiprocessors can be found in [6].

On shared memory multiprocessors, load balance for state determination in nonlinear analysis is achieved simply by assigning the processors to elements in a self scheduled manner - that is, elements are assigned to processors on the basis of first available, first served. This is of course possible because each processor can access the large shared memory of the system.

LINEAR AND NONLINEAR STATIC ANALYSES

In the following, the linear analysis is treated as a particular case of the nonlinear one.

Discrete equilibrium equations arising from finite element nonlinear formulations may be written in the general compact form

$$r(u, p, \theta) = 0 \qquad (2)$$

where u denotes the unknown vector of generalized displacements (rotations, temperatures, etc.) at the nodes of the discretized geometrical domain, p denotes a set of control parameters, θ is a functional of past history of the generalized deformation gradients, and r denotes the residual vector of out-of-balance generalized forces (moments, fluxes, etc.). Equation (2) covers all geometrical nonlinearities, material nonlinearities and several types of boundary condition non-linearities.

The Newton-Raphson method and its numerous variants, collectively known as *Newton-like* methods, are the most popular class of methods for the solution of (2) on conventional computers. These methods can embed either a direct or an iterative algorithm for the solution of a linearized system of equations.

Direct solution techniques have been popular among engineers, mainly because of two advantages they possess over iterative schemes:

- they are robust for ill-conditioned systems which often arise in the analysis of flexible space structures

- their execution time can be estimated for any given problem.

They can be sensitive to round-off error (matrices with high condition number), but their main disadvantage is that they suffer from having excessive storage requirements for large matrices, so that an out-of-core solution is often required. However, they are still attractive for supercomputers such as CRAY-2, where the main memory can store up to 256 million double precision words.

On the other hand, the use of iterative schemes for the solution, at each step, of the linearized system of equations has two desirable advantages:

- it efficiently exploits the sparsity of the involved matrices and therefore requires less storage than direct schemes

- it provides a means of controlling the accuracy of the solution.

The preconditioned conjugate gradient (PCG) has emerged over the last decade as a favorite algorithm for solving large sparse systems of linearized equations on sequential, vector [7, 8], and parallel [9, 10, 11] computers.

However, conventional *Newton-like* methods may behave poorly near bifurcation points and often fail to handle path-dependent problems such as plastic flow, where the stiffness matrix may oscillate wildly as the solution changes by small amounts. For such problems, explicit dynamic relaxation (DR) is a very robust iterative computational strategy.

Since the analyst would like to select the right strategy for the right problem, we have implemented within the same prototype parallel code *Newton—like* methods with iterative and direct solvers and explicit dynamic relaxation algorithms.

NEWTON-LIKE METHODS

Newton-like methods for solving nonlinear systems of equations having the general form $r(u, p, \theta) = 0$ are usually related to the following iteration scheme:

For $k = 0,1,2,...$ until convergence do:
$$\text{Solve } K(u_k)\delta u_{k+1} = -\delta f(u_k)$$
$$\text{Set } u_{k+1} = u_k + \delta u_{k+1}$$

where $K(u_k)$ is the stiffness matrix evaluated in the displacement state u_k, δu_{k+1} is the vector of unknown displacement increments, and $\delta f(u_k)$ is the vector of out-of-balance forces.

Next we overview both parallel direct and iterative methods for the solution at each iteration of the linearized system of equations. We emphasize only important parallel computation aspects and refer the reader to references [12, 13, 14, 15] for implementation details on local and shared memory multiprocessors.

- *Direct Method*

A 5-step direct algorithm for the solution at each iteration of the complete finite element system (see equation (1)) is given below in a symbolic form:

(S1)	Factor locally	$\mathbf{K}_{jj} = \mathbf{L}_j \mathbf{D}_j \mathbf{L}_j{}^T$	$(j = 1, N_p)$
(S2)	Eliminate	$\mathbf{K}_{ji}{}^T \mathbf{K}_{jj}{}^{-1} \mathbf{K}_{ji}$	$(j = 1, N_p)$
		and $\mathbf{K}_{ji}{}^T \mathbf{K}_{jj}{}^{-1} \delta \mathbf{f}_j$	$(j = 1, N_p)$
(S3)	Update interface	$\mathbf{K}_{ii} = \mathbf{K}_{ii} - \displaystyle\sum_{j=1}^{j=N_i} \mathbf{K}_{ji}{}^T \mathbf{K}_{jj}{}^{-1} \mathbf{K}_{ji}$	
		and $\delta \mathbf{f}_i = \delta \mathbf{f}_i - \displaystyle\sum_{j=1}^{j=N_i} \mathbf{K}_{ji}{}^T \mathbf{K}_{jj}{}^{-1} \delta \mathbf{f}_j$	
(S4)	Solve interface	$\mathbf{K}_{ii} \delta \mathbf{u}_i = \delta \mathbf{f}_i$	
(S5)	Back-solve locally	$\mathbf{K}_{jj} \delta \mathbf{u}_j = \delta \mathbf{f}_j - \mathbf{K}_{ji} \delta \mathbf{u}_i$	$(j = 1, N_p)$

Clearly, steps (S1), (S2) and (S5) can be carried out concurrently on all N_p processors without any interprocessor communication (local memory MIMD machines) or synchronization (shared memory multiprocessors). Step (S3) deserves special attention. On local memory MIMD multiprocessors, some message passing is required because $\mathbf{K}_{ii}$ is scattered column-wise among the processors [12]. On shared memory multiprocessors, it can be shown that step (S3) can be carried out in parallel without any memory conflict [13]. Finally, step (S4) is treated with a dedicated parallel profile equation solver [14].

Remarks

1. Even on shared memory multiprocessors, we have found that this parallel algorithm is faster than the one consisting of applying the parallel profile solver to the complete finite element system. On one hand, steps (S1), (S2), (S5) (and (S3) on shared memory systems) can be executed asynchronously. On the other hand, if n denotes the number of equations to be solved, the parallel profile equation solver requires n broadcasts on a local memory system and n synchronizations on a shared memory machine. In our strategy, $n = N_i$, where N_i denotes the number of interface equations (the reader should recall that the finite element decomposer attempts to minimize N_i). The alternative algorithm - that is, applying the parallel profile solver to the N_t complete equations resulting from the finite element formulation - would require $n = N_t \gg N_i$ broadcasts or synchronizations. In other words, the parallelization of the finite element direct solution at the equation level is less efficient than at the subdomain level, on both hardware architectures. See reference [15] for further details concerning the impact of hardware features on algorithm implementation.

2. At each iteration, after the generalized displacements are found, state determination is carried out in parallel as indicated earlier.

3. On multiprocessors with a limited amount of memory, the interface final system of equations (Schur's complement) may not fit in core. This can happen, for example, if the mesh size is small and the number of processors is relatively high. In this case, the program switches to a parallel profile solver for the complete finite element system. Other possibilities for overcoming this difficulty include an out-of-core strategy or a semi-iterative solution scheme.

• *Semi-iterative Methods*

Equation (1) has inspired several preconditioners for the conjugate gradient method. Results from domain decomposition theory suggest that the interface equations are better conditioned than the entire system of equations [16]. Consequently, it is more efficient to apply PCG only to the system of interface equations

$$(\mathbf{K}_{ii} - \sum_{j=1}^{j=N_i} \mathbf{K}_{ji}{}^T \mathbf{K}_{jj}{}^{-1} \mathbf{K}_{ji}) \delta \mathbf{u}_i = \delta \mathbf{f}_i - \sum_{j=1}^{j=N_i} \mathbf{K}_{ji}{}^T \mathbf{K}_{jj}{}^{-1} \delta \mathbf{f}_j \tag{3}$$

Note that the left hand side of equation (3) need not be formed and stored. During a conjugate gradient iteration, it is used only multiplicatively. See references [13, 16, 17] for examples of preconditioners and their performances.

EXPLICIT DYNAMIC RELAXATION

Dynamic Relaxation is a robust iterative method for solving highly nonlinear systems. Unlike the conjugate gradient method, it does not need to be embedded within a Newton outer loop. The algorithm solves the nonlinear discrete quasistatic finite element equations (4) by viewing them as the steady state solution of the second-order pseudo-dynamic problem

$$M\ddot{u} + C\dot{u} + S(u,p,\theta) - f = 0 \qquad (4)$$

where M and C are fictitious mass and damping matrices constructed in a way that achieves computational efficiency when integrating (2) with the central difference scheme. $S(u,p,\theta)$ denotes the internal forces. To preserve the explicit form of the central difference integrator M must be diagonal and C is chosen, for example, as $C = c\,M$. Parameter c and stepsize δt are selected to obtain the fastest convergence:

$$\delta t \le 2/(\omega_{max}^2 + \omega_{min}^2)^{(1/2)}$$
$$c = 2\omega_{min}/(1 + \omega_{min}^2/\omega_{max}^2)^{(1/2)}$$

where ω_{min} and ω_{max} are respectively the lower and higher pseudo-frequencies of the pseudo-dynamic problem. These quantities need not be computed exactly. Rough estimates are sufficient. Further details on M, C, c and δt may be found in reference [18].

Dynamic Relaxation may be slow in some cases. However, it presents two advantages: it is very robust and it is fully vectorizable and parallelizable. The vector form of (4) clearly demonstrates the suitability of the algorithm for vector processors. By eliminating the need for matrices, one saves not only storage but also the overhead associated with interprocessor communication that is usually required by algorithms manipulating two dimensional arrays. The explicit nature of the central difference method allows computations on different vector subcomponents to be performed in parallel without any processor synchronization and/or interprocessor communication, except for the evaluation of $S(u,p,\theta)$ which requires special attention [13]. The internal generalized force vector is obtained by accumulating the contributions of several connected elements. If u_k denotes the generalized displacement associated with the $k-th$ degree of freedom, then

$$[S(u,p,\theta)]_k = \sum_{el=1}^{el=N_k} [S(u,p,\theta)]_k^{(el)}$$

where $[S(u,p,\theta)]_k^{(el)}$ denotes the contribution of element el connected to the $k-th$ degree of freedom and is computed directly at the element level from a potential functional. On a local memory system, the processors which are assigned to adjacent subdomains communicate to exchange partial results at common interface nodes. On a shared memory multiprocessor, these partial results are accumulated via the coloring technique described earlier. See reference [13] for implementation details.

LINEAR AND NONLINEAR DYNAMIC ANALYSES

In this section, we summarize parallel methods for the solution of the equations of equilibrium for a finite element system in motion:

$$M\ddot{u} + C\dot{u} + Ku = f \qquad (5)$$

where M, C, and K are the real mass, damping and stiffness $n \times n$ matrices; f is the prescribed external time dependent load vector. First we focus on Rayleigh-Ritz solution schemes, then we look at direct time integration algorithms. Detailed implementations on parallel processors for both strategies may be found in [6, 19].

RAYLEIGH-RITZ SOLUTION

For a very large n, say over 10,000, it is common to use a reduction technique before determining the solution of equation (5). In this approach, the solution $u(s,t)$ is assumed to be a linear combination of a set of independent vectors $v_i(s)$:

$$u(s,t) = \sum_{i=1}^{i=q} \alpha_i(t) v_i(s) \qquad (6)$$

where s represents the space variables, t is the time variable and q is much smaller than n. If V is the $n \times q$ matrix whose columns are v_i and α is the vector of dimension q and with entries $\alpha_i(t)$, the matrix form of (6) is:

$$u(s,t) = V\alpha \qquad (7)$$

Premultiplying (5) by V^t and substituting (7) lead to the reduced equations of motions:

$$(V^t M V)\ddot{\alpha} + (V^t C V)\dot{\alpha} + (V^t K V)\alpha = V^t f \qquad (8)$$

These reduced equations of motion are of size $q \ll n$.

The Lanczos algorithm [20] and other Lanczos-like algorithms [21] generate a set of M-orthonormal vectors v_i, $i = 1, q$, at a fraction of the computational effort required for the calculation of exact mode shapes. The Lanczos vectors have the following properties:

- $V^t M V = I_q$
- $(V^t K V)^{-1} = T_q$

where I_q is the $q \times q$ identity matrix and T_q is an $q \times q$ symmetric tridiagonal matrix.

Using the Lanczos vectors and assuming a Rayleigh damping matrix $C = m M + k K$ transforms the original equations of motion (5) into the reduced tridiagonal system of second order differential equations:

$$T_q \ddot{\alpha} + (m T_q + k I_q)\dot{\alpha} + I_q \alpha = T_q V^t f \qquad (10)$$

The reduced tridiagonal equations above may be either integrated directly or the eigenvectors and eigenvalues of T_q may be calculated and the solution to (10) may be obtained by integrating the uncoupled equations. The latter approach is treated first. The direct integration of (10) is a particular case of the direct time integration of (5) and is treated next.

A Rayleigh-Ritz procedure based on a Lanczos-like algorithm [21] is summarized below.

•	$K = L^t D L$	$n \times n$ system
•	$K \bar{v}_1 = f$	solve for $\bar{v}_1$
	$b_1 = (\bar{v}_1^t M \bar{v}_1)^{(1/2)}$	M-normalization
	$v_1 = \bar{v}_1 * 1/b_1$	
	$K \bar{v}_i = M v_{i-1}$	solve for v_i
	$a^{i-1} = \bar{v}_i^t M v_{i-1}$	diagonal of T_q
	$c^j = v_j^t M \bar{v}_i$	compute for $j = 1,...,i-1$
	$\hat{v}_i = \bar{v}_i - \sum_{j=1}^{j=i-1} c_j v_j$	M-orthogonalization
	$b_i = (\hat{v}_i^t M \hat{v}_i)^{(1/2)}$	off-diagonal of T_q
	$v_i = \hat{v}_i * 1/b_i$	M-normalization

$$\left[T_r \right] = \begin{bmatrix} a_1 & b_2 & 0 & & & 0 \\ b_2 & a_2 & b_3 & \cdot & & \cdot \\ 0 & b_3 & a_3 & \cdot & \cdot & \cdot \\ 0 & 0 & \cdot & \cdot & \cdot & 0 \\ 0 & \cdot & \cdot & b_{r-1} & a_{r-1} & b_r \\ & & & & b_r & a_r \end{bmatrix}$$

$$
\begin{aligned}
\bullet \quad \mathbf{T}_q\, \mathbf{Z} &= \mathbf{Z}\, \Omega \\
[\,\omega^2\,] &= [\,1/\Omega\,]
\end{aligned}
$$

$$
\bullet \quad \mathbf{V}^* = \mathbf{V}\, \mathbf{Z}
$$

The above algorithm consists essentially of four phases:

- factorization of the stiffness matrix **K**
- generation of Lanczos vectors **v** and construction of $\mathbf{T}_q$
- solution of the reduced eigenvalue problem:
  $$\mathbf{T}_q\, \mathbf{Z} = \mathbf{Z}\, \Omega$$
- computation of Ritz vectors and corresponding frequencies:
  $$\mathbf{V}^* = \mathbf{V}\, \mathbf{Z}$$
  $$[\,\omega^2\,] = [\,1/\Omega\,]$$

After the computations above are performed, the uncoupled equations of motion are solved for $\alpha_i(t)$ and the solution to (5) is computed as $\mathbf{u} = \mathbf{V}^*\, \alpha = \mathbf{V}\, \mathbf{Z}\, \alpha$.

The parallel algorithms for the finite element static analysis are used to factor **K** and generate the Lanczos vectors. The construction of $\mathbf{T}_q$ involves vector manipulations consisting of inner products and trivial scaling. On multiprocessors with local memory, an inner product is carried out first locally, then the partial dot products are accumulated in $\log_2(N_p)$ stages, following a binary tree. On a shared memory multiprocessor, the coloring technique described earlier is invoked to avoid memory conflicts during the accumulation phase.

As noted earlier, $\mathbf{T}_q$ is an $q \times q$ tridiagonal symmetric matrix, and q is of the order of 100. On a local memory multiprocessor, T_q can be duplicated concurrently with a very low storage cost in each processor, as a set of two one dimensional arrays, storing respectively the main and the upper diagonals.

The solution of the reduced eigenvalue problem $\mathbf{T}_q\, \mathbf{Z} = \mathbf{Z}\, \Omega$ does not involve a significant amount of computational effort, since the system is tridiagonal. However, it can be further reduced by the use of parallel processing. A simple algorithm for a concurrent extraction of all of the eigenvalues of T_q has been described in [19]. Basically, the spectrum of eigenvalues is divided into $N_p + 1$ subintervals $[\,s_p\,, e_p\,]$ each containing an almost equal number of eigenvalues (see reference [6] for details on that splitting). The task of each processor p is to compute the eigenvalues of T_q which lie within its assigned interval $[\,s_p\,, e_p\,]$, as the roots of the polynomial $P(\lambda) = \det \mathbf{T}_q - \lambda \mathbf{I}$. A regula-falsi search method together with the Sturm Sequence Property is used for this purpose. The corresponding eigenvectors are obtained via Inverse Shifted Iteration. This procedure has two advantages:

- No overlapping between the frequency subdomains can occur, and hence none of the eigenvalues/eigenvectors is computed twice (by two different processors).

- All processors perform their task concurrently without any communication or synchronization.

The parallel implementation of all four phases on local memory multiprocessors can be found in [6, 19].

DIRECT TIME INTEGRATION

Whether for integrating equation (5) or its reduced form (10), for linear or nonlinear problems, direct integration methods can be implicit or explicit. They involve algebraic computations which are similar to those of static analysis (factoring matrices, performing matrix-vector products and inner products, ...). Implicit schemes are parallelized at the subdomain level, as for the nonlinear static analysis, and explicit schemes are parallelized at the element level, as for the case of Dynamic Relaxation. A parallel implementation of the *Wilson*–θ method on Intel's iPSC is described in [6]. Its conversion for shared memory multiprocessors such as CRAY-2 is straightforward.

SPACE STRUCTURES

Systems arising from the finite element discretization of flexible space and aerospace structures are inherently ill-conditioned. Among the sources of this ill-conditioning we note

• the system size, which is typically very large

• the hybrid nature of the structure

• the presence of mesh irregularities and mesh distortions

• the use of shell elements for modeling.

Consider the finite element discretization of the flexible aircraft shown in figure 1.a. A shell element is used for modeling the behavior of the structure under aerodynamic loading. The computed condition number $\kappa_2(K) = ||K||_2 ||K^{-1}||_2 = \frac{\lambda_{max}}{\lambda_{min}}$, of the resulting system of equations is reported in figure 4 shown below, for increasing mesh sizes. These results are contrasted with the condition number associated with a regular three-dimensional cubic mesh where a linear brick element is used for modeling.

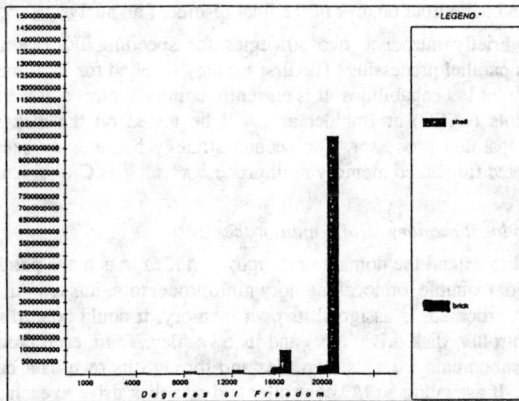

FIG. 4. Condition number of an aerospace structure

Clearly, for both meshes, the condition number increases dramatically with the mesh size. However, in all cases, the flexible shell element yields condition numbers which are four orders of magnitude higher than the ones resulting from the solid brick element.

It is known that the number of conjugate gradient (CG) iterations which are required for convergence is of the order of the square root of the condition number of the system. This means that for the mesh of figure 1, approximately 10,000 CG iterations are required in order to achieve convergence.

An important way around this difficulty is, of course, to precondition the system - that is, to find a preconditioner matrix P satisfying $\kappa(P^{-1}K) \rightarrow 1$. Then, one focuses on solving the preconditioned system $P^{-1}K\,u = P^{-1}f$ rather than on solving the original system of equations $K\,u = f$. Finding such a P which:

- is effective at reducing the condition number associated with very flexible structures
- requires little storage and little computer time to factor
- is vectorizable and parallelizable,

is still an active area of intensive research. Until such a **P** is found and well accepted, direct solvers will always find their golden place in commercial finite element packages.

PARALLEL I/O

Realistic finite element modeling of real engineering systems involves the handling of very large data spaces which can amount to several gigabytes of memory. To cope with this, many programs, in the general area of solid mechanics and structural analysis, use out-of-core data base management systems. However, I/O traffic between the disk and the processor memory slows down the computations significantly and can dominate the overall cost of the analysis.

In a typical finite element analysis, nodal and element data are retrieved from a storage disk before their processing, then stored back on the same storage disk after their processing has been completed. Examples include the transfer of nodal point coordinates, and of elemental mass and stiffness matrices in element-by-element computational procedures, and of history response arrays in time-stepping algorithms for linear and nonlinear dynamics. Other examples include the movement, into core and out of core, of blocks of an assembled stiffness or mass matrix in original or factored form, and the output on disk of the final results of an analysis.

Here, we briefly introduce two strategies for speeding I/O operations in finite element analysis through parallel processing. The first strategy is tuned for local memory MIMD multiprocessors with parallel I/O capabilities. It is currently being developed at the Center for Space Structures and Controls (CSSC) at Boulder and will be tested on the disk farming capabilities of NCUBE's 1024 parallel processor. The second strategy has already been designed, also at the CSSC. It is targeted for shared memory multiprocessors such as CRAY-2 and IBM 3090.

• *Parallel I/O on local memory MIMD multiprocessors*

It is very natural to extend the domain decomposition idea to achieve parallel I/O in the finite element analysis. For example, on local memory multiprocessors, it is tempting to imagine that in the same way that a processor is assigned its own memory, it could be attributed its own set of I/O devices (I/O controller, disk drive, *etc.*) and its own files. Then, each processor would read/write the data for its subdomain from its own files and through its own data base, in parallel with the other processors. If assigning an I/O controller and/or a disk drive to each processor is impractical and/or impossible, as is probably the case for a fine-grain system, it is possible for a cluster of processors.

After a given finite element domain is decomposed, it is grouped into regions R_i, each containing a cluster of subdomains $D^{R_i}{}_j$. A host processor p^{h_i} is uniquely mapped onto each region R_i. It is assigned the task of handling I/O manipulations associated with computations performed primarily in the cluster of subdomains within R_i. Host processor p^{h_i} directly transfers data from p_j's RAM, to its attached disk and vice versa. Each host processor $p^h{}_i$ is loaded with the same program driver, which we will call the *listener*, and the same copy of a data base manager, DBM. The main task of the listener is to listen to processor p_j's requests for I/O, and to activate DBM accordingly. These requests may be:

- receive data from p_j and store it in disk using DBM;
- retrieve data from disk through DBM and send it to p_j;
- retrieve data from disk through DBM, send it to another host processor $p^h{}_j$ together with the instruction of broadcasting it to a specified number of computational nodes that are directly connected to $p^h{}_j$; this particular operation implements potential exchange of data between subdomains.

• *Parallel I/O on shared memory MIMD multiprocessors*

Unlike the previous approach, a single executable version of a sequential DBM is stored in the global memory of the multiprocessor. Moreover, there is no need for a listener since all processors can access directly DBM, the I/O library and the disks. However, the core of the computational routines needs to be slightly modified to distinguish between global variables, which are shared by all the defined processes, and local variables, which have a single name to ease programming but a distinct value for each process. The essence of the strategy consists of:

- distinguishing between synchronous and asynchronous I/O requests
- distinguishing between shared and private data
- partitioning the data stream into a balanced number of contiguous subsets equal to the number of calling processes.

The design, implementation and performance of a simple parallel I/O manager, PIOM, operating with the above logic will be published elsewhere in detail.

SOFTWARE PORTABILITY

There is no standard parallel language that is currently available, even for a given class of multiprocessors. Consequently, a researcher in the area of applied parallel processing who is lucky enough to have access to a wide variety of parallel computers, may give up on experimenting with these after considering the amount of re-coding that he would need to do in order to transport his programs from one machine to another. The lack of a standard parallel language is also discouraging commercial finite element software developers from moving their products to parallel processors.

In the short term, one can find a temporary solution in portable parallel constructs. *The Force* [22] provides a FORTRAN style parallel programming language for shared memory MIMD multiprocessors, utilizing an extensive set of parallel constructs. It offers two desirable advantages:

- it insulates the programmer from process management, leaving him free to concentrate on the synchronization issues of parallel programming.
- it ensures the portability of the programmer's code to several different shared memory multiprocessors. Basically, the same code is run on any machine where *The Force* has been installed.

On ALLIANT computers, the compiler recognizes inherent parallelism at the DO loop level without the need for the programmer to invoke any explicit parallel construct. This may facilitate the work for the programmer. However, his code would not run on the CRAY-2, for example, because software support for multitasking on this supermultiprocessor is at the library level, where the user makes calls to ask the system for multitasking functions. For synchronization it is commonly necessary to wait until all processes have terminated a given task or to make sure that at any given time only one process modifies a variable. The procedures for these synchronizations are machine dependent. *The Force* relieves the programmer from the burden of modifying his code in order to port it to a new shared memory multiprocessor. Because only the *Force* constructs need be re-programmed from one multiprocessor to another, his precious code need not be modified. For example, if the desire is to request all processes to wait until the longest one has terminated, the same *Force* construct "*Barrier*" is invoked on any multiprocessor. On the CRAY-2 running under UNICOS, the *Force* preprocessor will read the simple "*Barrier*" statement and generate the following complex FORTRAN code:

```
CALL LOCKON(BARLCK)
IF (FFNBAR.LT.(NP - 1)) THEN
    FFNBAR = FFNBAR + 1
```

```
      CALL LOCKOFF(BARLCK)
      CALL LOCKON(BARWIT)
ENDIF
IF (FFNBAR .EQ. (NP-1))  THEN
ENDIF
IF(FFNBAR.EQ.0) THEN
  CALL LOCKOFF(BARLCK)
  ELSE
  FFNBAR = FFNBAR - 1
  CALL LOCKOFF(BARWIT)
ENDIF
```

which invokes the appropriate UNICOS multitasking software utilities.

The shared memory version of our prototype parallel code has been successfully implemented on ENCORE MULTIMAX, SEQUENT-BALANCE, ALLIANT FX/8 and CRAY-2, using the macros of *The Force*. Since we are not aware of any similar product for local memory machines, we had to write two other separate versions, one for Intel's iPSC and another for the CONNECTION MACHINE.

REALISTIC EXAMPLES - PERFORMANCE

Here we demonstrate the validity of our approach to linear and nonlinear, static and dynamic finite element parallel computations, and assess its performance on iPSC, ALLIANT FX/8, CRAY-2, and the CONNECTION MACHINE, which scan the various trends in today's parallel processing technology. We do not report absolute performance (MFLOPS rates) because the comparison of these multiprocessors is beyond the scope of this paper. For a given machine, we emphasize the measured speed-ups and demonstrate the benefits of parallel processing. The speed-up, S_{N_p}, is defined as the ratio $\frac{T_1}{T_{N_p}}$, where T_1 denotes the CPU time elapsed using only one processor and T_{N_p} denotes the CPU time elapsed using N_p processors. It is very important to note that T_1 measures the performance of a sequential version of the code which is different from the parallel one, in the sense that it does not contain any of the synchronization calls and it does not include the preprocessing phases such as domain decomposition and element coloring. Hence, the results we report herein account for the very little extra time spent in domain decomposition and element coloring.

Intel's iPSC

The Intel Personal SuperComputer is one of the first commercially available local memory parallel systems, with 32, 64, or 128 processors. Each processor is an Intel 80286 CPU with an Intel 80287 floating point accelerator, and has 512 Kbytes of local memory. The interconnection network is a hypercube (or n-dimensional cube, n = 5,6,7). Approximately 275 Kbytes of the 512 Kbytes RAM on each processor is available to the user's application, the rest being consumed by the operating system. The latter provides a simple message passing interface between the *nodes* (processors).

First, we report on our experience with solving finite element static problems on an iPSC-32. We seek the linear response of the flexible aircraft of figure 1.a to a prescribed aerodynamic loading. The mesh contains 1280 shell elements and 7680 degrees of freedom. We use the automatic decomposer of reference [2] to partition the finite element mesh into 32 balanced subdomains. The interface subdomain has 2030 equations. The speed-up of each of the phases of the static analysis is reported in table I below:

Problem set-up	31.4
Subdomain condensation	30.2
Interface solution	24.9
Subdomain solution	31.4
Stress computations	31.4
Overall	28.1

TABLE 1. Speed-up of a linear static analysis on iPSC-32

Clearly, relatively high speed-ups are achieved. Note however that the interface parallel solution is less efficient than the subdomain parallel solution.

Next, we consider the linear dynamic analysis of the same structure, using the Rayleigh-Ritz procedure described above. We perform the analysis using 2, 4, 8, 16, and 32 processors. For each case, we design the finite element mesh to completely fill the local memory of the activated processors. The measured speed-ups are reported in figure 6, and contrasted with the theoretical linear speed-ups. The displayed results demonstrate the efficiency of our approach to finite element parallel computations on local memory multiprocessors.

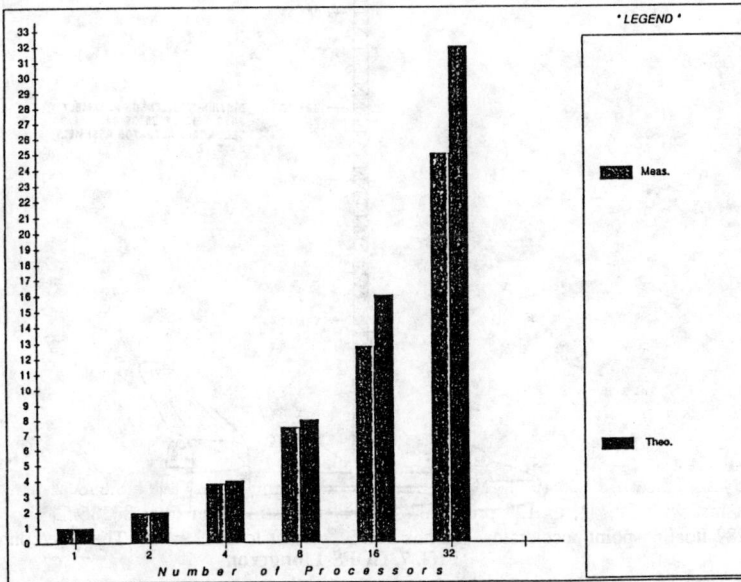

FIG. 6. Speed-up of a linear dynamic analysis on iPSC-32

Our experience with the iPSC is limited to linear finite element simulations, because of the lack of true I/O facilities on the *nodes* (a processor cannot open, read/write in a file).

ALLIANT FX/8

The ALLIANT FX/8 is a shared memory system consisting of up to 8 processors. Each processor supports the M68020 instruction set augmented with instructions for supporting floating point arithmetic, vector arithmetic, and concurrency.

Finite element computations on the ALLIANT FX/8 may be speeded up through a proper combination of vectorization and concurrency. In the following, we report separately each of the achieved speed-ups.

The nonlinear static analysis of NASA's COFS-I truss beam shown in figure 7 is performed twice on the ALLIANT FX/8, first using preconditioned conjugate gradient, then dynamic relaxation.

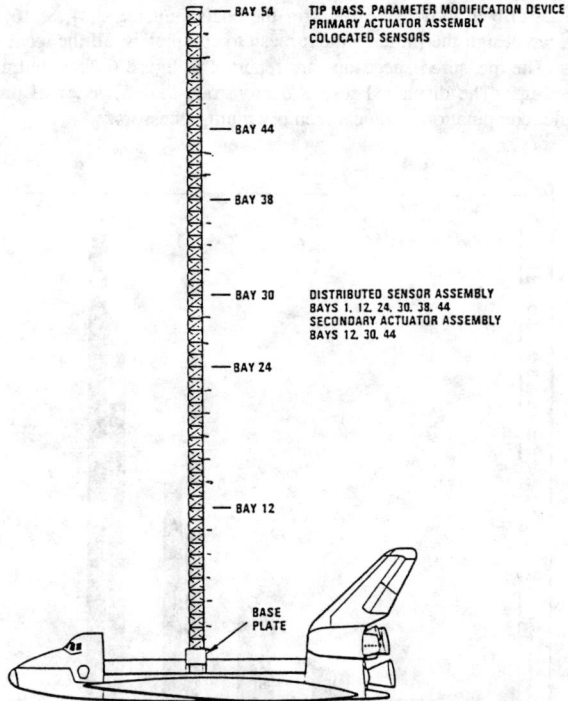

BAY 54 — TIP MASS, PARAMETER MODIFICATION DEVICE
PRIMARY ACTUATOR ASSEMBLY
COLOCATED SENSORS

BAY 44

BAY 38

BAY 30 — DISTRIBUTED SENSOR ASSEMBLY
BAYS 1, 12, 24, 30, 38, 44
SECONDARY ACTUATOR ASSEMBLY
BAYS 12, 30, 44

BAY 24

BAY 12

BASE PLATE

FIG. 7. COFS-I longeron

The finite element discretization of the longeron is shown in figure 8. It contains 1971 beam elements and 2016 degrees of freedom.

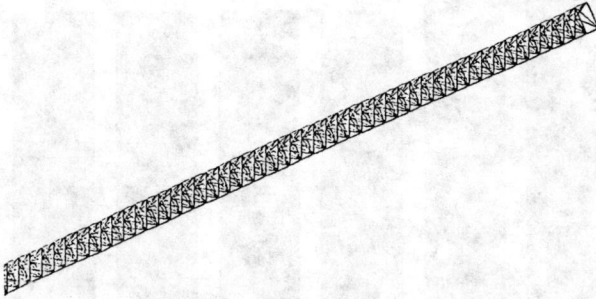

FIG. 8. Finite element discretization

The finite element mesh is first partitioned into 9 sets of disconnected elements, each represented by a color (see color figure 9).* Hence, at each round of explicit computations over the entire elements of the mesh, only 9 synchronization points are required. The elements of each set are further partitioned into subsets which are processed in parallel. Within each subset, vectorization is achieved by processing the elements in blocks of 32 at a time. The measured speed-ups due to vectorization and to concurrency are reported separately in figures 10 and 11. Here, the speed-up due to vectorization is defined as the ratio of the elapsed time using p processors with vectorization turned off, over the elapsed time using p processors with vectorization turned on.

*See Figure 9 in colour section, page 167.

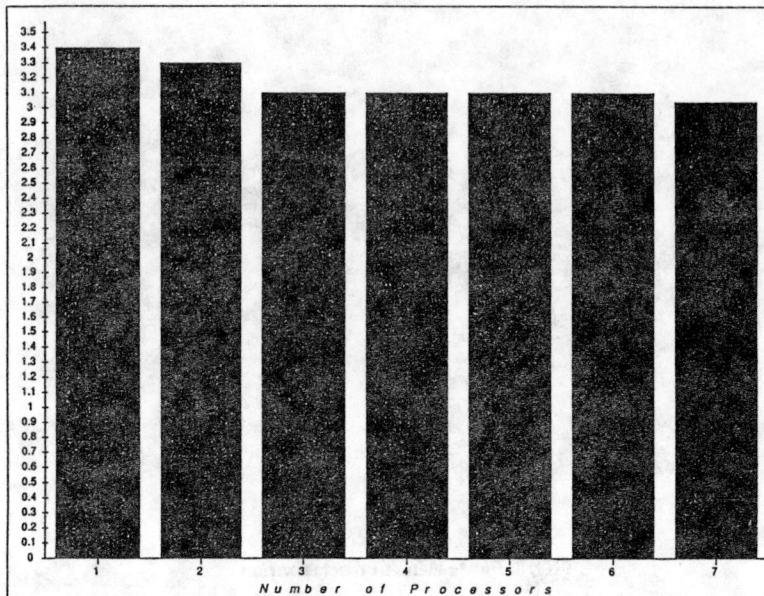

FIG. 10. Speed-up due to vectorization on ALLIANT FX/8

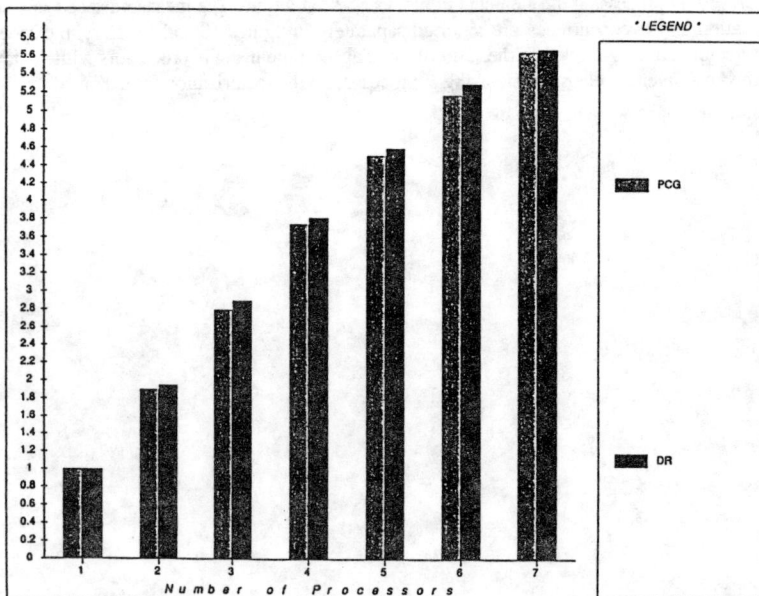

FIG. 11. Speed-up due to concurrency on ALLIANT FX/8

For a given finite element problem, the speed-up due to vectorization slightly decreases when the number of processors increases, because the length of a given vector within a process decreases. However, the total speed-up due to the combination of concurrency and vectorization remains high.

CRAY-2

The CRAY-2 supercomputer consists of up to 4 super vector processors sharing a large memory of 256 Mwords. Finite element computations on this system can be speeded-up through both vectorization and parallelization.

Figure 12 shown below displays a space station model.

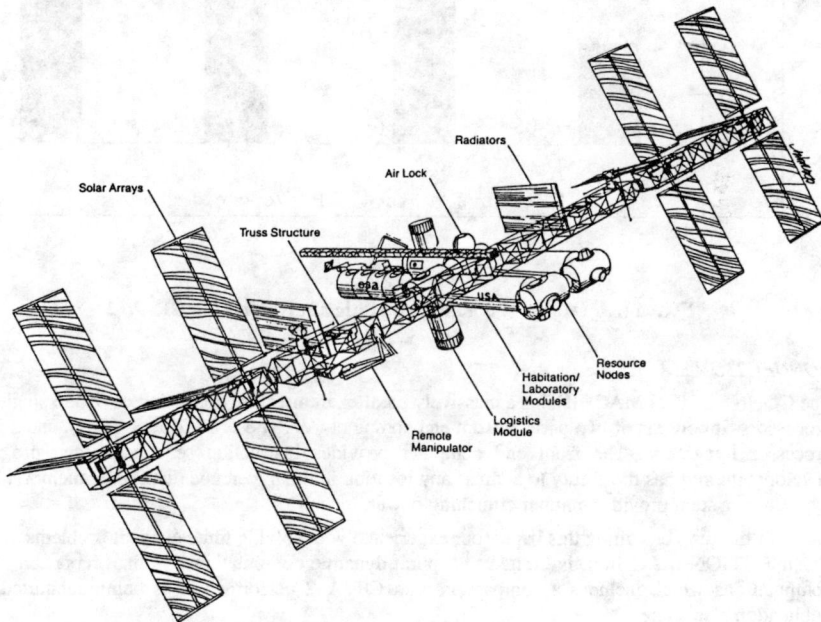

FIG. 12. Space station model

The finite element discretization of the truss box results in 11715 elements and 10800 degrees of freedom. The nonlinear static analysis of this structure is performed using a Newton-like algorithm with a direct method. The mesh is decomposed into 1, 2, 3, and 4 subdomains (see color figure 13)*corresponding to runs activating 1, 2, 3, and 4 processors. The coloring technique is invoked for explicit computations and interface processing. The corresponding speed-ups are reported in figure 14, shown below.

*See Figure 13 in colour section, page 168.

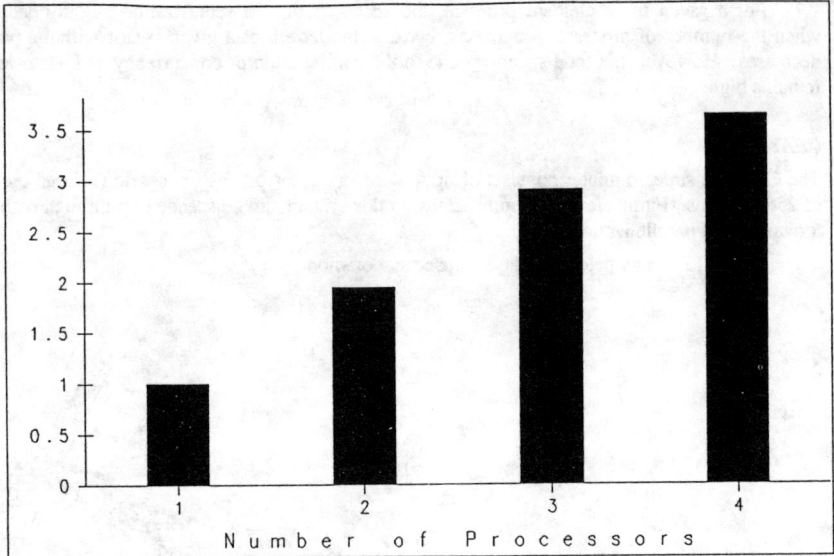

FIG. 14. Speed-up of a nonlinear static analysis on the CRAY-2

CONNECTION MACHINE

The CONNECTION MACHINE is a massively parallel architecture consisting of 65536 single bit processors. It consists of two parts, a front end computer (VAX, SYMBOLICS, SUN), and a 64K processor hypercube. The front end computer provides instruction sequencing and program development and has the ability to address any location in the hypercube distributed memory. The hypercube system provides number crunching power.

At the time of writing this paper, our experience with solving finite element problems on the CONNECTION MACHINE is limited to explicit dynamic computations. A final report on these computations which includes a comparison with CRAY-2 performances is being submitted for publication elsewhere.

CONCLUSION

A novel software architecture for finite element parallel computations is presented in this paper. It is flexible enough to be ported on both shared memory and local memory multiprocessors. It preserves vectorization and builds on top of it a layer of concurrency. Two kinds of parallelism are exploited:

- a *natural* parallelism which is inherent to finite element formulations
- an *artificial* parallelism which is introduced at the algebraic level.

Parallel data structures which accommodate both kinds of parallelism are also described. With these data structures, implicit static and dynamic computations are parallelized at the subdomain level. Explicit static and dynamic computations are parallelized at the element level, using a coloring technique. Strategies for speeding I/O bound problems are also described in this paper. A finite element prototype code based on the proposed software architecture has been implemented on iPSC-32, ALLIANT FX/8, CRAY-2 and the CONNECTION MACHINE. The performance of this code is assessed in this paper through several realistic finite element simulations. Very high speed-ups are demonstrated. It is hoped that the material presented in this paper can serve as a model for porting commercial finite element software to parallel processors.

REFERENCES

[1] Babb II, R. G. (Ed.). Programming Parallel Processors, Addison-Wesley Publishing, Inc., 1988.

[2] C. Farhat, A Simple and Efficient Automatic FEM Domain Decomposer, Computers & Structures, Vol. 28, No. 5, pp. 579-602, 1988.

[3] P. Berger, P. Brouaye and J. Syre, A Mesh Coloring Method for Efficient MIMD Processing in Finite Element Problems, Proc. Int. Conf. Par. Proc. , pp. 41-46, 1982.

[4] C. Farhat, Multiprocessors in Computational Mechanics, Ph. D Thesis, University of California at Berkeley, 1986.

[5] M. Hoit and E. Wilson, An Equation Numbering Algorithm Based on a Minimum Front Criteria, Computers & Structures Vol.16, No.1-4, pp.225-239, 1983.

[6] C. Farhat and E. Wilson, A New Finite Element Concurrent Computer Program Architecture, Int. J. Num. Meth. Eng., Vol. 24, pp. 1771-1792, 1987.

[7] G. Carey and B. Jiang, Element-by-Element Preconditioned Conjugate Gradient Algorithm for Compressible Flow, Proc. Int. Conf. Innovative Methods for Nonlinear Problems, W.Liu, T.Belytschko and K.Park (Ed.), Pineridge Press International Limited, Swansea, U.K., pp. 41-49, 1984.

[8] H. A. Van Der Vorst, The Performance of FORTRAN Implementations for Preconditioned Conjugate Gradients on Vector Computers, Parallel Computing, Vol. 3, No. 1, pp.49-58, 1986.

[9] K. Law, A Parallel Finite Element Solution Method, Computers & Structures, Vol. 23, No. 6, pp. 845-858, 1986.

[10] J. Kowalik and S. Kumar, An Efficient Parallel Block Conjugate Gradient Method for Linear Equations, Proc. 1982 Int. Conf. Par. Proc., pp. 47-52, 1982.

[11] R. Benner and G. Montry, Overview of Preconditioned Conjugate Gradient (PCG) Methods in Concurrent Finite Element Analysis, Internal Report from the Advanced Computer Science Project, Fluid and Thermal Sciences Department, Sandia National Laboratories, Albuquerque, New Mexico 87185, NTIS Report No. SAND85-2727, 15 pp., 1986.

[12] C. Farhat, E. Wilson and G. Powell, Solution of Finite Element Systems on Concurrent Processing Computers, Engineering With Computers, Vol. 2, No. 3, pp. 157-165, 1987.

[13] C. Farhat and L. Crivelli, A General Approach to Nonlinear FE Computations on Shared Memory Multiprocessors, Computer Methods in Applied Mechanics and Engineering, (in press)

[14] C. Farhat and E. Wilson, A Parallel Active Column Equation Solver, Computers & Structures, Vol. 28, No. 4, pp. 289-304, 1988.

[15] C. Farhat, C. Felippa and K. C. Park, Implementation Aspects of Concurrent Finite Element Computations, Parallel Computations and their Impact on Mechanics, ed. by A. K. Noor, ASME, New-York, pp. 301-316, 1987.

[16] P. E. Bjorstad and O. B. Widlund, Iterative Methods for the Solution of Elliptic Problems on Regions Partitioned Into Substructures, SIAM. J. NUMER. ANAL, Vol. 23, No. 6, December 1986.

[17] J. H. Bramble, J. E. Pasciak and A. H. Schatz, Preconditioners for interface problems on mesh domains, Dept. Math., Cornell Univ., Ithaca, NY, 1984.

132

[18] M. Papadrakakis, A Method for the Automated Evaluation of the Dynamic Relaxation Parameters, *Computer Methods in Applied Mechanics and Engineering*, Vol. 25, pp. 35-48, 1981.

[19] C. Farhat and E. Wilson, Modal Superposition Analysis on Concurrent Multiprocessors, *Engineering Computation*, Vol. 3, No. 4, pp. 305-311, 1986.

[20] B. N. Parlett, The Symmetric Eigenvalue Problem, Prentice Hall, Inc. Englewood Cliffs, 1980.

[21] E. Wilson, M. Yuan and J. Dickens, Dynamic Analysis by Direct Superposition of Ritz Vectors, *Earthquake eng. struct. dynam.*, Vol. 10, pp. 813-821, 1983.

[22] H. F. Jordan, Structuring parallel algorithms in an MIMD, shared memory environment, *Parallel Computing*, Vol. 3, No. 2, pp. 93-110, May 1986.

Some Comments On Structural Impact

Norman Jones

Impact Research Centre, Department of Mechanical Engineering, The University of Liverpool, P.O. Box 147, Liverpool L69 3BX, UK

SUMMARY

This article examines some aspects of the behaviour of structures subjected to large dynamic loads which cause an inelastic response. Particular emphasis is placed on several phenomena, which are incompletely understood, and yet have important implications for the development of numerical codes. It is concluded that further experimental work on ductile rupture and the scaling laws is required, which should be pursued in partnership with the architects of numerical codes.

NOTATION

h	depth of a double-shear beam specimen between the tips of the notches
ℓ_1	distance of the impact point from the nearest support
t	time
2B	width of a rectangular plate
E_a	actual energy absorbed in a full-scale prototype
E_p	energy absorption which is predicted for a full-scale prototype according to the geometrically similar scaling laws and the actual energy absorbed in a small-scale model.
G	impact mass
H	thickness of a structural member
2L	length of a beam or rectangular plate
M_o	fully plastic collapse moment for a perfectly plastic structural cross-section
P	axial crushing force in a thin-walled tube
V_o	impulsive velocity, or impact velocity
W_m	maximum permanent transverse displacement of a structural member
β	aspect ratio, B/L
δ	transverse displacement of a double-shear beam
δ_o	transverse displacement at the onset of cracking
δ_f	transverse displacement at complete severance
λ	geometric scale factor
ρ	density of material
σ_o	uniaxial yield stress
Δ	axial displacement of a thin-walled tube

1 Introduction

The term **'structural impact'** embraces a rich variety of engineering problems, ranging from the dynamic behaviour of elastic structures, such as beams, ships and buildings, to the structural crashworthiness of vehicles and the meteoroid impact of spacecraft. It includes ballistics and stress wave propagation, is concerned with the safety of nuclear power components and systems, and the protection of other engineering installations from various hazards. The structures may be made from a wide range of different materials, with the dynamic loads caused by dropped objects, collisions, explosions, or developed in dedicated energy absorbing systems which may be activated under various accident conditions.

A considerable body of research work has been published on various aspects of elastic impact, including stress wave propagation, which, therefore, is not discussed further. Thus, we shall focus our attention on the effect of large dynamic loads which produce plastic strains, and examine only structures made from ductile materials. Moreover, ballistics is not discussed because it has developed into a quite separate and distinct subject [1,2]. Despite these exclusions, the dynamic plastic behaviour of structures, which are subjected to dynamic loads producing large plastic strains, is important for a broad range of engineering problems.

Quite often, any considerations of impact loads are an afterthought in design. Frequently, structural systems are designed and built with no attention whatsoever paid to their performance under severe impact loads. For sufficiently small structures, the regulatory bodies require, in some cases, impact tests to be performed on the final design, such as for nuclear transportation casks. In many other cases, (for example, the response due to an aircraft crashing on a nuclear power station, or a ship collision), the impact event may only be modelled with small-scale model tests, or using theoretical idealisations. Unfortunately, the response of a structure under an impact load may be quite different from the corresponding static behaviour, as shown in [3] for the axial impact of small-scale model motor coaches. Thus, the structural design and performance under impact loads may be inefficient when the structure has been designed from a static viewpoint.

Impact loadings are now featuring more prominently in design, because economic pressures are demanding that structures become more efficient and lighter, and travel at greater speeds. This means that the consequences of impact loads are more severe. For example, railway coaches were built sturdily in the past, so that they could withstand heavy buffing loads. However, modern designs of lightweight coaches may be damaged more easily by buffing loads which would, therefore, require more maintenance. Thus, impact loads are now becoming of primary importance to designers and railway operators who are seeking to reduce the lifetime costs of rolling stock [4].

References [5-8] contain a general introduction to the structural impact field, together with many citations for further work. Moreover, a summary of recent studies on the dynamic plastic behaviour of structures is published in [9], while [10] discusses some trends in structural impact. It is not the object of this article to review the extensive published literature, which has been done adequately in the cited references, but rather to discuss some current trends in structural impact research and, in particular, to show how these impinge on the development of numerical codes.

Some aspects of theoretical and numerical methods are discussed in the next section, together with various phenomena which arise in this field. Section 3 examines the failure of structures when the ductility of the material is exhausted, while Section 4 discusses experimental modelling and the observed departures from the laws of geometrically similar scaling. Some conclusions complete the article in Section 5.

2   Theoretical and Numerical Predictions

Many articles have been written over the past 40 years since the pioneering work in this field during the Second World War. Considerable theoretical understanding has been achieved and progress made with the aid of the simple rigid-plastic methods of analysis [8,11]. Figures 1 to 3 illustrate the agreement which may be achieved between experimental results and the simple rigid-plastic methods of analysis for some basic structural shapes.

Elementary numerical schemes have been developed for more complex structures [20, 21], while numerical codes are available for complex systems. Finite difference codes have been developed by Witmer et al. [22], and others, while finite-element methods have been developed by many authors, and applied to a wide range of engineering applications, as noted in [10]. The ABAQUS finite-element programme, for example, has been used to examine some beam impact problems [16, 23, 24], as shown in Figure 2.

The predictions of many computer programmes have been compared with experimental test results which have been conducted on a wide range of structural shapes. However, Symonds et al. [25, 26], have discovered an unexpected instability in a transversely loaded beam, which is a normally stable structure. They examined ten well-known computer codes, and observed significant differences betwen the displacement-time histories and permanent displacements. It transpired that the numerical solutions are very sensitive to the physical parameters within a certain range of values, and small changes in the solution strategy may lead to disproportionate changes in the output.

It is evident, from this body of published work, that material strain rate sensitivity must be retained in the analysis when the material is strain rate sensitive, as indicated in Figure 4. The influence of finite-displacements, or geometry changes, is also important in most practical cases, because the displacements are much larger than wholly elastic ones. A comparison between the theoretical predictions from infinitesimal and finite-displacement theories is given in Figures 1, 2 and 4. Transverse shear effects may be more important for a dynamic loading problem than for the corresponding static loading case [29, 30]. Thus, it is necessary, for a certain class of dynamic problems, to include the influence of transverse shear forces in the yield condition which controls plastic flow. However, rotatory inertia effects are not significant [31, 32].

Dynamic plastic buckling develops in a certain class of structural members when the dynamic loads are large enough to cause material plastic behaviour and important inertia effects. This phenomenon may occur in axially loaded beams, plates and cylindrical shells, and various shells subjected to external pressure pulses and impact loads [33]. The solution of these problems may cause numerical difficulties, as remarked in [33] and [34]. For example, it transpires that the theoretical solution for the direct buckling of an axial impacted elastic-plastic column is sensitive to the time step, while columns with other parameters may develop indirect buckling after relatively long times. Thus, different numerical strategies are required for the two cases, although it is not known which type of buckling would develop for a particular problem. In fact, incorrect numerical solutions have been published for some problems because the calculations were terminated prematurely.

Another phenomenon, which is known as dynamic progressive buckling, develops at small impact velocities (typically, up to tens of meters/sec), when the inertia forces may be neglected, and the problem taken as quasi-static [7]. However, the phenomenon of strain rate sensitivity must be retained in the analysis if the material is strain rate sensitive. Dynamic progressive buckling has been studied for circular, square and rectangular thin-walled tubes subjected to axial impact loads which may arise in various structural crash-worthiness problems [35]. The axial load-axial displacement characteristics are cyclic for a circular tube, as shown in Figure 5, and two adjacent peaks are associated with the formation of each axisymmetric wrinkle, or buckle. This particular geometry is an efficient energy-absorber, [36, 37], with the total dynamic energy absorbed given by the mean value of the load (P_m) times the stroke (Δ).

Simple methods of analysis [35, 38] and empirical equations [39], have been developed for the dynamic progressive buckling of thin-walled members, and have proved valuable for design purposes. This class of problems is still difficult and expensive for many current numerical codes because of the large plastic strains and severe changes in geometry.

The experimental results which have been published on dynamically loaded beams [12, 27], frames [40, 41], circular plates [42], rectangular plates [18], shells [22, 43, 44], shell intersections [45], axially loaded tubes with circular and square cross-sections [35], etc., provide test cases for the calibration of current and future numerical codes.

3   Failure Modes

The structural impact problems in the previous sections may suffer large permanent displacements which hinder the continued safe operation of an engineering system, even though the material remains ductile throughout the entire response. This is a type of structural failure. Moreover, dynamic plastic buckling and dynamic progressive buckling may be regarded as further types of structural failure, although the phenomenon of dynamic progressive buckling is sometimes utilised for improving the energy-absorbing capacity of a system [7, 37]. Again, the behaviour is assumed to remain wholly ductile.

Another type of structural failure occurs when the ductility of the material is exhausted. Unfortunately, most experimental studies have focused on the dynamic response of ductile structures, and have not explored the conditions required for rupture. Nevertheless, it is important for a designer to assess the requirements for material failure, so that a realistic margin of safety might be estimated for a structure, when subjected to unusual impact events.

The first systematic study on the dynamic inelastic failure of a structure was reported by Menkes and Opat [46], who identified three failure modes for a metal beam, which was fully clamped at both supports, and subjected to an impulsive velocity distributed uniformly over the entire span. The failure modes were characterised as large permanent ductile deformations (Mode 1), tensile tearing (Mode 2), and transverse shear failure (Mode 3).

It was shown, in [47], that simple rigid-plastic methods of analysis which incorporate the influence of finite-displacements, or geometry changes [14], gave good agreement with the experimental values recorded on the aluminium 6061T6, strain rate insensitive, beams, which suffered a Mode 1 response. It was also found in [47], that the threshold impulsive velocity, required to produce a Mode 2 tensile tearing failure, was also predicted fairly accurately, using the same theoretical procedure, but with the maximum tensile strain equal to the rupture strain recorded in a static uniaxial tensile test. A theoretical method was developed, in [47], for the threshold impulsive velocity required for a Mode 3 transverse shear failure. This estimate was obtained using a rigid-plastic analysis with infinitesimal displacements, but incorporating the transverse shear force, as well as a bending moment, in the yield condition [29, 30]. Transverse shear sliding may develop in a beam when the associated transverse shear force lies on the yield curve, just as, more familiarly, a bending hinge may be associated with a bending moment.

Virtually no experimental or theoretical studies have been published in this important area for almost a decade, since [46] and [47], except for several applications (e.g. [48-50]). Articles summarising the published work on the dynamic inelastic failure of beams and shells are presented in [51] and [52], respectively.

Recently, experimental tests have been reported on the failure of ductile beams struck by masses at various locations across a fully clamped span [15, 53]. The dropped masses cause large transverse displacements before a tensile tearing failure, in some cases; while, in others, the displacements remain small, and transverse shear failures occur. The experimental tests were conducted on aluminium alloy (strain rate insensitive) and mild steel (strain rate sensitive) beams, having several different thicknesses with a constant span. A comparison between the two sets of experimental results allows an estimate to be made of the importance of material strain rate sensitivity. In addition, tests were conducted on beams with flat or enlarged ends in order to assess the importance of different clamped end conditions.

It appears that the dynamic failure of strain rate insensitive aluminium alloy beams, which are subjected to impact loads, is more complicated than for impulsive velocities. Again, at low impact velocities, the Mode 1 ductile response is obtained, and good agreement is found between the theoretical predictions of rigid-plastic methods [17] and the corresponding experimental results, as shown in Figure 2. As the impact velocity is increased, all the beams with flat ends, except one, failed due to tensile tearing (Mode 2), either at the supports or underneath the striker [53]. The location of the Mode 2 failure tended to change from the impact point to a support with a reduction in the value of ℓ_1, where ℓ_1 is the distance of the impact point from the nearest support. However, a tensile tearing (Mode 2) failure occurred always at the support for the aluminium alloy beams with enlarged ends. Thus, it appears that the dynamic failure of aluminium alloy beams due to impact loads is governed largely by the Mode 2 failure mode which was introduced, in [46] and [47], for impulsive velocity loadings.

The impact experiments were repeated on beams made from mild steel, which is a highly strain rate sensitive material. It transpires that, for sufficiently large impact energies, the beams with flat ends failed in a shear mode at the impact point. However, this shear failure is more complex than the Mode 3 response for an impulsive loading which was found only when the mass struck close to a support. It turns out that the steel beams with enlarged ends failed due to tensile tearing (Mode 2) when struck near a support, and in a shear mode when struck near the mid-span, as shown in Figure 6.

The experimental results in [15] emphasise the sensitivity of dynamic failure to the material properties and support conditions, and show the value of simple rigid-plastic methods of analysis and illustrate our poor understanding of this important practical topic.

Experimental tests are vital in this general area in order to reveal the dynamic inelastic failure modes of structures. However, current experimental methods alone are unable to provide the criteria which may be used by a designer for the failure of a wide range of practical problems. Simple analytical methods also have limitations at this stage of their development.

The beam impact problem in [15] and [17] has been modelled using the ABAQUS finite-element numerical code [16], as shown in Figure 7. This figure gives the equivalent plastic strains throughout the thickness of a beam, as well as detailed distributions at different times in the region where failure was observed in the experimental tests. In particular, the numerical code was used to calculate the behaviour of a beam which had just cracked in an experimental test when motion ceased. Thus, there was no need to introduce any failure criteria into the numerical code, since the beam remained intact throughout the entire response. It was the object of this study [16], to interpret these numerical calculations in order to develop the criteria of failure which may be used by designers in analytical, or numerical methods.

This example illustrates the potential of a powerful partnership between experimental studies and numerical calculations. At the present stage of development of this highly non-linear field, both experimental test programmes and numerical schemes have severe limitations. For example, it is difficult to record the detailed strain distributions within the highly strained and localised regions of impacted metal structures. The numerical codes require information on the conditions which must be satisfied for a strain rate sensitive material to start cracking in a dynamically loaded structure. Of course, one could use a limit forming diagram [52] or the crude assumption that the equivalent strain equals the static uniaxial rupture strain. However, the maximum impact load which a structure may withstand is sensitive to the assumptions used for the initiation of failure. It was observed, in [16], for example, that the location of failure may change from underneath the stricker to the support region when the fully clamped end of a beam is changed from a flat end to an enlarged one. The difference between the strain distributions which are associated with these two support conditions is illustrated in Figure 8 for otherwise identical problems.

4   Experimental Modelling

Several articles have been written, in recent years, drawing attention to the difficulties associated with the geometrically similar scaling laws which relate the dynamic response of small-scale models and the impact behaviour of full-scale prototypes [7, 54-56]. The discrepancies may be significant, and full-scale prototypes, with overall deformations about twice as large as those expected from theoretical predictions based on one-quarter scale model test results, have been observed in several investigations [54-56]. It appears that this discrepancy may arise only when the impact loadings cause tearing [55, 57].

The resolution of these difficulties is of paramount importance to numerical analysts because a computer programme may predict the dynamic response of a small-scale model quite accurately, but it could give misleading results for the behaviour of a full-time prototype. To overcome this problem it is necessary to incorporate the distortion of the scaling laws into a computer programme. However, what are the laws which give this distortion for the dynamic failure of ductile strain rate sensitive structures? Unfortunately, precious little data is available for guidance.

A recent experimental study has been conducted into the dynamic inelastic failure of double-shear beams having a wide range of thicknesses [58]. The results reveal that the transverse displacement of a beam at the initiation of cracking, δ_o, was almost a constant proportion of the corresponding specimen thickness, h, and could be expressed in the form;

$$\delta_o/h = 0.157 \qquad \text{and} \qquad \delta_o/h = 0.165 \qquad \text{(1a, b)}$$

for aluminium alloy and mild steel, respectively. This behaviour satisfies the geometrically similar scaling laws and presents, therefore, no potential difficulty for structural designs or numerical calculations.

Additional experimental impact tests in [58] examined the margin of safety between the transverse displacement at the initiation of cracking, δ_o, and at complete severance, δ_f, of a double-shear beam. These experimental tests on beams having a wide range of thicknesses show that the transverse displacement immediately before complete severance is a greater proportion of the beam thickness for the thinnest specimens, or;

$$\delta_f/h = 0.16 + 1.19/h, \qquad \delta_f/h = 0.27 + 0.94/h \qquad \text{(2a, b)}$$

for aluminium alloy and mild steel, respectively, where h is measured in mm. This behaviour does not obey the elementary laws of geometrically similar scaling. In other words, the difference between the transverse displacements at the initiation of cracking and complete severance of a beam, $\delta_f - \delta_o$, is less for the thickest specimens. The response is consistent with the observations in [54] and [55], in which it is noted that the full-scale prototypes suffered a greater amount of tearing and larger associated permanent displacements than in the geometrically similar tests on small-scale models.

The test results in [58] may be interpreted in terms of energies rather than displacements. It may be shown that the impact energy, which is required for complete severance, is proportional to $h^{2.24}$ and $h^{2.5}$ for the aluminium alloy and mild steel double-shear beam specimens, respectively. This behaviour contrasts with the geometrically similar scaling laws, which requre a variation of energy with h^3 while, by comparison, linear elastic fracture mechanics and area scaling would vary with h^2. Figure 9 illustrates this non-scaling phenomenon for the strain rate insensitive aluminium alloy double-shear beam specimens in [58].

Further discussion on the scaling of the double-shear beam impact experiments in [58] is reported in [59] and [60], together with some theoretical developments.

It is interesting to note that the results of some recent static tearing tests on thin aluminium alloy sheet specimens [61] may also be used to show that the laws of geometrically similar scaling are not satisfied. The bending energy, which is consumed in deforming plastically the test specimens, scales with the volume of material (h^3), as expected from the laws of geometrically similar scaling. However, the energy required to tear these specimens varies as $h^{2.61}$, which does not satisfy the geometrically similar scaling laws.

As noted in Section 3, it is possible that progress may be made more rapidly in this area by using numerical codes in partnership with experimental test programmes. In fact, the experimental test results in [54, 56, 57] provide well documented test cases for the calibration of numerical programmes. Clearly, any reliable numerical scheme must be capable of predicting the non-scaling behaviour of the simpler wedge-plate and double-shear beam specimens in [56, 58], respectively, as well as reproducing the dynamic response of the more complex geometrically similar structural tests which are reported in [54]. Incidentally, it is important to take account of any energy losses into the impact rigs which are used for the dynamic testing of structures. These energy losses may be significant for high impact forces in some test rigs, as noted in [62].

It was noted earlier that the departure from the elementary laws of geometricallly similar scaling occurs, possibly, only when tearing occurs. However, a small-scale model may respond without any tearing, while a geometrically similar full-scale prototype may fracture. This phenomenon occurs when a ductile-brittle failure transition is encountered as the dimensions of a structure are increased [55].

5 Conclusions

This article discusses some aspects of the behaviour of structures subjected to large dynamic loads which cause an inelastic response. The topic is a large one, and has been reviewed more fully elsewhere. Particular emphasis is placed, in this article, on several phenomena, which are incompletely understood, and yet have important implications for the development of numerical codes.

It is shown that the failure criteria for the tearing and rupture of structures is complex and not well understood, which is further complicated by the paucity of experimental data on the variation of fracture strain with strain rate, even for uniaxial behaviour. The scaling laws for dynamically loaded small-scale structures have not been established, and existing evidence reveals important departures from the elementary geometrically similar scaling laws.

It is clear that further experimental work is required to provide a sound understanding of the various phenomena for the development of reliable numerical codes. Progress is likely to be made most rapidly by a partnership between the architects of numerical codes and those responsible for organising experimental test programmes. Currently, the predictions from numerical codes for many practical problems of interest in this article may only provide rough approximations to the actual response, no matter how sophisticated the numerical scheme. This situation prevails because of uncertainties in the dynamic loading (e.g., shape of pressure-time characteristics), lack of adequate human injury criteria for the collision protection of passengers in various transportation modes, and the paucity of material properties under high rates of strain and large plastic strains.

Acknowledgements

The author wishes to thank the Science and Engineering Research Council for their suppport of this study through grant GR/B/89737. The author is also indebted to the Impact Research Centre and the Department of Mechanical Engineering at the University of Liverpool, in particular, Dr. R.S. Birch, Dr. W.S. Jouri and Dr. Jilin Yu for their assistance with the figures, Mrs. M. White for her secretarial assistance, and Mr. H. Parker and Mrs. A. Green for their assistance with the tracings.

References

1. Backman, M.E. and Goldsmith, W.; The Mechanics of Penetration of Projectiles into Targets, International Journal of Engineering Science, Vol. 16, pp. 1-99, 1978.

2. Proceedings of the Symposium on Hypervelocity Impact; (Guest Editor Anderson,

C.E.), International Journal of Impact Engineering, Vol. 5, Nos. 1-4, pp. 1-760, 1987.

3. Lowe, W.T., Al-Hassani, S.T.S. and Johnson, W.; Impact Behaviour of Small-Scale Model Motor Coaches, Proceedings of the Institution of Mechanical Engineers, Vol. 186, pp. 409-419, 1972.

4. Scott, G.A.; The Development of a Theoretical Technique for Rail Vehicle Structural Crashworthiness, Proceedings of the Institution of Mechanical Engineers, Vol. 201 (D2), pp. 123-128, 1987.

5. Johnson, W.; Impact Strength of Materials, Edward Arnold, London and Crane Russak, New York, 1972.

6. Johnson, W. and Mamalis, A.G.; Crashworthiness of Vehicles, Mechanical Engineering Publications, London, 1978.

7. Jones, N.; Structural Impact, Cambridge University Press, (In Press).

8. Jones, N.; A Literature Review of the Dynamic Plastic Response of Structures, The Shock and Vibration Digest, Vol. 7, No. 8, pp. 89-105, 1975.
Recent Progress in the Dynamic Plastic Behaviour of Structures, Part 1, The Shock and Vibration Digest, Vol. 10, No.9, pp. 21-33, 1978.
Part 2, Vol. 10, No. 10, pp. 13-19, 1978.
Part 3, Vol. 13, No. 10, pp. 3-16, 1981.
Part 4, Vol. 17, No. 2, pp. 35-47, 1985.

9. Jones, N.; Recent Studies on the Dynamic Plastic Behaviour of Structures, Impact Research Centre Report Number ES/40/88, Applied Mechanics Reviews, University of Liverpool, Department of Mechanical Engineering, (In press).

10. Jones, N.; Some Trends in Structural Impact. Colloquium on the Future of Structural Testing, Bristol University, Impact Research Centre Report Number ES/38/88, University of Liverpool, Department of Mechanical Engineering, 1988.

11. Jones, N.; Response of Structures to Dynamic Loading, Mechanical Properties at High Rates of Strain (Ed. Harding, J.), pp. 254-276, Institute of Physics Conference Series No. 47, London, 1979.

12. Jones, N., Griffin, R.N. and Van Duzer, R.E.; An Experimental Study into the Dynamic Plastic Behaviour of Wide Beams and Rectangular Plates, International Journal of Mechanical Sciences, Vol. 13, pp. 721-735, 1971.

13. Symonds, P.S. and Mentel, T.J.; Impulsive Loading of Plastic Beams with Axial Constraints, Journal of the Mechanics and Physics of Solids, Vol. 6,pp. 186-202, 1958.

14. Jones, N.; A Theoretical Study of the Dynamic Plastic Behaviour of Beams and Plates with Finite-Deflections, International Journal of Solids and Structures, Vol. 7, pp. 1007-1029, 1971 .

15. Liu, J.H. and Jones, N.; Experimental Investigation of Clamped Beams Struck Transversely by a Mass, International Journal of Impact Engineering, Vol. 6, No. 4, pp. 303-335, 1987.

16. Yu, J. and Jones, N.; Numerical Simulation of a Clamped Beam Under Impact Loading, Computers and Structures, (In Press).

142

17. Liu, J.H. and Jones, N.; Dynamic Response of a Rigid Plastic Clamped Beam Struck By a Mass at any Point on the Span, International Journal of Solids and Structures, Vol. 24, No. 3, pp. 251-270, 1988.

18. Jones, N., Uran, T.O. and Tekin, S.A.; The Dynamic Plastic Behaviour of Fully Clamped Rectangular Plates, International Journal of Solids and Structures, Vol. 6, pp. 1499-1512, 1970.

19. Jones, N. and Baeder, R.A.; An Experimental Study of the Dynamic Plastic Behaviour of Rectangular Plates, Symposium on Plastic Analysis of Structures, Ministry of Education, Polytechnic Institute of Jassy, Civil Engineering Faculty, Rumania, Vol. 1, pp. 476-497, 1972.

20. Symonds, P.S.; Finite Elastic and Plastic Deformations of Pulse Loaded Structures by an Extended Mode Technique, International Journal of Mechanical Sciences, Vol. 22, pp. 597-605, 1980.

21. Symonds, P.S. and Mosquera, J.M.; A Simplified Approach to Elastic-Plastic Response to General Pulse Loads, Journal of Applied Mechanics, Vol. 52, pp. 115-121, 1985.

22. Witmer, E.A., Balmer, H.A., Leech, J.W. and Pian, T.H.H.; Large Dynamic Deformations of Beams, Circular Rings, Circular Plates and Shells, AIAA Journal, Vol. 1, pp. 1848-1857, 1963.

23. Symonds, P.S. and Fleming, W.T.; Parkes Revisited: On Rigid-Plastic and Elastic-Plastic Dynamic Structural Analysis, International Journal of Impact Engineering, Vol. 2, No. 1, pp. 1-36, 1984.

24. Reid, S.R. and Gui, X.G.; On the Elastic-Plastic Deformation of Cantilever Beams Subjected to Tip Impact, International Journal of Impact Engineering, Vol. 6, No. 2, pp. 109-127, 1987.

25. Symonds, P.S. and Yu, T.X.; Counterintuitive Behaviour of a Problem of Elastic-Plastic Beam Dynamics, Journal of Applied Mechanics, Vol. 52, No. 3, pp. 517-522, 1985.

26. Symonds, P.S., McNamara, J.F. and Genna, F.; Vibrations and Permanent Displacements of a Pin-Ended Beam Deformed Plastically by Short Pulse Excitation, International Journal of Impact Engineering, Vol. 4, No. 2, pp. 73-82, 1986.

27. Symonds, P.S. and Jones, N.; Impulsive Loading of Fully Clamped Beams with Finite Plastic Deflections and Strain Rate Sensitivity, International Journal of Mechanical Sciences, Vol 14, pp. 49-69, 1972.

28. Humphreys, J.S.; Plastic Deformation of Impulsively Loaded Straight Clamped Beams, Journal of Applied Mechanics, Vol. 32, pp. 7-10, 1965.

29. Symonds, P.S.; Plastic Shear Deformations in Dynamic Load Problems, Engineering Plasticity (Ed. Heyman, J. and Leckie, F.A.), pp. 647-664, Cambridge University Press, 1968.

30. de Oliveira, J.G. and Jones, N.; Some Remarks on the Influence of Transverse Shear on the Plastic Yielding of Structures, International Journal of Mechanical Sciences, Vol. 20, pp. 759-765, 1978.

31. Jones, N. and de Oliveira, J.G.; Dynamic Plastic Response of Circular Plates with Transverse Shear and Rotatory Inertia, Journal of Applied Mechanics, Vol. 47, pp. 27-34, 1980.

32. Jones, N. and de Oliveira, J.G.; Impulsive Loading of a Cylindrical Shell with Transverse Shear and Rotatory Inertia, International Journal of Solids and Structures, Vol. 19, No. 3, pp. 263-279, 1983.

33. Jones, N.; Dynamic Elastic and Inelastic Buckling of Shells. Chapter 2, Developments in Thin Walled Structures, (Ed. Rhodes, J. and Walker, A.C.), Vol. 2, pp. 49-91, Elsevier Applied Science Publishers, 1984.

34. Jones, N. and dos Reis, H.L.M.; On the Dynamic Buckling of a Simple Elastic-Plastic Model, International Journal of Solids and Structures, Vol. 16, pp. 969-989, 1980.

35. Abramowicz, W. and Jones, N.; Dynamic Progressive Buckling of Circular and Square Tubes, International Journal of Impact Engineering, Vol. 4, No. 4, pp. 243-270, 1986.

36. Ezra, A.A. and Fay, R.J.; An Assessment of Energy Absorbing Devices for Prospective Use in Aircraft Impact Situations. Chapter in Dynamic Response of Structures (Ed. Herrmann, G. and Perrone, N.), pp. 225-246, Pergamon Press, 1972.

37. Johnson, W. and Reid, S.R.; Metallic Energy Dissipating Systems, Applied Mechanics Reviews, Vol. 31, pp. 277-288, 1978 also Applied Mechanics Update, pp. 315-319, 1986.

38. Wierzbicki, T.; Crushing Behaviour of Plate Intersections. Chapter 3, Structural Crashworthiness, (Ed. Jones, N. and Wierzbicki, T.), pp. 66-95, Butterworths, London, 1983.

39. Thornton, P.H., Mahmood, H. F. and Magee, C.L.; Energy Absorption by Structural Collapse. Chapter 4, Structural Crashworthiness, (Ed. Jones, N. and Wierzbicki, T.), pp. 96-117, Butterworths, London, 1983.

40. Mosquera, J.M., Kolsky, H. and Symonds, P.S.; Impact Tests on Frames and Elastic-Plastic Solutions, proc. ASCE, Journal Engineering Mechanics Division, Vol. 111, No. 11, pp. 1380-1401, 1985.

41. Lindberg, B. and Pedersen, J. B.; Plastic Deformation of Impact Loaded Frames, International Journal of Impact Engineering, Vol. 6, No. 2, pp. 101-108, 1987.

42. Bodner, S.R. and Symonds, P.S.; Experiments on Viscoplastic Response of Circular Plates to Impulsive Loading. Journal of the Mechanics and Physics of Solids, Vol. 27, pp. 91-113, 1979.

43. Jones, N., Dumas, J.W., Giannotti, J.G. and Grassit, K.E.; The Dynamic Plastic Behaviour of Shells. Chapter in Dynamic Response of Structures, (Ed. Herrmann, G. and Perrone, N.), pp. 1-29, Pergamon Press, 1972.

44. Jones, N., Giannotti, J.G. and Grassit, K.E.; An Experimental Study into the Dynamic Inelastic Behaviour of Spherical Shells and Shell Intersections, Archiwum Budowy Maszyn (Archives of Mechanical Engineering), Vol. 20, No. 1, pp. 33-46, 1973.

45. Summers, A.B. and Jones, N.; Some Experiments on the Dynamic Plastic Behaviour of Shell Intersections, Nuclear Engineering and Design, Vol. 26, pp. 274-281, 1974.

46. Menkes, S.B. and Opat, H.J.; Broken Beams, Experimental Mechanics, Vol. 13, pp. 480-486, 1973.

144

47. Jones, N.; Plastic Failure of Ductile Beams Loaded Dynamically, Trans., ASME, Journal of Engineering for Industry, Vol. 98(B), pp. 131-136, 1976.

48. Pettersen, E. and Valsgard, S.; Collision Resistance of Marine Structures. Chapter 12, Structural Crashworthiness, (Ed. Jones, N. and Wierzbicki, T.), pp. 338-370, Butterworths, 1983.

49. Wierzbicki, T., Chryssostomidis, C. and Wiernicki, C.; Rupture Analysis of Ship Plating Due to Hydrodynamic Wave Impact, pp. 237-256, SNAME, Ship Structure Symposium '84, Arlington, Virginia, 1984.

50. Wiernicki, C. J.; Damage of Ship Plating Due to Wave Impact Loads, pp. 151-179, SNAME, STAR Symposium, Portland, Oregon, 1986.

51. Jones, N.; On the Dynamic Inelastic Failure of Beams. Chapter in Structural Failure, (Ed. Wierzbicki, T. and Jones, N.), John Wiley and Sons, New York, 1988.

52. Duffey, T.A.; Dynamic Rupture of Shells. Chapter in Structural Failure, (Ed. Wierzbicki, T. and Jones, N.), John Wiley and Sons, New York, 1988.

53. Liu, J. H. and Jones, N.; Plastic Failure of a Clamped Beam Struck Transversely By a Mass, University of Liverpool, Department of Mechanical Engineering, Impact Research Centre Report ES/31/87, 1987. Presented at 27th Polish Solid Mechanics Conference, Rytro, Poland, August 1988.

54. Booth, E., Collier, D. and Miles, J.; Impact Scalability of Plated Steel Structures. Chapter 6, Structural Crashworthiness, (Ed. Jones, N. and Wierzbicki, T.), pp. 136-174, Butterworths, 1983.

55. Jones, N.; Scaling of Inelastic Structures Loaded Dynamically. Chapter 2, Structural Impact and Crashworthiness, (Ed. Davies, G.A.O.), Vol. 1, pp. 45-74, Elsevier Applied Science Publishers, 1984.

56. Jones, N. and Jouri, W.S.; A Study of Plate Tearing for Ship Collision and Grounding Damage, Journal of Ship Research, Vol. 31, No. 4, pp. 253-268, 1987.

57. Dallard, P.R.B. and Miles, J.C.; Design Tools for Impact Engineers, Structural Impact and Crashworthiness, (Ed. Morton, J.), vol. 2, pp. 369-382, 1984.

58. Jouri, W.S. and Jones, N.; The Impact Behaviour of Aluminium Alloy and Mild Steel Double-Shear Specimens, International Journal of Mechanical Sciences, Vol. 30, Nos. 3/4, pp. 153-172, 1988.

59. Jouri, W.S. and Jones, N.; Scaling and Impact Behaviour of Ductile Double-Shear Specimens, University of Liverpool, Department of Mechanical Engineering, Impact Research Centre Report ES/36/88, 1988.

60. Atkins, A.G.; Scaling in Combined Plastic flow and Fracture, International Journal of Mechanical Sciences, Vol. 30, Nos. 3/4, pp. 173-191, 1988.

61. Yu, T.X., Zhang, D.J., Zhang, Y. and Zhou, Q.; A Study of the Quasi-Static Tearing of Thin Metal Sheets, International Journal of Mechanical Sciences, Vol. 30, Nos. 3/4, pp. 193-202, 1988.

62. Birch, R.S., Jones, N. and Jouri, W.S.; Performance Assessment of an Impact Rig, Proceedings of the Institution of Mechanical Engineers, part C, Vol. 202, No. 4, pp. 275-285, 1988.

FIGURE TITLES

Figure 1.

Comparison of experimental and theoretical maximum permanent dimensionless transverse displacements for impulsively loaded, fully clamped, axially restrained, aluminium 6061 T6 beams.

o, ◇, △, □ :	Experimental results [12]
————(1) :	Simple bending or first order theory [13]
————(2) :	Upper bound according to equation (60a) of [13]
————(3) :	Lower bound corresponding to 2 above
— —— —(4) :	Exact yield surface [14]
– – – – –(5) :	Circumscribing yield surface [14]
– – – – –(6) :	(5) with σ_o replaced by $0.618\sigma_o$, inscribing yield surface
(2)—(6)	Include the influence of finite-displacements or geometry changes

Figure 2:

Variation of dimensionless maximum permanent transverse displacements with dimensionless external impact energy of a mass striking the mid-span of a fully clamped aluminium alloy beam

⊗, ▲, ▼, ⊕, △, ▽, o, • :	Experimental results [15]
a-e:	Numerical finite-element results using ABAQUS [16]
——— — — :	Simple bending only theory
————(1) :	Equation (1) of Reference [15] for a circumscribing yield curve. Retains the influence of finite displacements, or geometry changes [17]
————(2) :	Same as (1) but for an inscribing yield curve

146

Figure 3:

Comparison of experimental and theoretical maximum permanent dimensionless transverse deflections for impulsively loaded, fully clamped, aluminium 6061 T6 rectangular plates.

∘, Δ, ∇, +, ◇ :	Experimental results [18,19]
————①:	Exact yield surface [14]
— — — —②:	Circumscribing yield surface [14]
— — — —③:	Same as ② for an inscribing yield surface

Figure 4:

Comparison of experimental and theoretical permanent transverse dimensionless deflections for impulsively loaded, fully clamped, axially restrained, hot rolled, mild steel beams.

Δ, ∇ X, ◇, □ :	Experimental results [27]
⊗H :	Experimental results [28]
oa-e:	Numerical finite-difference predictions [22]
————①:	Simple bending only or first order theory [13], perfectly plastic material
————②:	Upper bound according to equation (60a) of [13], perfectly plastic material
————③:	Circumscribing yield surface, perfectly plastic material [14]
————④:	Equations (11) and (29) of Reference [27] for beams with H = 0.09in. (2.29mm) and made from a rate-sensitive material
————⑤:	Same as ④ above but for beams with H = 0. 19in (4.83mm)

Figure 5:

Axial crushing force versus axial displacement of a thin-walled mild steel circular tube [7,35].

Figure 6:

Failure of a mild steel beam which originates from the corner of an indentation at the impact point [10,15].

Figure 7:

Equivalent plastic strain contours at the impact point (on the right hand side of the upper surface) and at the support (on the left hand side of the lower surface) of an aluminium alloy beam with fully clamped flat ends subjected to an impact load near the one-quarter point.

(a) t = 0.419 ms,   (b) t = 0.948 ms,

(c) t = 1.439 ms,   (d) t = 2.321 ms

The strain levels are 2; 0.04, 3; 0.08, 4; 0.12, 5; 0.16, 6; 0.20, 7; 0.24, 8; 0.28, and 9; 0.32 and are predicted using the ABAQUS numerical code [16].

Figure 8:

Equivalent plastic strain contours near the supports of fully clamped aluminium alloy beams struck at the mid-span with (a) flat ends at t = 2.25 ms, and (b) enlarged ends at t = 2.38 ms.

The strain levels are defined in the legend in Figure 7, and are predicted using the ABAQUS numerical code [16].

Figure 9:

Ratio of the actual energy absorbed in a full-scale prototype (E_a) to the energy predicted using the laws of geometrically similar scaling from a small-scale model to a full-scale prototype (E_p) versus the inverse scale factor for aluminum alloy double-shear beam

148

specimens with a scaled transverse displacement of $\delta\lambda = 6$ mm [59].

——————————————— :   Predictions for a full-scale prototype
                                           with h = 40.735 mm

— — — — — — :   Predictions for a full-scale prototype
                                           with h = 30.68 mm

— – — – — – :   Predictions for a full-scale prototype
                                           with h = 20.245 mm

— – – — – – — – – :   Predictions for a full-scale prototype
                                           with h = 10.55 mm

Experimental results

∇ :   h = 40.735 mm,   o :   h = 30.68 mm

Δ :   h = 20.245 mm, $\square$:   h = 10.55 mm

$\Diamond$:   h = 5.09 mm

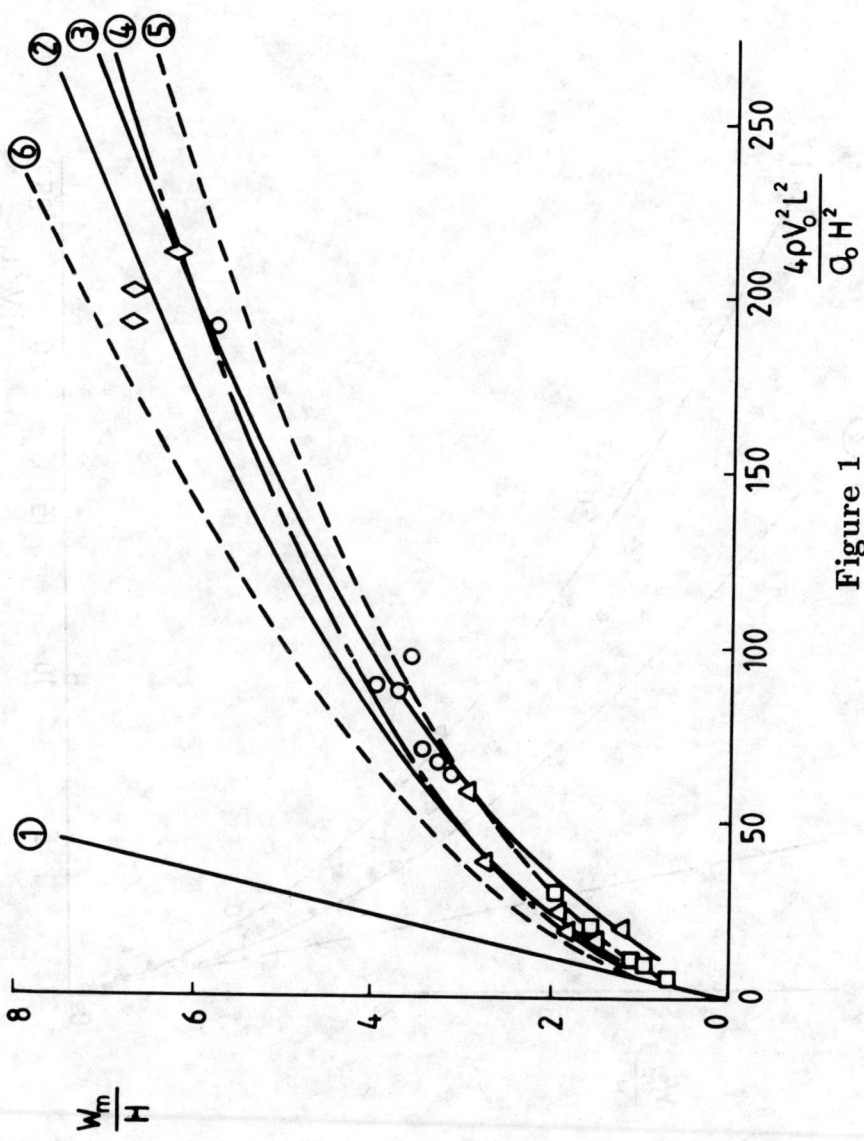

Figure 1

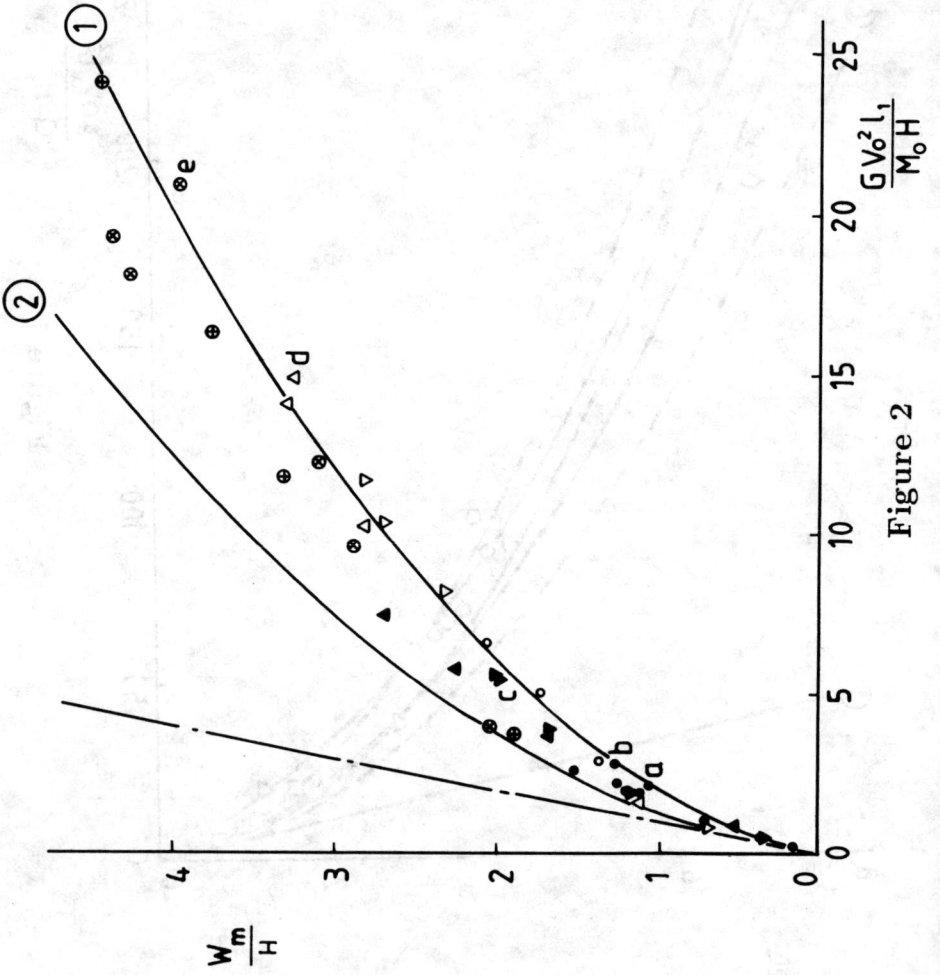

Figure 2

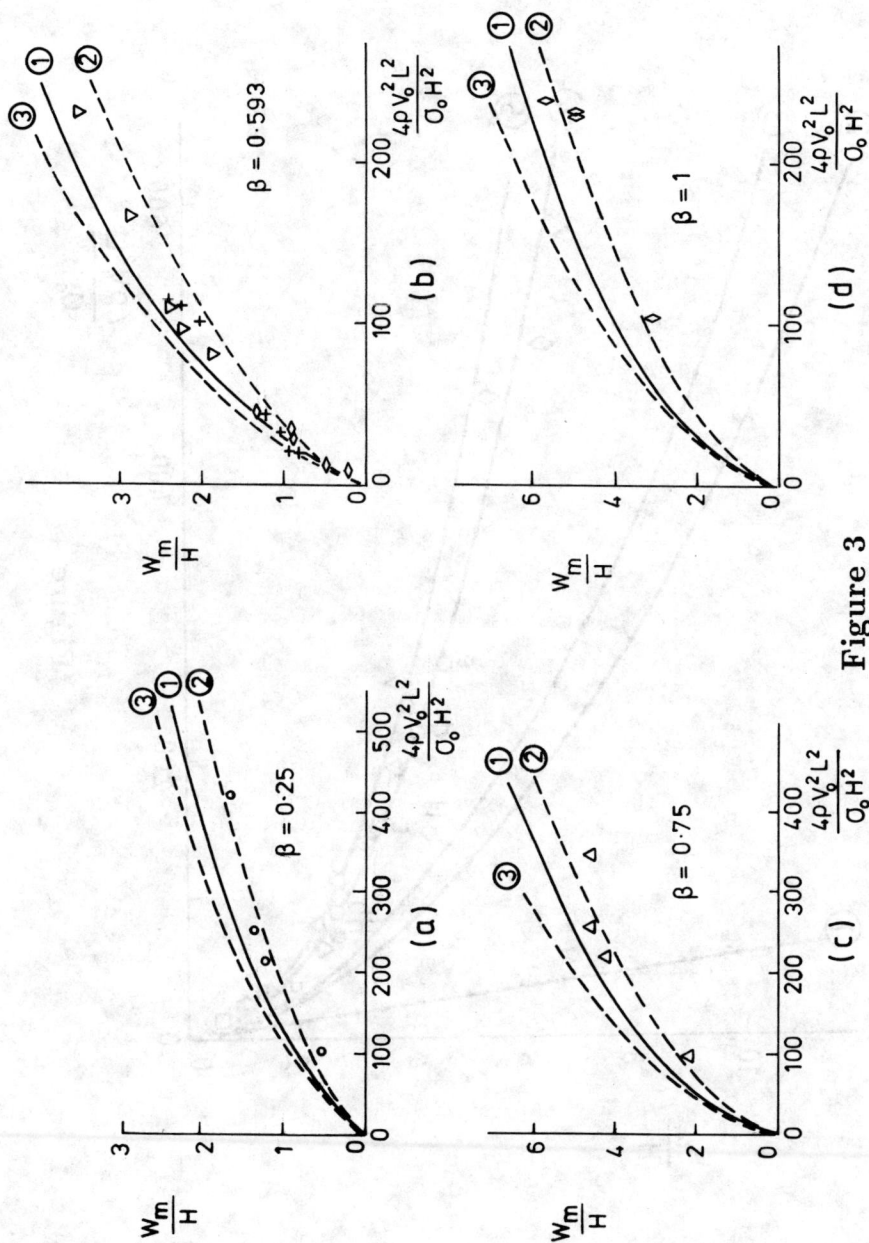

Figure 3

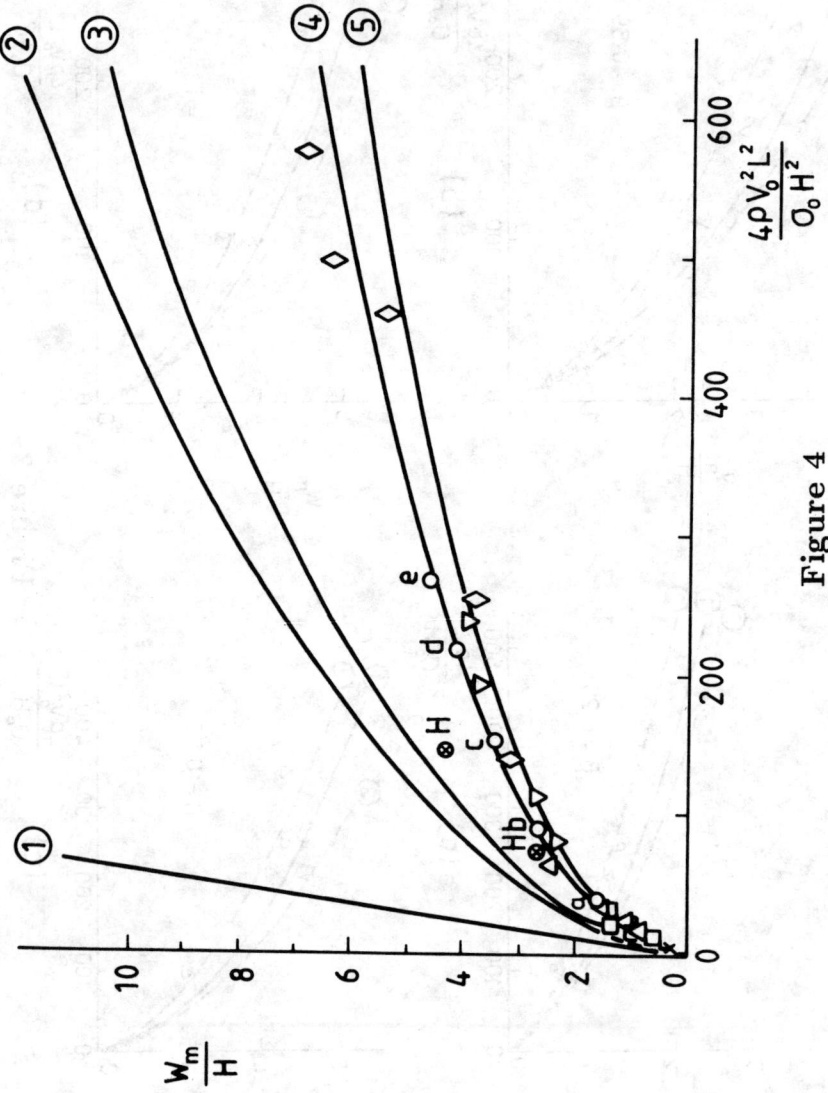

Figure 4

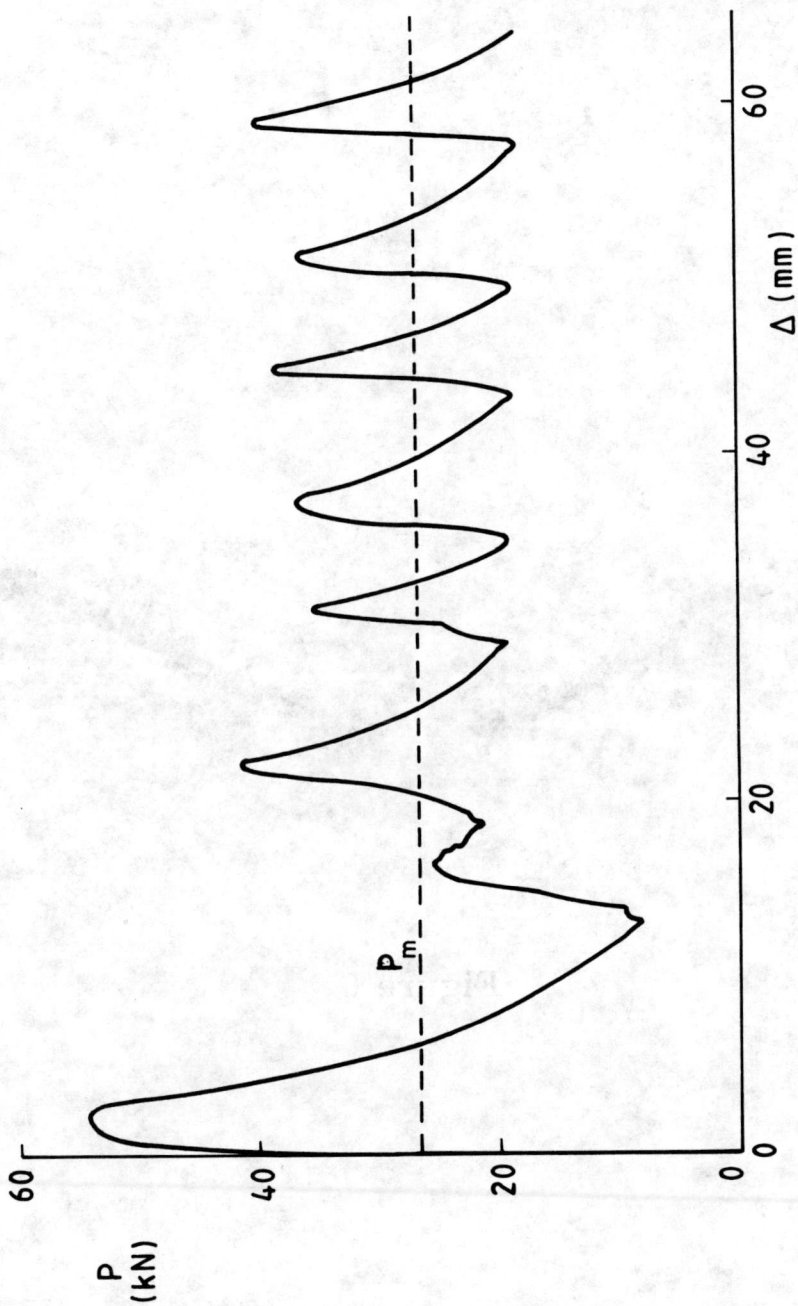

Figure 5

Figure 6

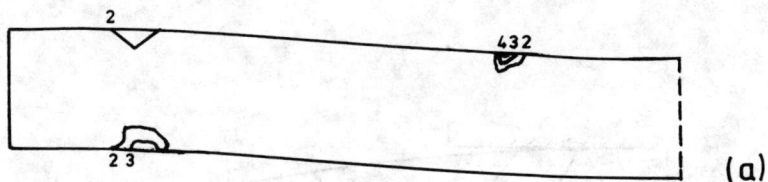

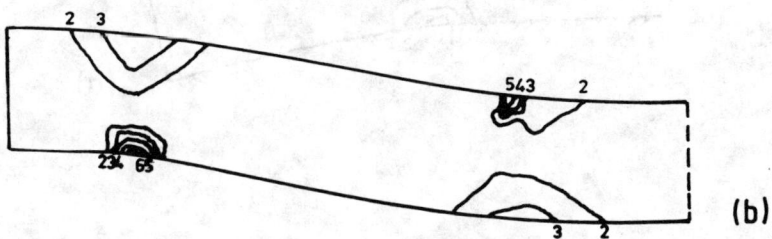

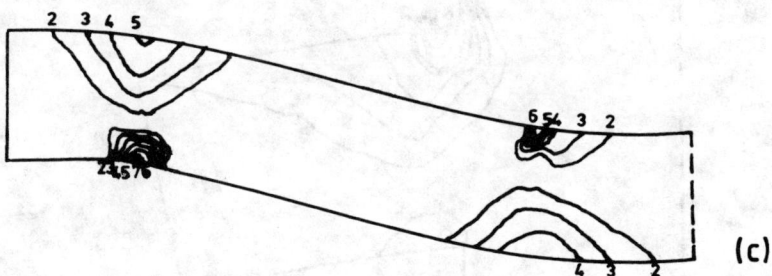

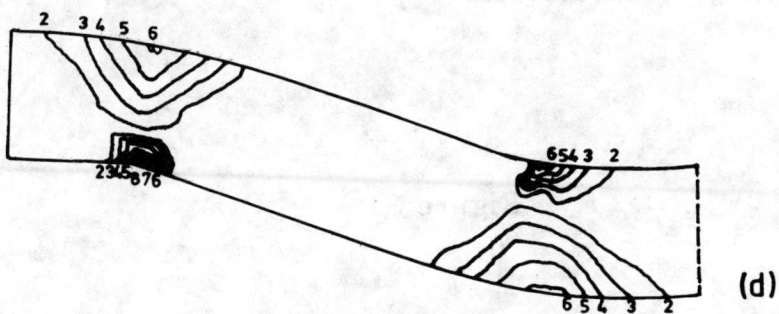

Figure 7

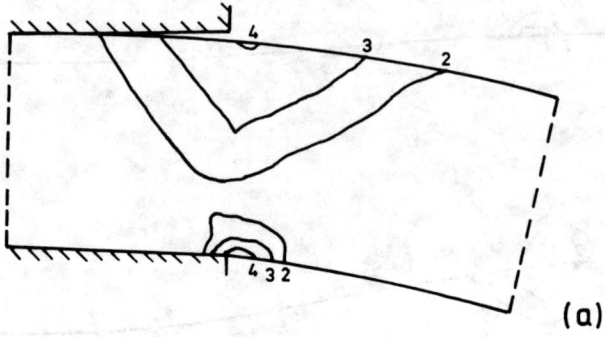

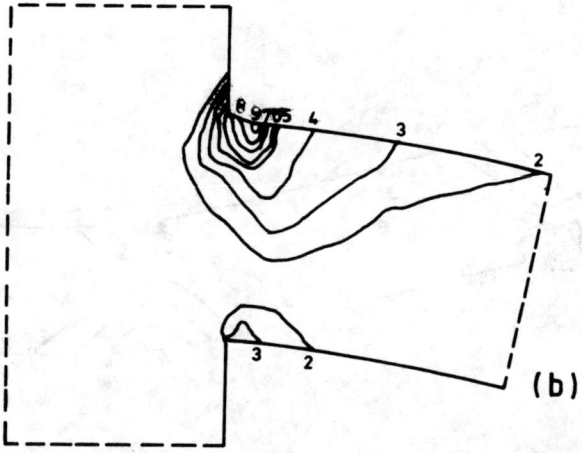

Figure 8

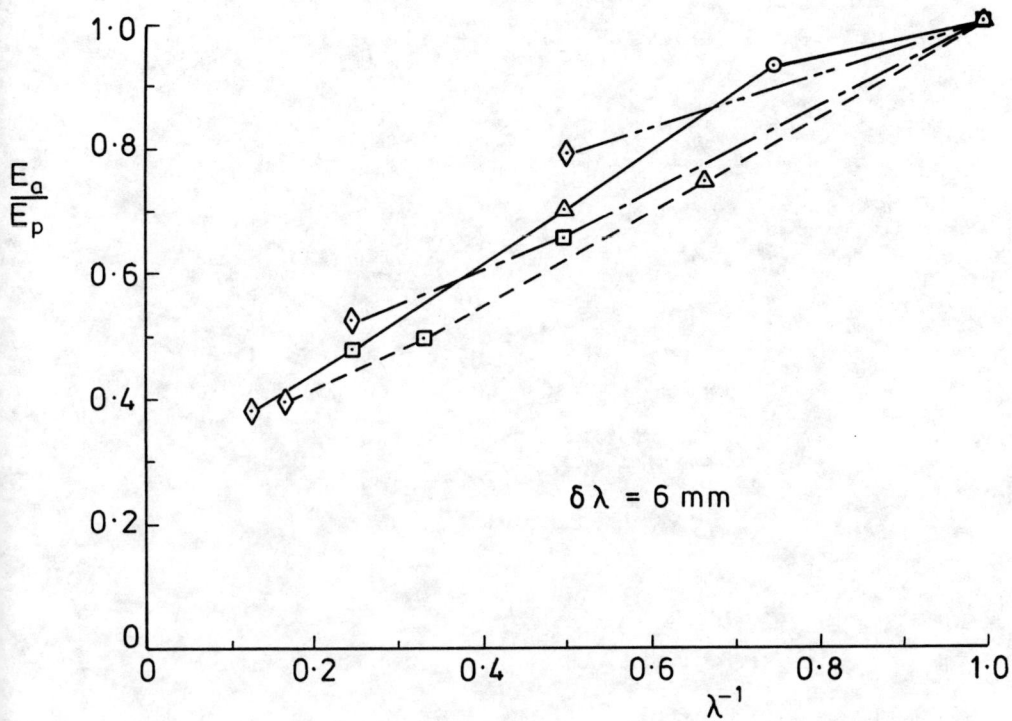

Figure 9

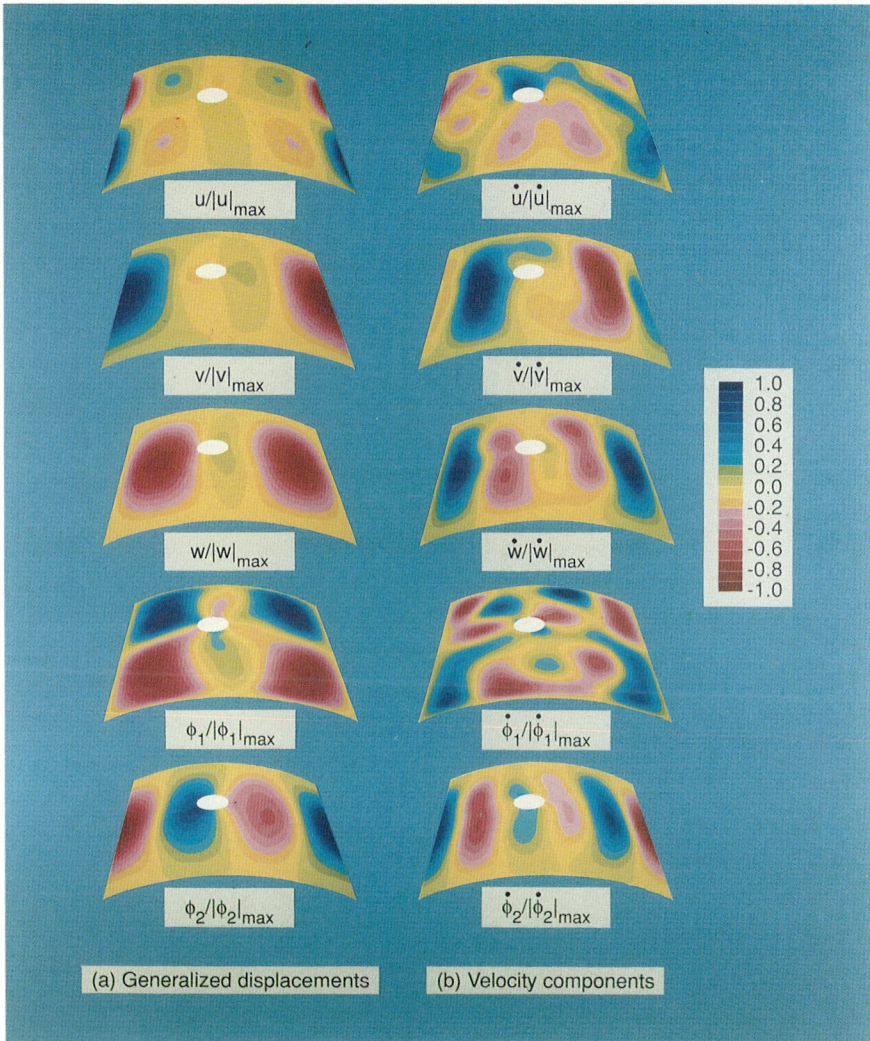

Figure 15 - Normalized contour plots for generalized displacements and velocity components at t=3.0 msec. Laminated anisotropic composite panel with an off-center circular cutout subjected to uniform normal loading p_o = -50,000 Pa (see Fig. 14).

Refer to Noor, page 34.

160

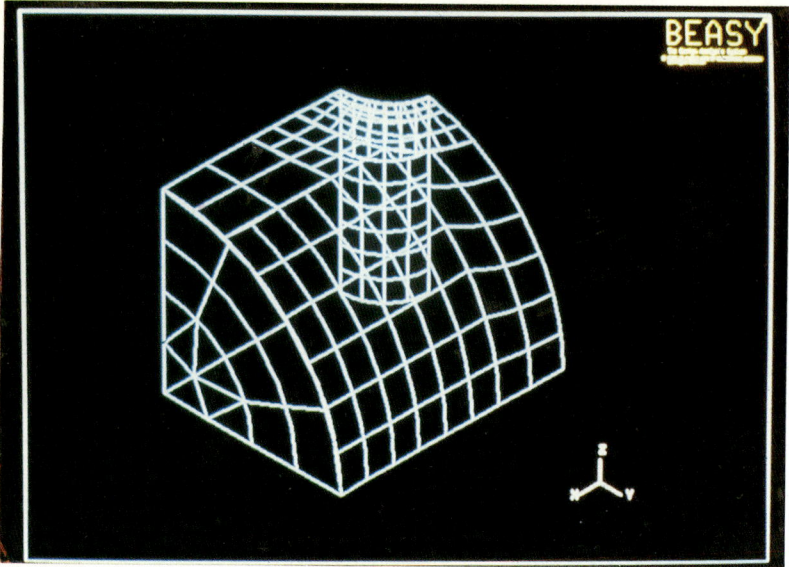

Figure 1   Part of a cylinder with cylindrical hole discretized using
            Discontinuous Boundary Elements

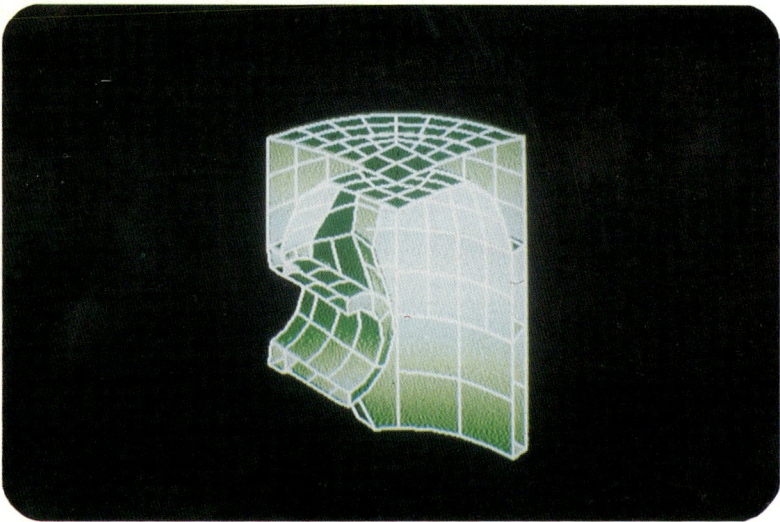

Figure 2   Part of a piston discretized using Boundary Elements

Refer to Brebbia, page 72.

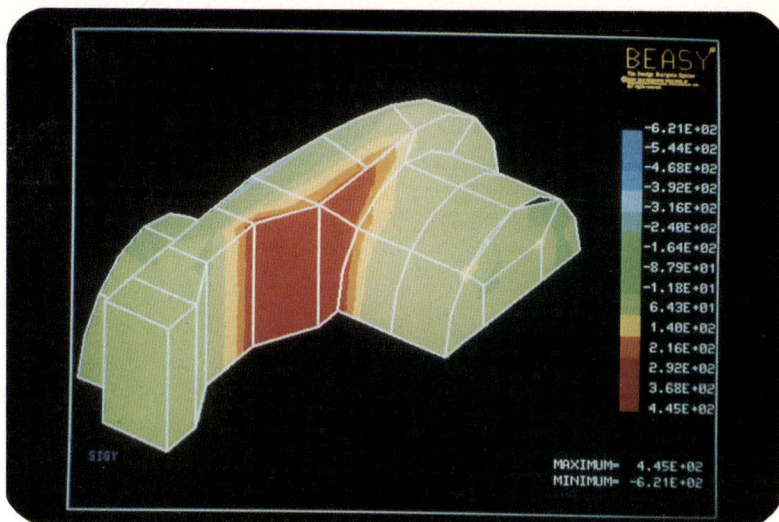

Figure 3   Crankshaft discretization

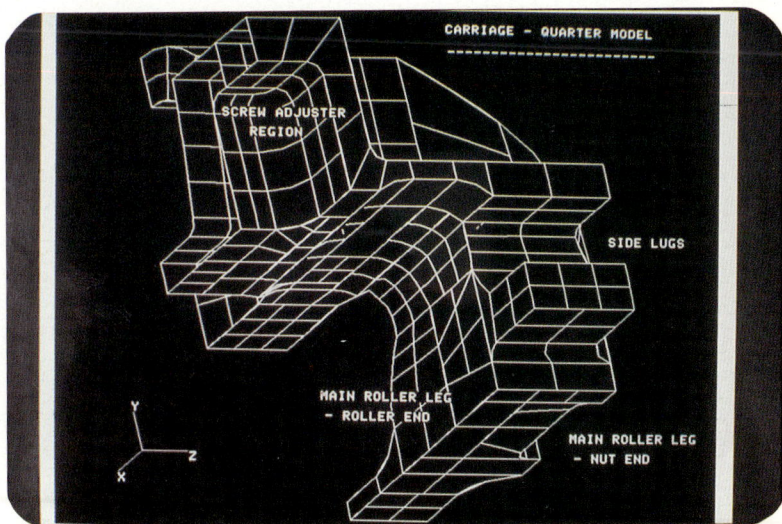

Figure 4   Discretization of a support bracket in an aircraft

Refer to Brebbia, page 72.

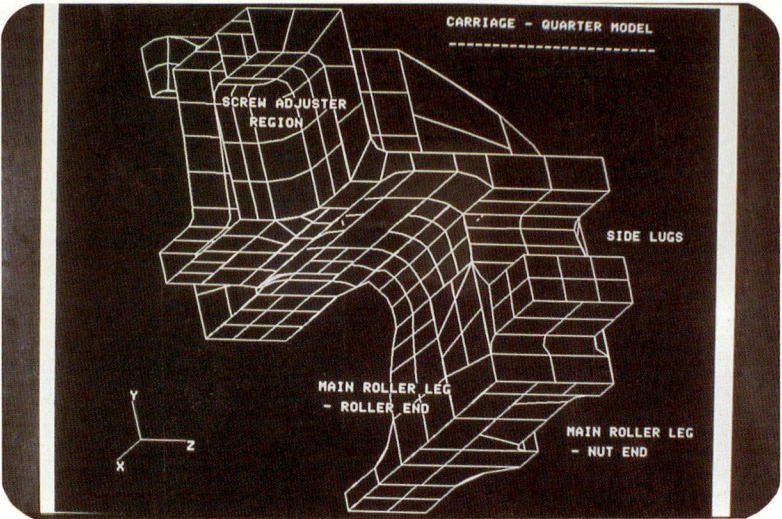

Figure 3 - BEASY Model of Part of a Modern Airliner *(Courtesy of British Aerospace)*

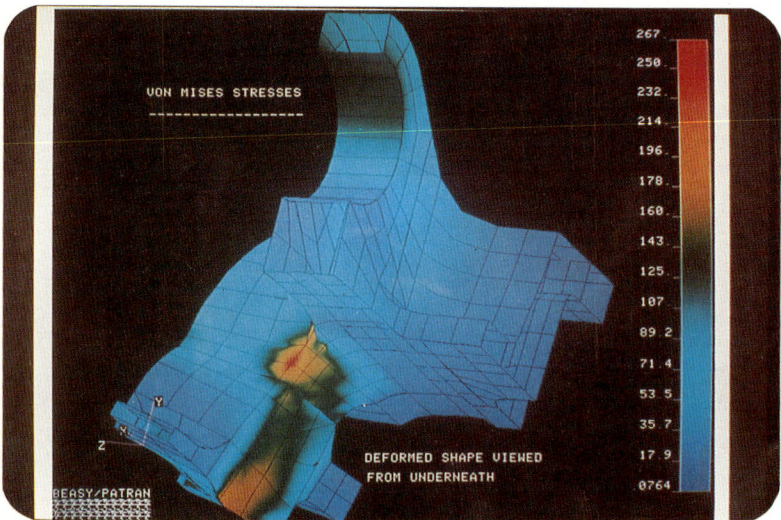

Figure 4 - Von Mises Stresses Predicted by BE Model

Refer to Adey, page 93.

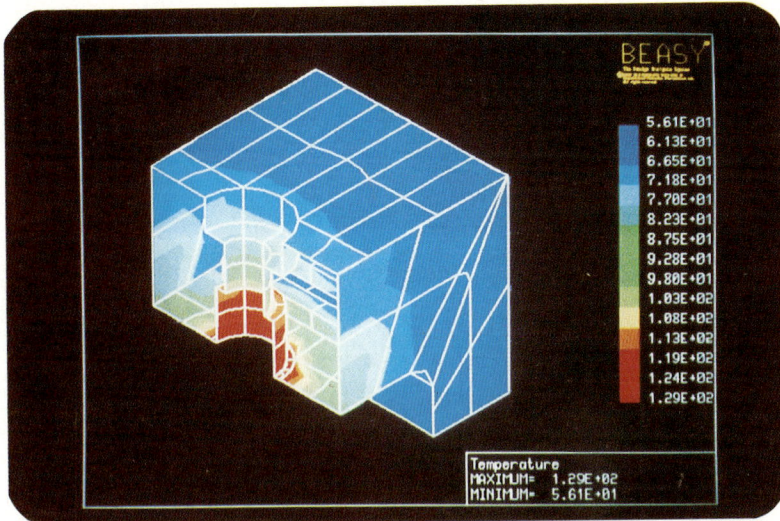

Figure 5 - Thermal Contours on Model of Casting Mould

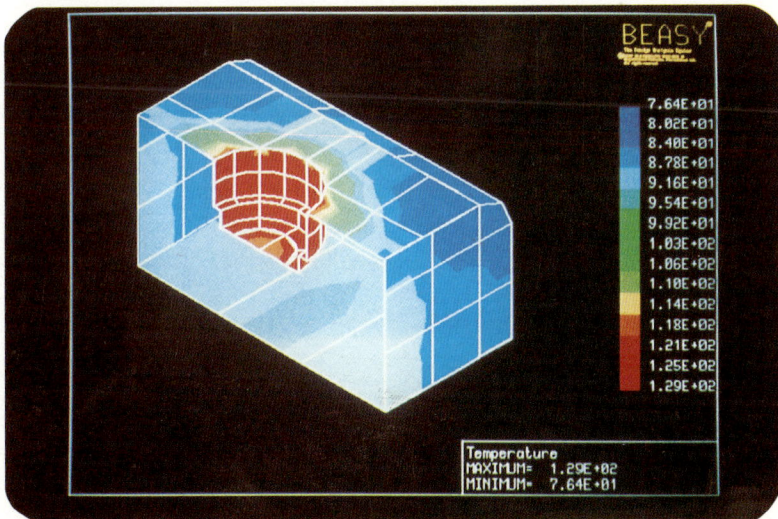

Figure 6 - Temperature Contours on Lower Part of Mould Clearly
Showing Elements Only on the Surface

Refer to Adey, page 93.

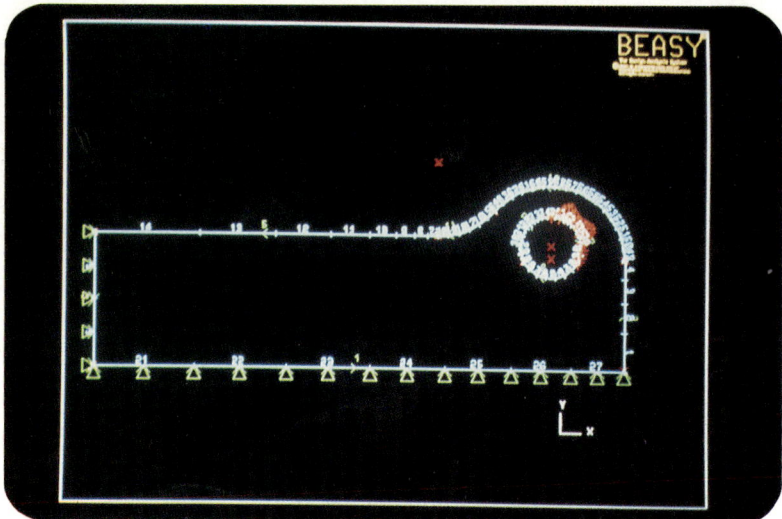

Figure 7 - Two Dimensional Stress Analysis Model of Lifting Lug

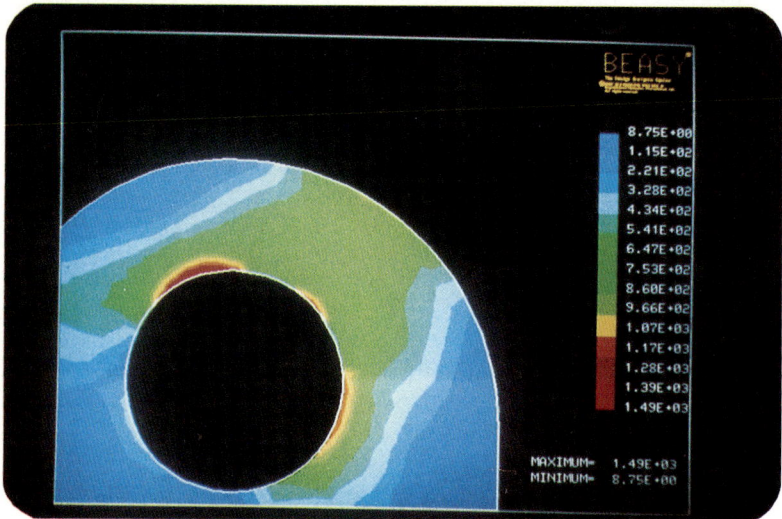

Figure 8 - Stress Contours Predicted Near the Lug Hole

Refer to Adey, page 94.

DECOMPOSITION INTO 8 BALANCED SUBDOMAINS

for parallel processing on ETA-10

Figure 1b

Refer to Farhat, page 109.

165

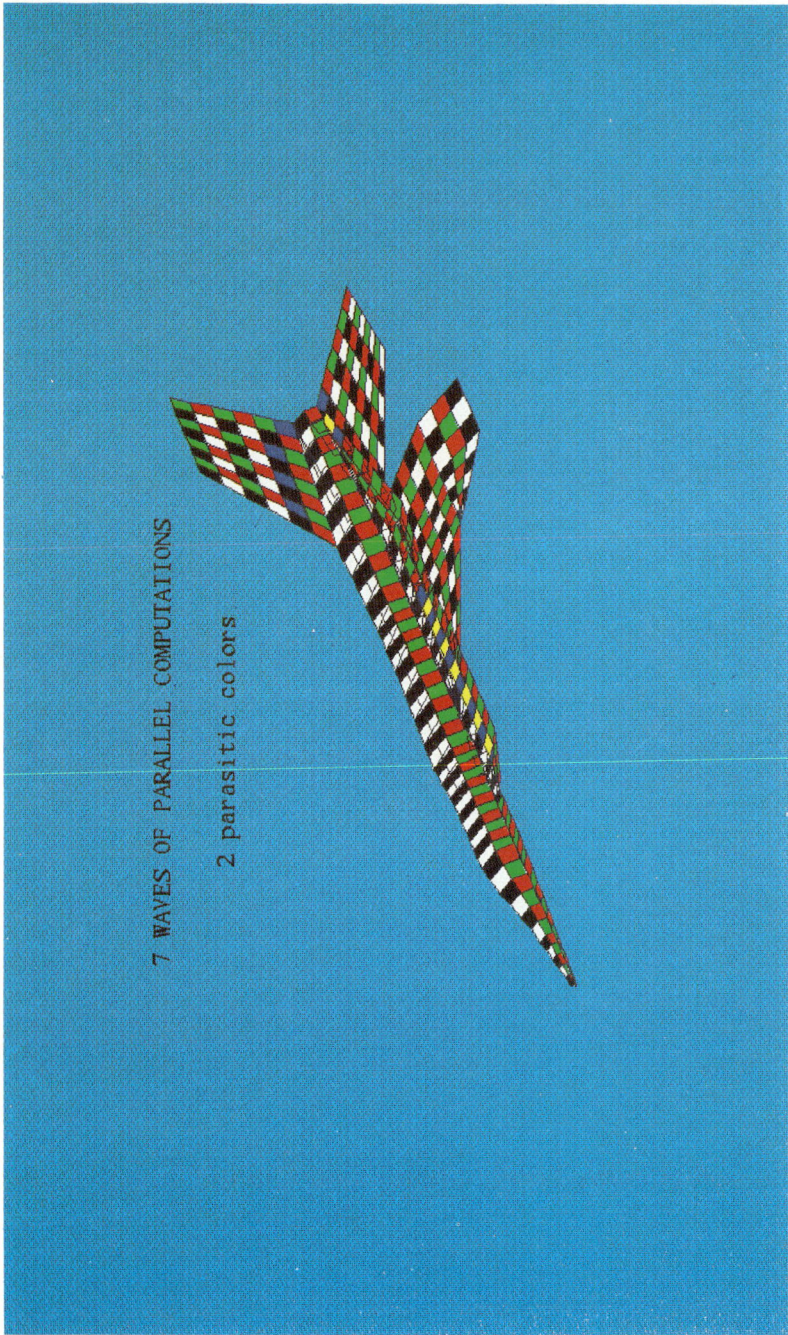

7 WAVES OF PARALLEL COMPUTATIONS

2 parasitic colors

Figure 2

Refer to Farhat, page 114.

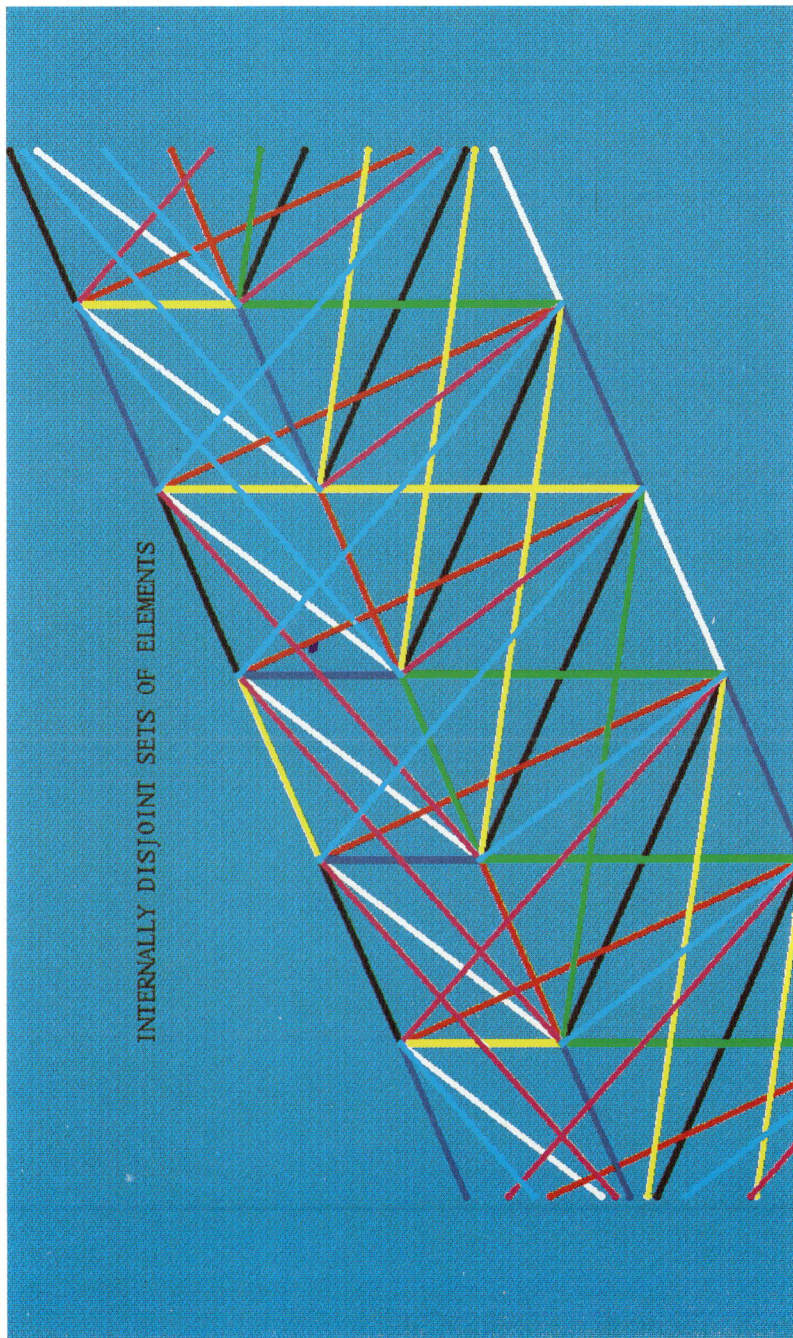

INTERNALLY DISJOINT SETS OF ELEMENTS

Figure 9

Refer to Farhat, page 127.

168

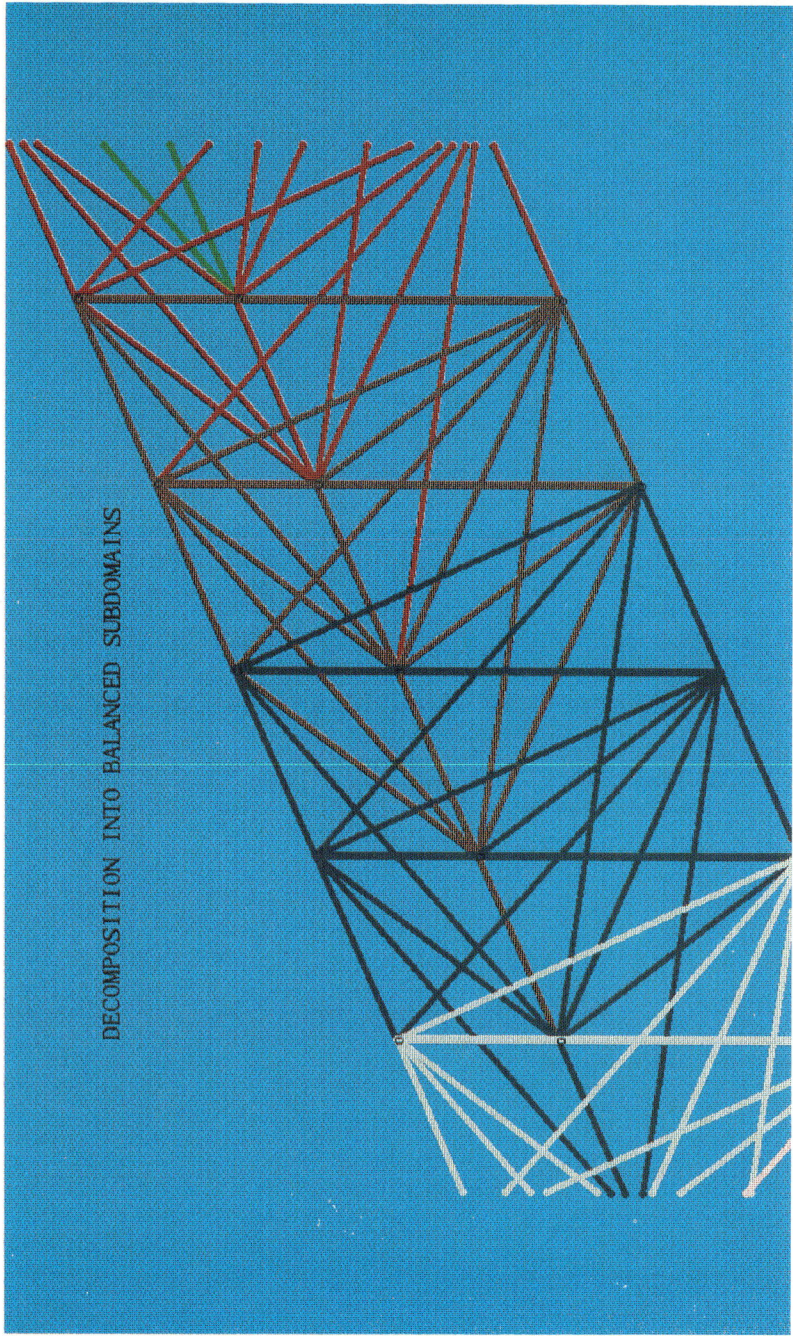

DECOMPOSITION INTO BALANCED SUBDOMAINS

Figure 13

Refer to Farhat, page 129.

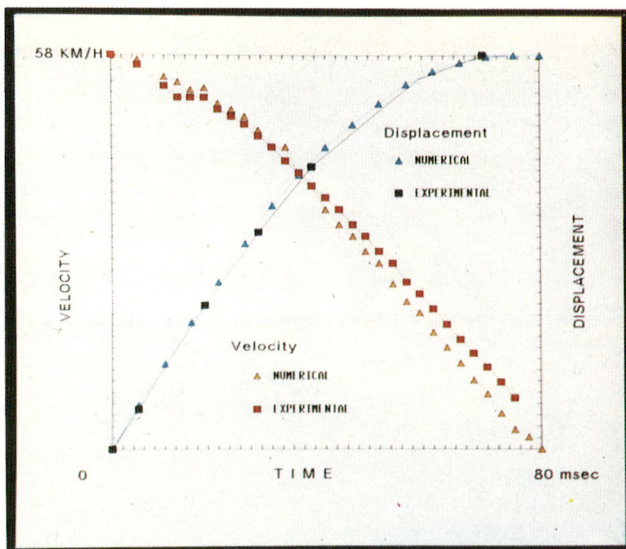

Frontal Crash on Rigid Wall.

Refer to Haug et al., page 171.

Credit: Citroën BX car model courtesy of PSA (Peugot SA) Co.

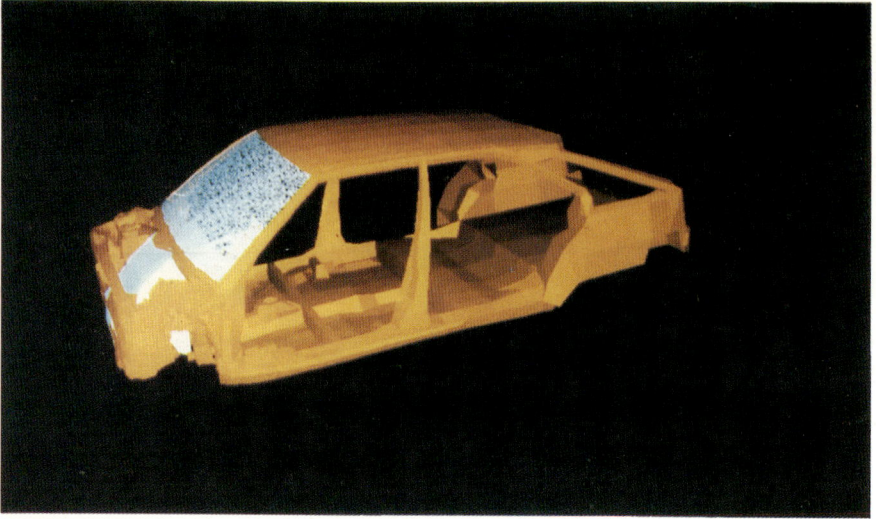

Full Car Crash Simulation

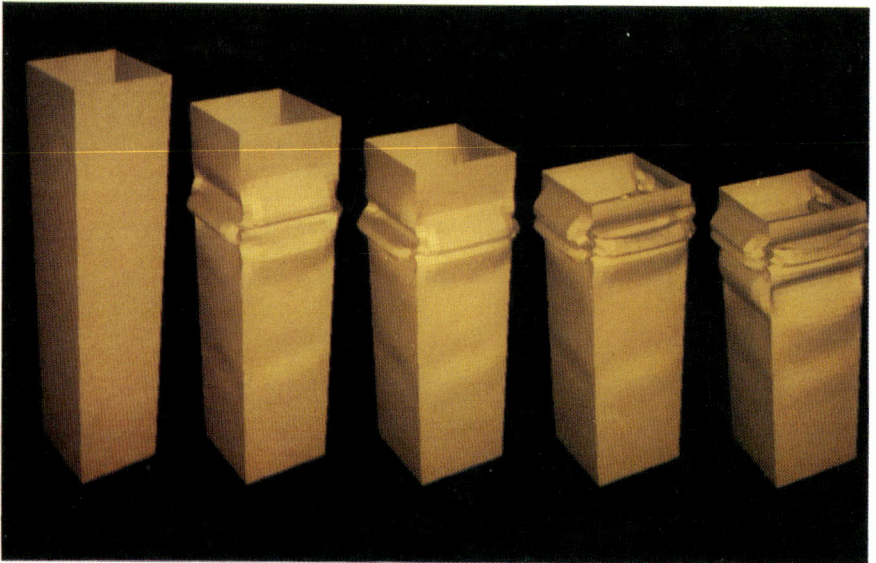

Box Beam Simulation Axial Crash (25 km/h)

Refer to Haug et al., page 171.

Credit: Citroën BX car model courtesy of PSA (Peugot SA) Co.

Industrial Crash Simulations using the PAM-CRASH Code

E. Haug, A. Dagba, J. Clinckemaillie, F. Aberlenc
Engineering Systems International S.A., 20 Rue Saarinen, Silic 270, 94578 Rungis-Cedex, France
A. Pickett, R. Hoffmann, D. Ulrich
Engineering Systems International GmbH, Frankfurterstr. 13-15, D-6236, Eschborn, West Germany

ABSTRACT

The present paper focuses on the industrial crash simulation of engineering structures, ranging from complete passenger car crash studies, to the simulation of helicopter crash, ship collision and railway car impact events, which involve steel, composite and fabric materials. Most of the analyses are carried out using the PAM-CRASH™ code which ESI developed from research prototype codes into an integrated industrial finite element software package with pre- and post-processors PRE-3D and DAISY.

INTRODUCTION

Over the past fifteen years ESI has been involved in the development of industrial nonlinear computational mechanics software packages for the detailed numerical simulation of a vast variety of complex engineering problems, which comprise nonlinear static and pre- and post-buckling analyses, nonlinear dynamic high-speed impact/penetration, crashworthiness, fluid flow and combustion, metal forming, stamping and composite stress, fracture, damage and failure analyses.

In particular, industrial crashworthiness simulations have gained considerable momentum over the past five years. The sudden growth of this field was made possible with the steadily increasing in-house availability of super-computers to the end user. Due to the use of supercomputers numerical crashworthiness calculations became industrially feasible and credible, which created the pressure needed to develop prototype simulation codes into industrial design tools, such as PAM-CRASH.

See pages 169 and 170 in colour section.

The validation of the PAM-CRASH code is demonstrated on the simulation of simple steel component crushing and on the results of full car crash analyses, juxtaposed with experimental results. Code developments in the areas of improved and new material models, contact schemes, rigid body options, solution schemes and convenient user interfaces are still ongoing and the code optimization on vector and parallel machines has not yet found an end. An integration of the code into industrial CAD/CAE software environments, e.g. based on ESI's DAMES relational data base management systems, and IBM's supported CATIA, CAEDS is in progress.

ESI's involvement with crash simulations dates back to 1976 (helicopter crash studies) and early approaches and results are published in Reference [1]. At that time supercomputers were either not available or at best centralized in a few computer centers per country and the early studies were carried out on CDC 7600 machines or later on an in-house VAX 11/780 mini-computer.

The early quasi-static crashworthiness simulations of thinwalled metal structures (steel, aluminum) were performed using the implicit solution technique available in ESI's general purpose nonlinear static and dynamic software package PAM-NL. This program is superseded today by the ARGUS/ESI package, and certain quasi-static or low velocity dynamics moderate deformation crash related safety tests (e.g., floor panel permanent deformation due to seat belt pulling action ; roof snap through ; etc.) are still today, and may continue to be in the future simulated advantageously using the implicit solution technique. The major advantage of the implicit solution technique over the explicit technique is its relative numerical stability (time steps orders of magnitude larger than the stable explicit time step may be used, as permitted for accurate tracking of the studied physical phenomenon) and the fact that it easily yields static end states (e.g. residual shape after elastic spring-back due to stamping deformation of a sheet of metal into a thinwalled profile), which are more difficult to achieve with the explicit approach.

The explicit PAM-CRASH code evolved from academic prototype software, developed mainly at US national defense laboratories and at universities in the years 1960 and 1970, into a strongly growing industrial software package, geared to the simulation of crash events. In 1974 ESI installed the finite difference (FD) explicit wave code, HEMP, written by Mark Wilkins [2], used for the simulation of impact/penetration and blast events in defense applications. After transcoding, validation and the addition of material models and of a rezoner, ESI added in 1976 the convected axisymmetric nonlinear thin shell finite element (FE) of Belytschko's WHAM2D-code [3] to the FD, HEMP code, which permitted the simulation of hypothetical core accidents in the french Superphenix fast breeder reactor by modelling the shell parts with thin shell finite elements and the fluid parts with finite differences, both parts being connected via sliding interface contact surfaces, Ref. [3bis].

One of the first genuine crash simulations was carried out in 1976 with the so created hybrid FE/FD code HEMP/ESI on the subject of an airplane impact on the outer hull of a nuclear power plant (c.f. [1]), where a thinwalled steel tube was shot at about 900 km/h against a concrete wall and where the code realistically simulated the formation of 30 axisymetric buckles due to the crimpling of the tube in the so called concertina buckling mode of failure. Meanwhile ESI performed several helicopter crash studies (c.f. [1]) using the implicit code PAM-NL. Due to the interest of the automotive industry, ESI performed between 1978 and 1983 several moderate deformation quasi-static component crashworthiness studies using the implicit technique [1].

The implicit line of action was abandoned for severe large deformation full scale three-dimensional automobile crash simulations in favor of the more efficient explicit approach. In 1976 and 1981 ESI acquired the 2D and 3D explicit non- linear FE codes WHAM-2D and WHAM-3D, respectively, written by T. Belytschko [4], and in 1978 DYNA3D, written by J.O. Hallquist [5]. After numerous validations and additions to these codes, ESI synthesized during the years 1980-1983 from the HEMP, WHAM and DYNA3D codes and from its own additions and modifications, its own industrial general purpose explicit FE wave code EFHYD-3D, applied mainly to national defense problems. The EFHYD-3D code derives its basic code architecture and input/output section from the (1981) DYNA-code, the under-integrated 2D and 3D solid, beam and thin shell elements go back to the WHAM-codes and the contact-impact algorithms are based on the HEMP and DYNA codes.

In 1984 ESI extracted the special purpose thin-walled metal structures crash analysis code, PAM-CRASH, from the EFHYD-3D code and replaced the three-noded Belytschko thin shell element with the computer time efficient under-integrated four-noded Belytschko thin shell element with hourglass control, based on Mindlin plate theory. The so created numerical tool made large scale thin-walled metal structure crash simulations industrially feasible at a time where none of the comparable competing codes had this capacity.

Since 1985 the PAM-CRASH package has been widely installed in the automotive industry, and ESI carried out in 1985 and 1986 the first successful full frontal car crash analyses, and the package continues to be developed intensively.

THE PAM-CRASH™ CODE

Code definition
PAM-CRASH™ is a three dimensional Lagrangean finite element explicit vectorized/multi-tasked code for the nonlinear dynamic analysis of structures. The code is streamlined for the dynamic crashing analysis of structures made of assemblages of thinwalled shells, plates, beams, solids, bars and rigid bodies.

Code description

In the following paragraphs a short description of some important existing and new features of the PAM-CRASH code is given.

Spatial discretization The finite element approach is chosen as a spatial discretization scheme, which adds modeling flexibility with respect to the finite difference discretization scheme used in older wave codes.

Time discretization The explicit central difference time integration scheme is chosen to integrate the nonlinear discrete equations of motion established via the spatial FE discretization. These equations, given by the matrix expression

$$Ma = F_{ext} - F_{int} \tag{1}$$

become uncoupled when the mass matrix M is diagonal and the process of equation solving for the accelerations a becomes trivial. F_{ext} are the externally applied loads and F_{int} are the internal resisting forces. At time t_n, accelerations a_n follow from (1) and velocities $v_{n+1/2}$ and displacements u_{n+1} follow from the central difference operator as follows

$$a_n = M^{-1} (F_{ext} - F_{int})_n \tag{2}$$
$$v_{n+1/2} = v_{n-1/2} + a_n \, \Delta t_n \tag{3}$$
$$u_{n+1} = u_n + v_{n+1/2} \, \Delta t_{n+1/2} \tag{4}$$

where subscripts $n-1/2$, n, $n+1/2$, $n+1$ denote present and future times at half and full time intervals with time steps, Δt_n and $\Delta t_{n+1/2}$, respectively, see Figure 1 [6]. The solution in time is advanced in time steps, Δt, the magnitude of which is limited for numerical stability.

Stable time step In the simplest case, the stable time step of the central difference explicit time integration scheme is proportional to the smallest discrete distance, d_{min}, in the FE-model and inversely proportional to the solid elastic stress wave speed,

$$c = (E_o/\rho)^{1/2} \tag{5}$$

where E_o is the initial elastic modulus of the material and ρ is the mass density. I.e.,

$$\Delta t \simeq d_{min} \, (\rho/E_o)^{1/2} \tag{6}$$

The cost of an explicit analysis is directly proportional to the number of time steps needed. It is therefore an important cost consideration to establish FE-meshes which avoid small values d_{min}, especially when only a few elements are much smaller than the bulk of the elements.

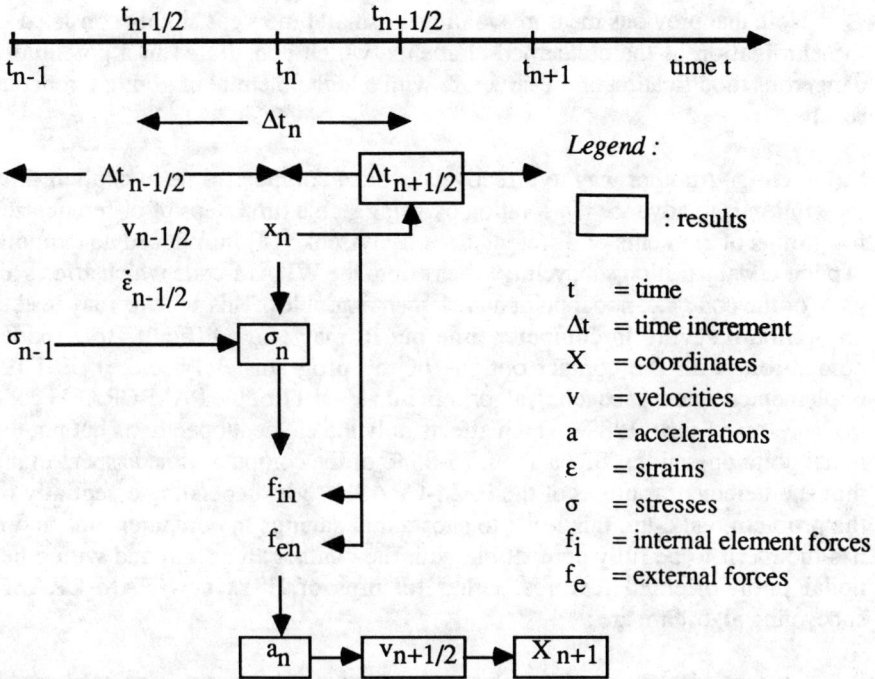

Fig. 1. PAM-CRASH flow chart

Further, the distance d_{min} may decrease during a crash analysis when certain limited regions of the structure undergo very severe plastic deformations and the number of stable time steps needed to further advance the solution will increase correspondingly. In order to limit the smallness of the time step, PAM-CRASH has a built-in time step control scheme [7] which permits the user to never let the stable time step fall below a specified minimum, Δt_{min}, Figure 2.

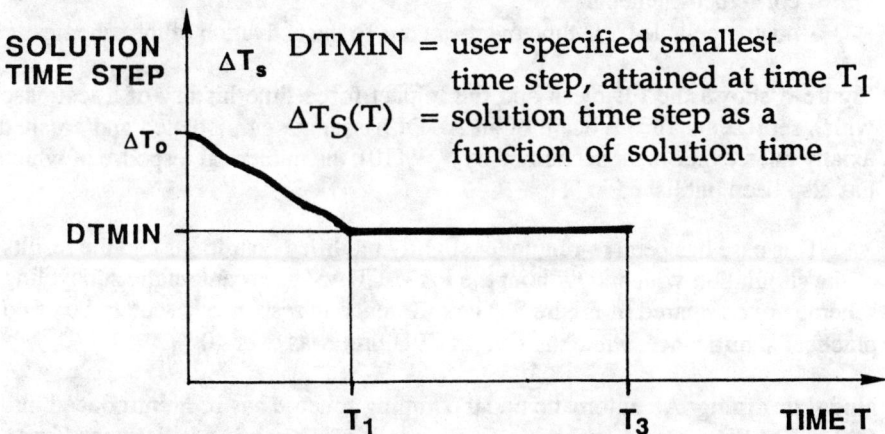

Fig. 2. PAM-CRASH time step control option

Note that previous measures to prevent small time steps simply consisted in the elimination of the concerned elements, which constituted in a potentially dangerous modification of the structure with a high potential of giving erroneous results.

Subcycling Another way to effectively reduce computer time in explicit time integration is to advance the solution by using stable time steps of different sizes for groups of elements of different sizes. Belytschko [8] introduced an explicit-explicit nodal partition subcycling scheme into the WHAM-code which affects all parts of the code, i.e. nodal point and element variables. This scheme may lead to an optimum saving in computer time but it may be difficult to render compatible with the contact options of the program (Belytschko). ESI [9] implemented a new subcycling algorithm into level 11 of the PAM-CRASH code (to be released early 1989), which affects only the element operations but not the nodal point operations. Since about 75-85 % of the computer time is spent in the thin shell element routines of the PAM-CRASH code, depending essentially on the problem make-up, this leads to substantial savings in computer time and it has the merit to be fully compatible with the contact algorithm and with other nodal point oriented features. Other features of ESI's new PAM-CRASH subcycling algorithm are :

- works with any mesh topology, i.e. elements of the same size need not be adjacent to each other
- automatic grouping of elements falling into the same stable time step group, each integrated with the same local stable time step
- no appreciable loss of solution accuracy
- total compatibility with all other program options
- extremely simple input : the user simply sets a flag
- if, during analysis, elements change time step groups due to deformation, the algorithm recognizes this fact and it automatically redistributes the concerned elements
- no appreciable CPU time overhead due to the activation of the scheme.

Figure 3 shows the FE-mesh and the impact force time history of a test case which represents a front beam of an AUDI passenger car, isolated and crushed axially in an experiment performed by VW [10], the numerical response of which has also been published in [1].

This case has been re-run under slightly modified conditions and the results of the simulation with and without the PAM-CRASH new automatic subcycling scheme are compared in Figure 3. The difference in response is seen to be of no practical significance, while the gain in CPU time was over 40 %.

Nodal damping An automatic nodal damping scheme has been introduced into PAM-CRASH which permits damping out residual dynamic oscillations after an achieved state of post-buckling or post-collapse. A typical example may be the

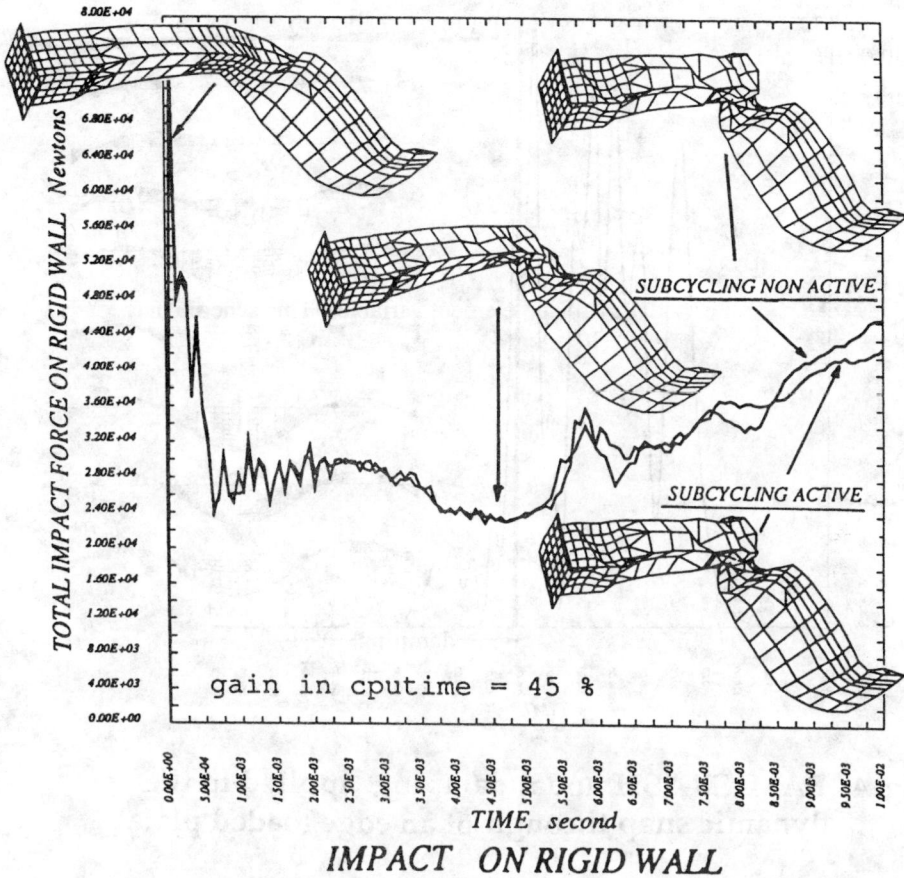

Fig. 3. PAM-CRASH subcycling scheme ; applied to a passenger car front beam member (VW/AUDI)

elimination of residual oscillations after a dynamic snap-through event, in order to evaluate the final static configuration after snap-through, or the evaluation of the residual deformed position of a sheet metal after metal forming deformations due to stamping or stretching (elastic spring-back). To this end mass proportional nodal dashpots the damping coefficients of which are calculated for a critical damping of the lowest residual vibration modes are introduced.

These dashpots may be activated at any time and for any duration. Figure 4 shows the calculated dynamic snap-through of a partly clamped plate [11], under the action of in-plane edge loads, which exhibits a dynamic snap-through from an initial lower order lateral buckling mode (order 2) into a final higher order buckling mode (order 3). The figure shows that the lateral displacement variation in time is calmed down to nearly zero amplitude after activation of the automatic nodal damping scheme of PAM-CRASH at 400 ms for a duration of 300 ms.

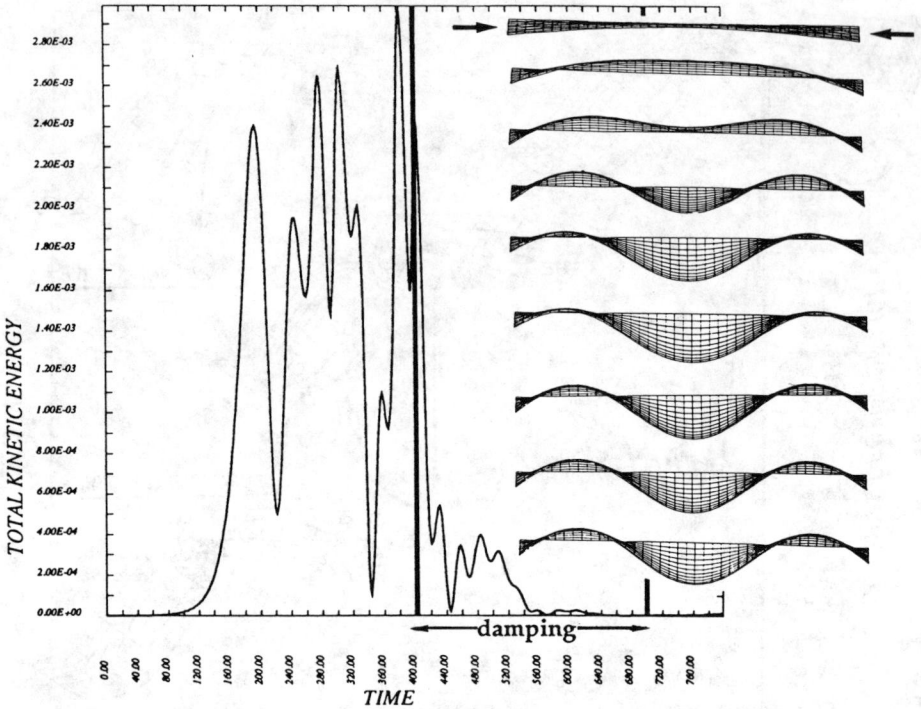

Fig. 4. PAM-CRASH nodal damping, applied to the dynamic snap through of an edge loaded plate

Dynamic relaxation While the just described nodal damping option may effectively dissipate residual elastic oscillations, a more severe damping or dissipation of dynamic effects is needed in order to effectively carry out certain quasi-static simulations with explicit codes. The well introduced schemes of dynamic relaxation [12] may achieve the described goal and the feasibility of their incorporation into the PAM-CRASH code is currently investigated.

Composite fracturing material Rather than by plasticity, composites typically fail in a fashion which is closer to a fracturing material behaviour, where Young's modulus is "damaged" via the opening of microcracks. A fracturing material model with a bi-linear modulus damage function and with strain-softening has been evaluated within the PAM-CRASH code, which permits simulation of the crash behaviour of composite parts. For the bending, shear and tensile fracture simulation of random fiber composite components, frequently used in automobiles, an isotropic fracturing law is sufficient. For the fracture simulation of layered long fiber reinforced composites an anisotropic or bi-phase fracturing law is required, such as implemented in ESI's implicit damage analysis code PAM-FISS. Fig.5 summarizes the used fracturing material law and the bi-phase approach, where the rheologies of the fiber and matrix phase of the composite can be treated separately [13]. In the bi-phase approach

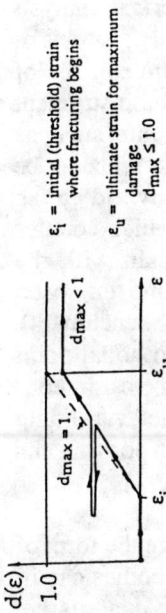

UD-COMPOSITE (UD) = FIBERS (F) + MATRIX (M)

UD : undirectional
f : fiber
m : matrix

Stress-Strain law:
$$\sigma^{UD} = C^{UD} \epsilon^{UD}$$
$$C^{UD} = C^f + C^m$$

Known material properties:
$E^{UD}_{11}, E^{UD}_{22}, G^{UD}_{12}, \nu^{UD}_{12}$ = in-plane UD material constants
E^f_{true} = true fiber modulus
α = fiber volume fraction

Calculated quantities:
$$\nu^{UD}_{21} = \nu^{UD}_{12} E^{UD}_{22} / E^{UD}_{11}$$
$$N^{UD} = 1 - \nu^{UD}_{12} \nu^{UD}_{21}$$
$$E^f_{11} = \alpha E^f_{true}$$

Derived orthotropic matrix material constants:
$$E^m_{11} = E^{UD}_{11} - E^f_{11}$$
$$E^m_{22} = E^{UD}_{22} / (1 + \nu^2_{12} (E^{UD}_{22} / E^{UD}_{11}) (E^f_{11} / (E^{UD}_{11} - E^f_{11})))$$
$$\nu^m_{12} = \nu^{UD}_{12}$$
$$\nu^m_{21} = \nu^{UD}_{21} / (1 - E^f_{11} N^{UD} / E^{UD}_{11}) \neq \nu^{UD}_{21}$$
$$G^m_{12} = G^{UD}_{12}$$

Program PAM-FISS/Bi-Phase rheological model

ϵ_i = initial (threshold) strain where fracturing begins

ϵ_u = ultimate strain for maximum damage
$d_{max} \leq 1.0$

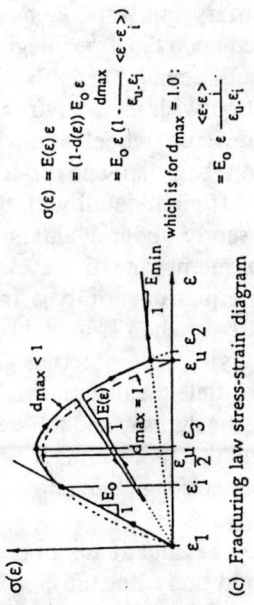

(a) Linear fracturing damage law

$$E(\epsilon) = (1-d(\epsilon)) E_o$$
$$E_{min} = (1-d_{max}) E_o$$
for $d_{max} = 1.0 : E_{min} = zero$

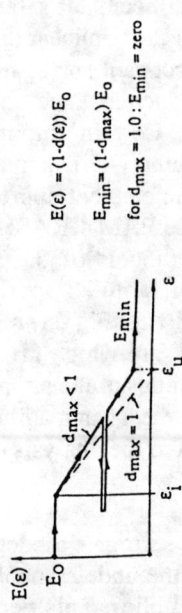

(b) Evolution of the axial modulus with damage

$$\sigma(\epsilon) = E(\epsilon) \epsilon$$
$$= (1-d(\epsilon)) E_o \epsilon$$
$$= E_o \epsilon (1 - \frac{d_{max}}{\epsilon_u - \epsilon_i} <\epsilon - \epsilon_i>)$$

which is for $d_{max} = 1.0$:
$$= E_o \epsilon (1 - \frac{<\epsilon - \epsilon_i>}{\epsilon_u - \epsilon_i})$$

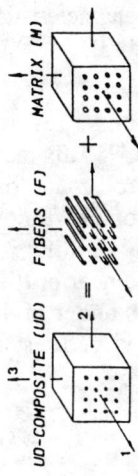

(c) Fracturing law stress-strain diagram

Fracturing material law in PAM-FISS

Fig. 5. PAM-CRASH composite rheological models (first developed in PAM-FISS)

the introduced fracturing law can be applied safely to the subcritical damage spread in the matrix phase, the micro-cracking of which is confined by the fibers. If applied to the overall composite (matrix plus fibers), strain localisation effects [27] can be overcome by the introduction of "critical damage over a characteristic volume" damage mechanics fracture criterion, described in detail in [13].

Viscous damping In strain softening materials fracture can occur under severe localization of strain with very little or no dissipation of energy [14]. This fact can render numerical simulations, e.g., of a step-loaded tensile rod made of fracturing material difficult as soon as the maximum resistance of one finite element is overcome and when this critical element fails rapidly. Unlike in plasticity, damage will remain local and stress waves caused by the near brittle failure of the material may trigger violent quasi-elastic vibrations under high frequencies in the remaining structure. In order to attenuate these vibrations, a viscous material damping with its one- dimensional form of

$$\sigma = E_O \, \varepsilon + \mu \, d\varepsilon/dt \tag{7}$$

has been introduced into the PAM-CRASH code. The damping coefficient, μ, can be calibrated such that the highest frequencies in a finite element model, which are caused by stress waves after sudden local fracturing, will be attenuated effectively, whereas the lower frequencies which are excited due to the loading will not be affected.

Occupant safety analysis Conventional occupant safety numerical analysis tools are based upon the time integration of the nonlinear differential equations of motion of the kinematic assemblies of occupant surrogates (dummies), and of their interaction with their environment, such as knee bolsters, the instrument panel, the steering wheel and airbags. Certain important events such as out-of-position (bag slap) and off-center dummy-airbag interactions can better be analyzed in sufficient detail with the finite element approach. A hybrid airbag model is presently under evaluation in the PAM-CRASH code which couples thin shell or membrane finite element models of the airbag skin with the gasdynamic equations of bag inflation, applied to the volume enclosed instantaneously by the FE-model of the airbag, Fig.6 (preliminary benchmark). The coupling yields the internal gas pressure which gives rise to nodal point forces which inflate the bag model. Interactions of the airbag with the car interior environment, the dummy and the steering wheel can readily be treated via sliding interfaces, and the approach is well suited to the analysis of out-of-position and off-axis dummy airbag interactions.

Also under evaluation is an occupant surrogate model, either in the form of connected rigid body linkages, or where the undeformable rigid bodies can in part be replaced by deformable FE assemblies. This permits the quantitative analysis of, e.g., chest deformation and of the parameters leading to face

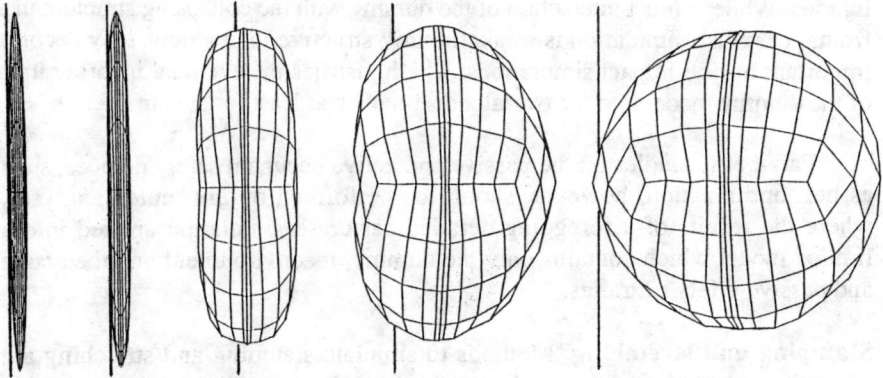

(a) inflation without contact

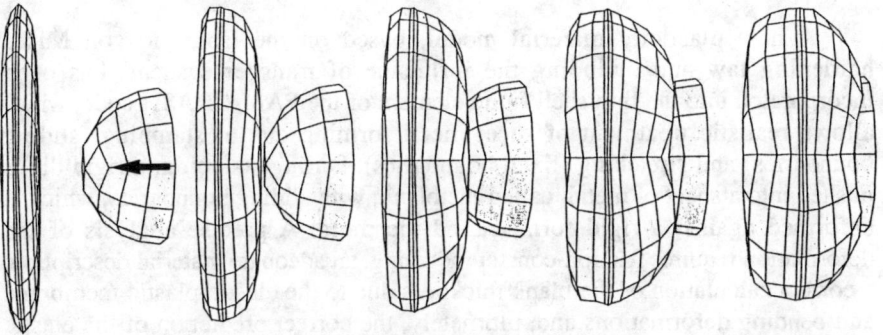

(b) contact during inflation (bag-slap)

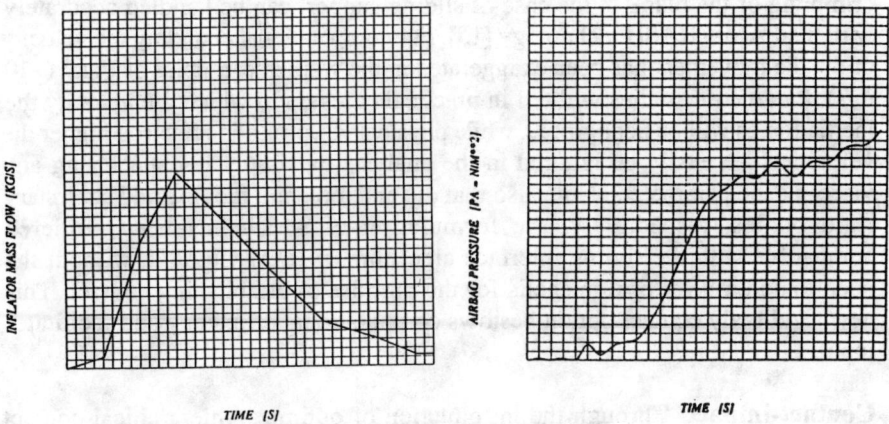

(c) inflator mass flow and bag pressure vs. time

Fig. 6. PAM-CRASH hybrid FE/gasdynamics airbag model (preliminary benchmarks)

injuries. While a direct interaction of the dummy with the collapsing structure in a frontal car crash simulation is weak, dummy-structure interactions may become important in side-impact simulations, which justifies the eventual incorporation of the dummy model into the overall crash model, at least for side impact.

Parametric studies on the passive and active occupant safety devices, such as belt optimization, however, should be performed by uncoupled analyses, where the results of a foregoing detailed car crash simulation are fed into a limited model which contains only the dummy, its environment and the active and passive safety restraints.

Stamping and stretching Methods to simulate stamping and stretching are currently evaluated. One approach uses the numerical capabilities implemented in PAM-CRASH, for which preliminary results have been obtained, as described below.

A new plasticity material model, based on the isotropic von Mises hardening law and including the influence of transverse shear, has been incorporated into the thin shell finite element of the PAM-CRASH code, which allows realistic treatment of sheet metal forming due to stamping (sliding boundaries) and stretching (fixed boundaries). During stamping, a rigid die is pushed into a sheet of metal, called the blank, with sliding boundaries, which is deformed against a rigid form, called the mold. A precise analysis of the deformation requires friction-contact modeling, an adequate material description, a correct calculation of the blank thickness due to the elastic-plastic membrane and bending deformations and, ultimately, the correct prediction of the elastic spring-back. Preliminary benchmark tests, Fig.7, show that large die movements, large blank deformations, friction contact and circumferential crimpling of the blank in the case of sliding support can be handled accurately with the PAM-CRASH technology [15]. For computational economy the velocity of the die movement has been exaggerated in the worked examples of Fig.7 (~10 m/s). Parametric studies will tell in practical cases up to which die velocity the inertia forces will be negligible, while physical visco-plastic effects at higher die velocities can easily be ignored in the analysis by simply not specifying any visco-plastic parameters. Note also that die and mold are moving and stationary rigid bodies, respectively, the formulation of which has been rendered compatible with the sliding interface algorithm of PAM-CRASH, without the need to add a layer of thin shells for the purpose of establishing contact. This new rigid body contact option bestows computational efficiency and facilitates input.

Contact-impact Through the introduction of optimized hierarchical contact search and contact evaluation techniques, the self-contact algorithm of PAM-CRASH has been rendered extremely fast and user friendly. For example, all of the 8000 facettes of a full frontal car crash FE model could now be specified in one single group of self-contact segments with a CPU time overhead significantly below 100 %. Of course, the user is encouraged to specify several

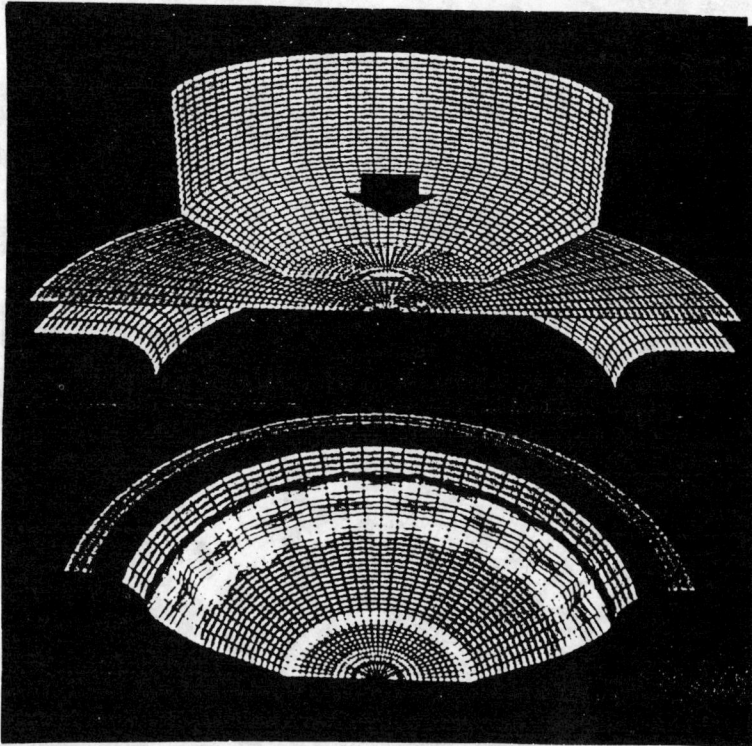

Fig. 7. PAM-CRASH stamping simulation (benchmark test)

distinct groups of self-contact segments, possibly with overlap, which will reduce CPU time overheads to a few percent. For the segments specified in self-contact groups the program verifies continually whether any node attached to this group of segments will penetrate any of the segments belonging to the group. If yes, the penetration is penalized and the concerned node will move along the surface of the segment under specified zero or nonzero friction, until separation.

The impact of an individual node onto a segment is fully plastic by default. Recently a partly plastic or an elastic impact with perfect rebound has been made available in order to permit the occasional study of elastic-plastic rebound of pointlike objects.

A typical benchmark simulation for self-contact problems is shown in Fig.8, where a hollow, thin-walled, metallic tube is crushed axially with the PAM-CRASH code. The analysis shows the formation of "dog bone" antisymmetric elasto-plastic buckles in which the insides and the outsides of the tube walls go into multiple self-contact.

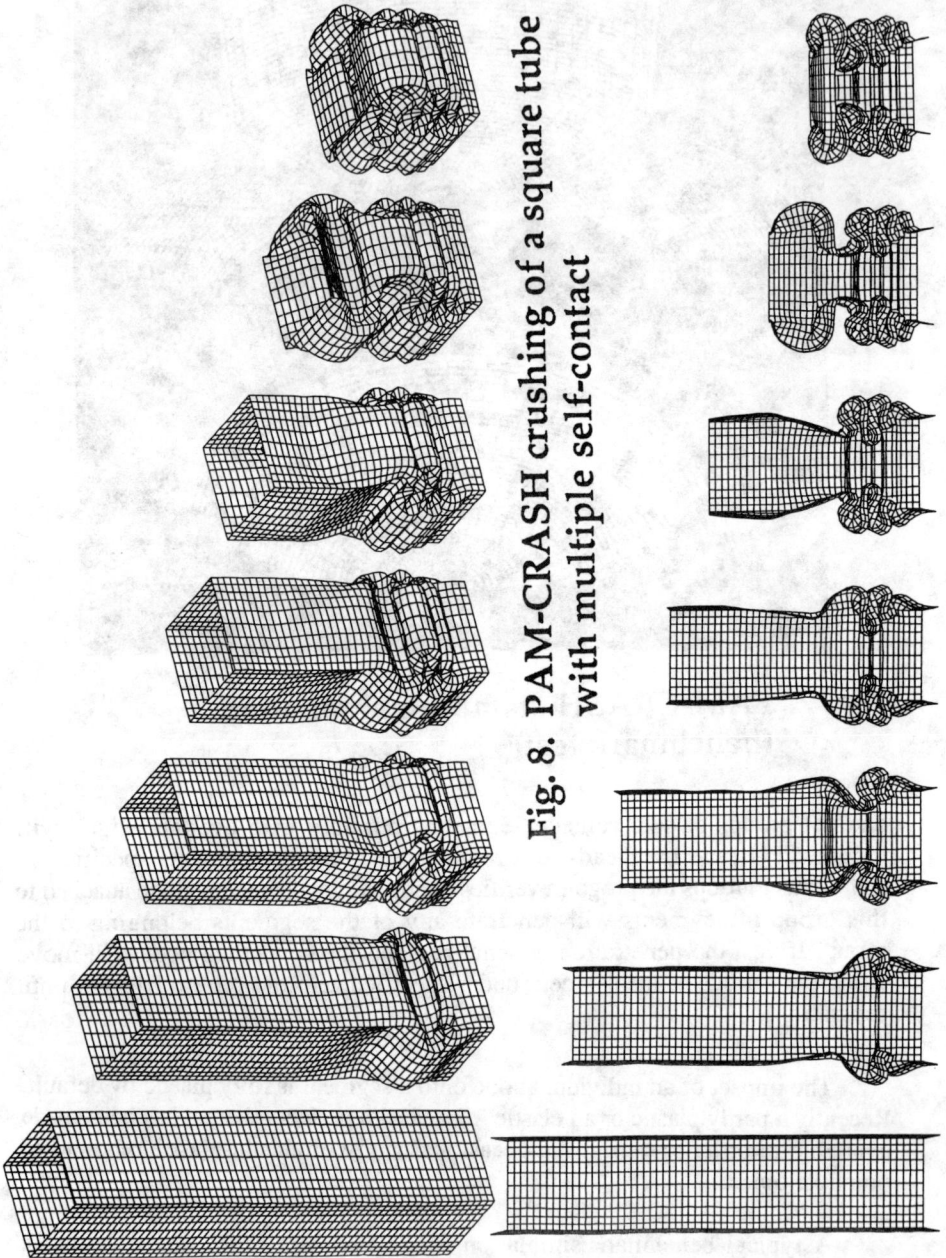

Fig. 8. PAM-CRASH crushing of a square tube with multiple self-contact

Rigid walls The rigid wall option permits study of impacts against undeformable stone walls. This option has been extended from an infinite non-moving rigid wall into a rigid obstacle specified via bricks and cylinders of arbitrary size. The so defined obstacle can be contacted over its entire surface and it can translate in space. An infinite or a finite mass can be assigned to the moving obstacle. Fig.9 shows a benchmark test of a moving cubic finite rigid wall obstacle with finite mass, which is shot at some initial velocity into a sheet of flexible material. The motion pictures show how the hit membrane progressively wraps around the brick shaped "bullet". Another application of this type has been the modeling of an obstacle in a hypothetical railway car incident, Fig.17.

APPLICATIONS

Previous studies
In one of the earliest ESI publications [16] on industrial crashworthiness simulations, based on a 1979 Euromech Conference presentation in Jablonna, Poland, solutions of airplane crash simulations on the protective hulls of nuclear power plants, and of nuclear fast breeder reactor rapid bubble expansion hypothetical core accidents have been reported, which already at that time made use of the explicit time integration technique for some examples. This technique was implemented in the mixed FD (solids) and FE (thin shells) code HEMP-ESI. Other examples have been run on the implicit nonlinear axisymmetric thin shell FE code (with non-axisymmetric loading via Fourier series expansion), PAM-AX3D. This publication also contains implicit studies of quasi-static crushing events related to helicopter hard landings, bus frame roll-over tests and a first attempt to solve an axial crushing event of a passenger car front beam component. The latter simulations were carried out with the implicit 3D nonlinear general purpose FE analysis code PAM-NL.

A comprehensive review concerning ESI's work on the subject of structural crash simulation appeared in 1983 [1], which included the topics of the implicit quasi-static incremental collapse thin shell FE analysis of the A-pillar and of a frontal assembly of a passenger car, and the explicit thin shell FE simulation of a front beam component of a passenger car and of a car frame modeled by beams. This publication also contains a first "super-element" approach to the helicopter crash test simulation described in [16], and the first detailed, complex, axisymmetric "concertina" mode collapse simulation of a hollow tube shot against a concrete slab, which correctly predicted the formation of 30 buckles with self-contact under due consideration of the variable plate thickness.

An overview paper [17] summarized the lagrangean, eulerian and arbitrary lagrange-eulerian (ALE) solution techniques used in ESI's numerical codes. It further gives an overview on the sophisticated concrete and steel material models used in impact-penetration studies, and of the composite material model used for the damage analysis in composites. The paper gives numerous examples on impact, penetration and explosion which range from medium to hypervelocities

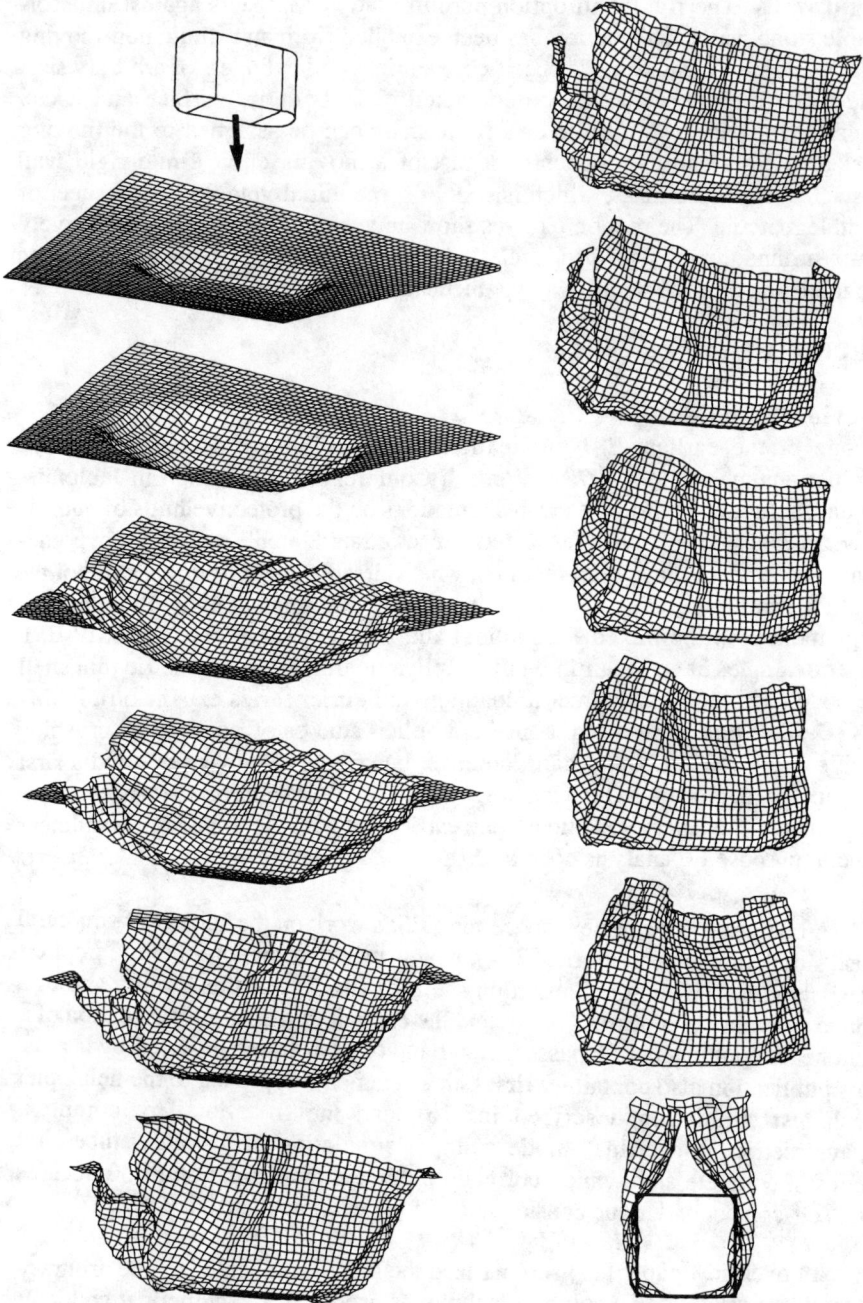

Fig. 9. PAM-CRASH movable finite rigid wall contact body hitting a flexible sheet

impact, penetration and explosion which range from medium to hypervelocities in the fields of armor plate penetration, missile impact, airplane impact, micrometeorite impact on the Giotto satellite dust protection system, and the paper summarizes earlier crashworthiness simulations for the transportation industry.

These selected papers reflect ESI's deep and broad industrial involvement and experience in impact/penetration and crash related topics, which created the background for the continual and ongoing development of the crash simulation program PAM- CRASH.

Recent car crash studies.
As soon as supercomputers were available, the field of industrial crashworthiness simulation literally exploded. The following examples reflect the state of the art in full frontal car crash simulation, where recently published results of studies carried out by clients in the automotive industry and by ESI are summarized and compared to experimental results, if available.

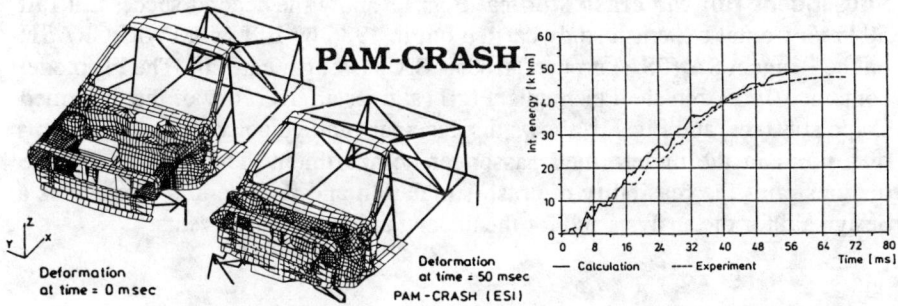

(a) FE mesh and deformed shape

(b) internal energy vs. time

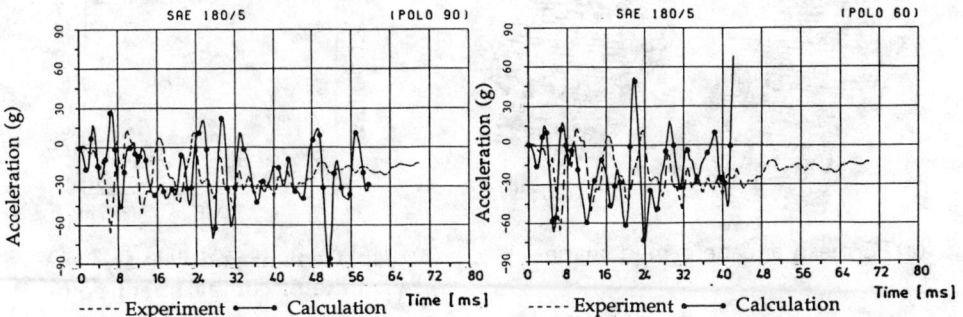

(c) compartment acceleration vs. time (primary analysis without engine support break)

(d) compartment acceleration vs. time (secondary analysis with engine support break at 7 msec)

Fig. 10. VW POLO first successful frontal car crash simulation (PAM-CRASH, 1985)

First full frontal car crash simulation The first successful full frontal car crash simulation at 50 km/h has been carried out in 1985 by ESI on the VW POLO car [18], [19]. Fig10a shows the PAM-CRASH FE model with 5555 thin shell and 106 nonlinear beam elements. The internal energy vs. time diagram of a first analysis is shown in Fig.10b, and the predicted and the measured curves are in good agreement. The calculated acceleration vs. time diagram of a point in the undeformed passenger compartment, Fig.10c, however, shows discrepancies if compared to the measured curve [18]. In fact, in the full scale crash test carried out in the VW lab, the right hand side front engine support broke at an unknown time. The rupture of this component has considerable influence on the more sensitive acceleration and impact force responses. This is seen in Fig.10d, where the acceleration vs. time diagram of a second analysis compares far better with the measured diagram. This result has been achieved when the right hand side front engine support has been eliminated at an elapsed crash time of 7 milliseconds, at which time the support member had reached its greatest stress maximum during the first analysis, and was likely to break.

Subsequent full car crash studies Figs.11 show the second successful full 50 km/h frontal car crash study carried out in 1986 by ESI with PAM-CRASH on the Peugeot SA (PSA) group Citroën - BX passenger car [20]. The FE model contained 7900 thin shell elements. Fig.11a gives an overview of the deformed shape at 80 ms, and Fig.11b shows displacement and velocity vs. time diagrams of a point in the undeformed passenger compartment. The shown example demonstrates the feasibility of crash simulation and its potential to serve as a design aid for the analysis and for the understanding of crash events.

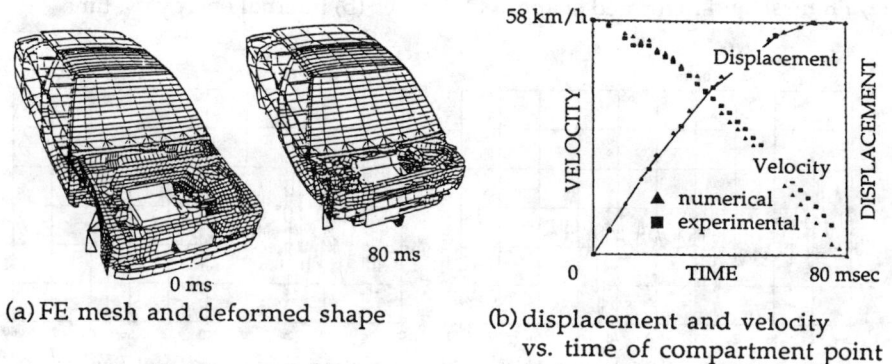

(a) FE mesh and deformed shape

(b) displacement and velocity
vs. time of compartment point

Fig. 11. CITROEN-BX (PSA) frontal car crash simulation (PAM-CRASH, 1986)

Figs.12 [21] show the results of a 50 km/h passenger car frontal crash study carried out at OPEL with PAM-CRASH. The displacement and velocity vs. time diagrams, Fig.12b, are in good agreement with the experimental values.

Fig.13a [22] shows the final deformed shape of a GM80 passenger car FE model after simulation of a 30 mph frontal crash by General Motors, using PAM-CRASH. The barrier force vs. rocker displacement diagram, Fig.13b, closely coincides with the experimentally determined curve.

Finally, Figs.14 [23] show a detailed 20 km/h impact study, performed with PAM-CRASH on a frontal assembly of a FIAT Thema passenger car without the engine, Fig.14a. The resulting displacement and velocity vs. time diagrams coincided closely with the measured curves. Fig.14b shows the absorbed internal energy vs. time diagrams of several parts of the frontal assembly. The latter results are difficult to achieve from experiments and the possibility that such results can be obtained easily in a crash simulation further underlines the utility of crash simulations.

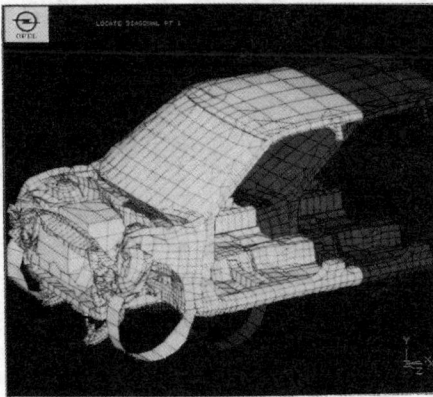

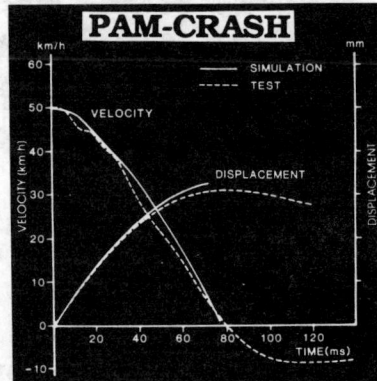

(a) undeformed and deformed   (b) displacement and
    shape   velocity vs. time

Fig. 12. OPEL Ascona frontal car crash simulation
(with PAM-CRASH, 1986)

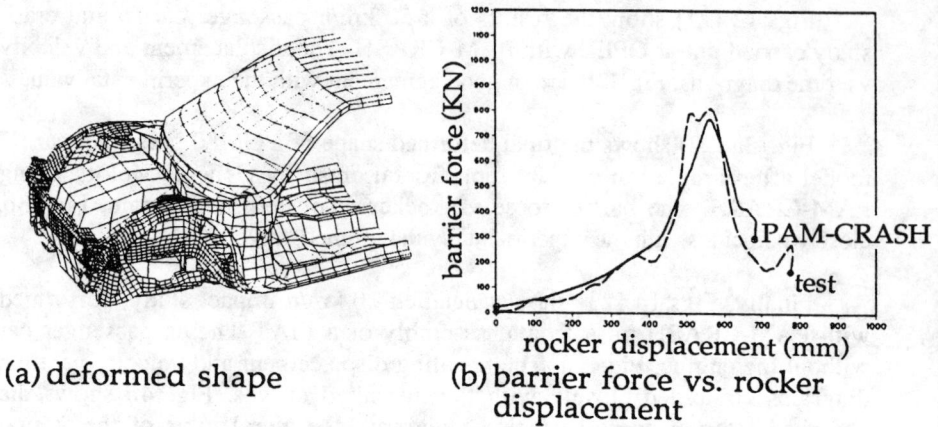

(a) deformed shape

(b) barrier force vs. rocker displacement

Fig. 13. General Motors GM80 frontal car crash simulation (with PAM-CRASH, 1987)

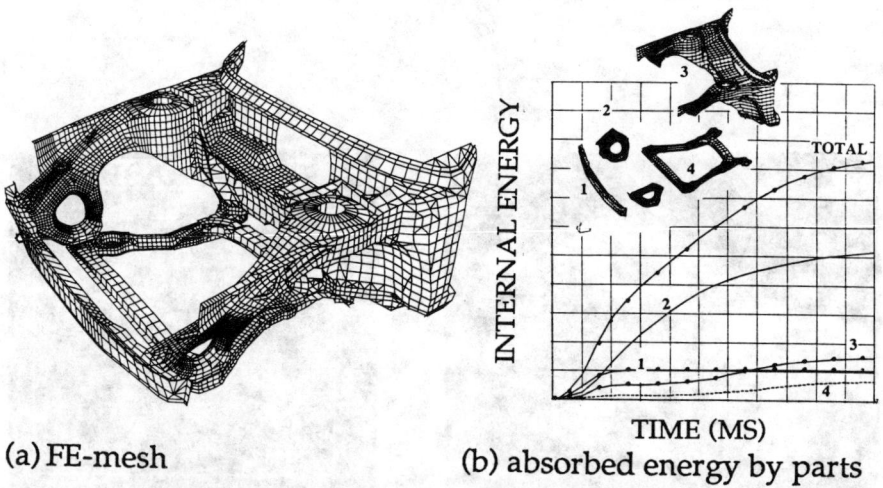

(a) FE-mesh

(b) absorbed energy by parts

Fig. 14. FIAT Thema frontal crash simulation (with PAM-CRASH, 1988)

Further studies.

The following paragraphs show the utility of PAM-CRASH for the simulation of crash events in transportation industries other than the automotive industry.

Helicopter crash The failure of the engine power of a helicopter provokes a hard landing at about 10 m/s vertical velocity. The corresponding energy must be absorbed in part by the stiffened panel fuel cell box structure beneath the occupant space. The crashworthiness of these components has been tested and simulated with PAM-CRASH [1], [16], [17], [24], [25]. Fig.15 shows the

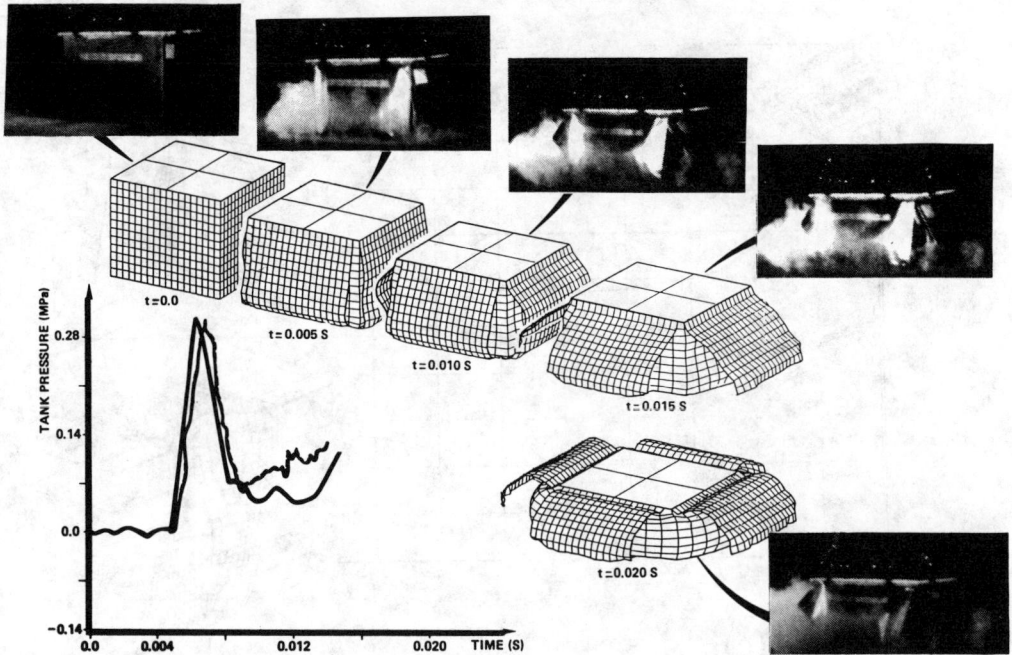

Fig. 15. Helicopter fuel cell drop test simulation (EFHYD-3D / PAM-CRASH)

simulation of a drop test where one weighted stiffened panel fuel cell box is dropped from a certain height. The panel walls are seen to be crushed upon impact and their rivet connections near the corners of the box are seen to fail. The fuel, which is stored in the leak proof fully filled up interior flexible rubber coated nylon reinforced tissue tank, is incompressible, and its pressure deforms the inner tank, which in turn exerts a contact pressure on the outer panel walls. The top weight is stopped by the resistance of the stiffened panels and by the resistance of the flexible tank. The simulation with EFHYD-3D (the program from which PAM-CRASH has been extracted and specialized for crash simulations) uses elasto-plastic thin shell finite elements for the panel walls, which can tear upon rivet failure, and membranes for the flexible tank. The fuel stored in the flexible tank has been modeled by lagrange brick elements. The fuel bricks interact with the wall insides of the flexible tank via sliding interface contact surfaces and the wall outsides of the flexible tank interact with the wall insides of the stiffened panels via the same contact option. Fig.15 shows excellent agreement between the measured and the calculated tank pressure.

Ship collision The 90° lateral collision between a rigid impacting and a deformable impacted ship, Fig.16a, has been simulated with PAM-CRASH [25]. The impacted ship absorbs energy in part via membrane stretching of its outer hull and in part by crushing of its cruciform deck structure. Fig.16b shows the

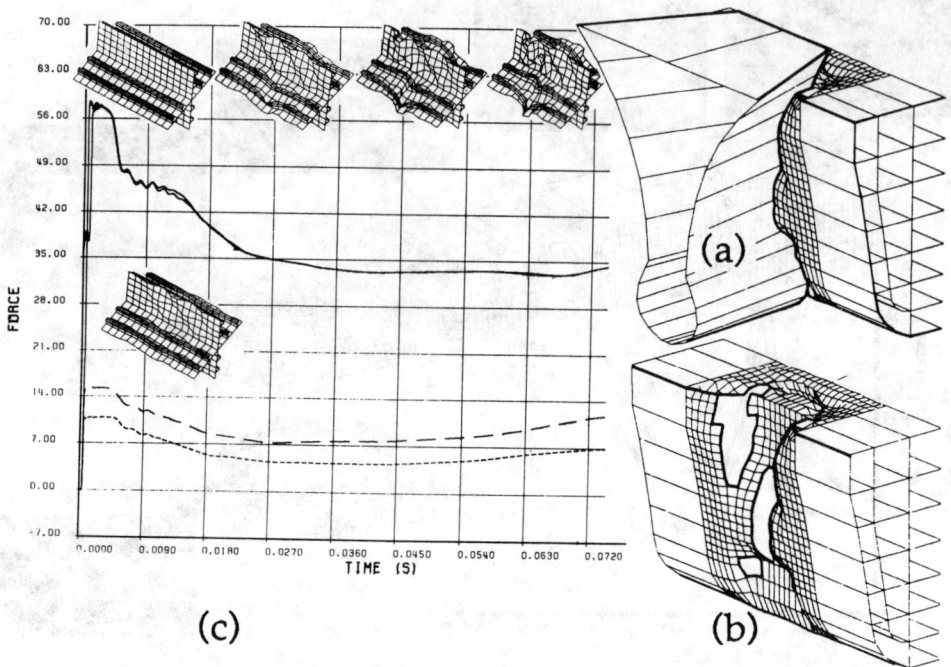

Fig. 16. Ship collision, (a) rigid bow against deformable boat,
(b) deformed shape, (c) typical cruciform crushing
force vs. time (PAM-CRASH)

deformed shape with the impacting ship removed and reveals extensive tearing of
the outer hull. After tearing, the energy absorbing mechanism is shifted to the
crushing of the deck structure. Fig.16c shows a typical repetitive stiffened panel
cruciform element of the deck structure, which has been modeled by 500 thin
shell finite elements, and crushed axially. The resulting crushing force vs. time
curve for the finely modeled cruciform panel assembly serves as a target result
for a much coarser model, in which each panel of each cruciform is represented
by only 4 x 2 plate elements, i.e. each cruciform contains only 32 elements
instead of 500. The plastic material parameters of these elements were adjusted
such that the folding of the so created "macro-mesh" produced the same mean
crushing force or energy absorbing characteristics as the target. This permitted to
model the deck structure of the ship more economically using the coarse
macro-mesh cruciforms, which had the correct energy absorbtion capacity.

Railway car crash Fig.17 shows a FE model of a railway car segment which
has been prepared for a numerical simulation with PAM-CRASH. The picture
demonstrates the great value of the code in all sectors of transportation. In the
context of a research program of the french national railway system, SNCF, the
program is presently applied to the study of impact phenomena on railway cars.

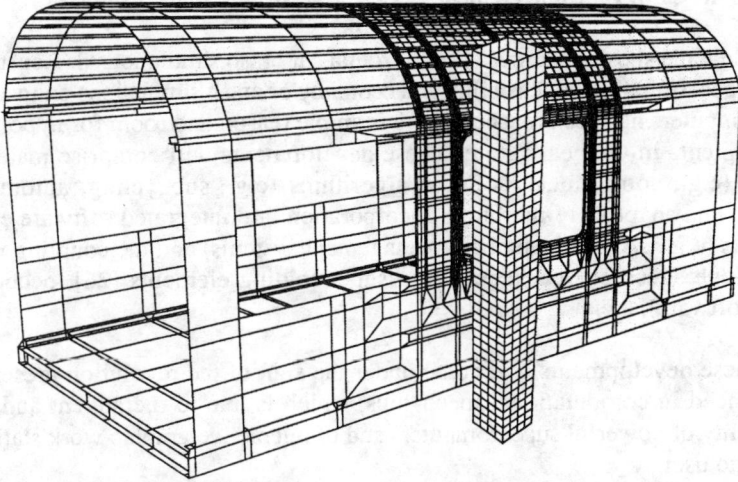

Fig. 17. SNCF railway car and "finite rigid wall" obstacle (PAM-CRASH)

SUMMARY AND CONCLUSIONS

The paper traced the history of ESI's long standing involvement, experience and software development, in sophisticated impact/penetration and blast simulations and in crashworthiness studies. These activities created the background for the ongoing important development of the specialized industrial crashworthiness analysis program PAM-CRASH. The synthesis of this program from the HEMP, WHAM and DYNA public domain softwares and from ESI's own developments has been described and some important and more recent features of the code have been highlighted, which contribute new analysis capabilities, precision, robustness and user friendliness, as well as considerable CPU performance gains. Certain of the described features are still under development in a large and active development effort carried out presently at ESI.

The final goal of automotive crashworthiness simulation is to construct cars, apt to warrant occupant safety. Great efforts will therefore go into the detailed modeling of active and passive safety restraints and of their interactions with regulatory occupant surrogates. Currently available dummy simulation software cannot, for example, give answers to important questions, such as "bag slap" and "off-axis" dummies in airbag simulations. The hybrid FE/gasdynamics airbag model, presently developed and incorporated into PAM-CRASH, can give answers and it leads to the development of detailed rigid and deformable linkage FE-type occupant surrogate models. The PAM-CRASH airbag model can also be interfaced with standard occupant surrogate models found in currently available occupant safety analysis programs, such as CAL3D and MADYMO. Furthermore, the inclusion of FE occupant safety models into PAM-CRASH will permit coupled crash simulations to be carried out, with occupant safety models added to the car body finite element models, e.g., in side impact studies.

While the shown examples of full frontal car crash simulations demonstrate that the goal of industrial feasibility and economy of crash simulations on modern supercomputers has been reached, there is ample reason and room for important developments in the near future. These developments will comprise material models (e.g., composites), solution algorithms (e.g., subcycling, automatic meshing and adaptive re-meshing), incorporation into integrated software environments (e.g. CAD and data base management systems) and the coupling with stand-alone accessory packages (e.g., super-folding elements [26], occupant safety software), etc.

These developments will come under the sign of the revolution presently experienced in computational mechanics, which is due to the advent and the availability of powerful supercomputers and of interactive graphic work stations to the end user.

ACKNOWLEDGEMENTS

Numerous personalities and collaborators contributed to the success of the PAM-CRASH code. Mark Wilkins of the Lawrence Livermore National Laboratory (LLNL) and Ted Belytschko of the University of Chicago were key consultants to ESI in the years of fact finding. J.O. Hallquist of LLNL let ESI have early versions of the DYNA-3D code. Drs. T. Scharnhorst (Volkswagen Werke), W.Dirschmid (AUDI), K. Hieronimus and E.J. Nalepa (OPEL), as well as Dr. E. Hase (Daimler-Benz), Mr. I. Raasch (BMW) and Dr. H. Goldstein (FORD) had the courage and the vision to believe in the feasibility of crashworthiness simulation, proposed to them by ESI and performed elsewhere. Dr. A. Deshpande (General Motors) fostered early installation in the US. Dr. D. Bigi (FIAT) supervised the Thema car simulation. Mr. J. Mens (Aérospatiale) was in charge of the helicopter crash studies, Mssrs. Galbe and Auroire, STCAN-MSN (french navy) of the ship collision and Mr. Lagneau, (french national railway, SNCF) of the railway car simulation.

Drs. P. Melli and F. Angeleri (IBM NIC Rome) tuned the program to IBM 3090 vectorized/multi-tasked supercomputers, with the help of Drs. S. Katoh and S. Suda (IBM NIC Tokyo). Vectorized CRAY versions of the program originated at CRAY Research Inc. under Dr. C. Marino. Recent advances were developed on the CONVEX mini-supercomputer installed at ESI.

At ESI, Drs. J. Dubois, A. de Rouvray (ESI) instigated the development of the PAM-CRASH code. Mr. J.C. Bianchini piloted the helicopter and railway studies. Mr. S. Aïta gave valuable advice on recent developments. H. Charlier, C. Pauquet, F. Wijnant, F. Sanchez and J.F. Lefebvre (ESI) interfaced the program with ESI's pre- and post-processors (PRE-3D, DAISY).

All efforts, encouragements and contributions are gratefully acknowledged.

REFERENCES

1. Haug, E., Arnaudeau, F., Dubois, J., de Rouvray, A. and Chedmail, J.F., Static and Dynamic Finite Element Analysis of Structural Crashworthiness in the Automotive and Aerospace Industries, Chapter 7 in Structural Crashworthiness, Eds. N.Jones & T.Wierzbicki, pp.175-217, Butterworths, 1983.

2. Wilkins, M.L., Calculation of Elastic-Plastic Flow, Report UCRL-7322, Lawrence Radiation Laboratory, California, January 1969.

3. Belytschko, T.B. and Hsieh, B.J., Nonlinear - Transient Analysis of Shells and Solids of Revolution by Convected Elements, Journal AIAA 12, 1031- 1035, 1974.

3bis. Dubois, J., et al., A Coupled Model for the Finite Difference - Finite Element Analysis of Hydrodynamic and Elasto-Plastic problems in the HEMP-ESI Code, Int. Conf. on Innovative Methods in Engineering, Versailles, France, May 1977.

4. Belytschko,T.B., Lin, J.I. and Tsay, C.S., Explicit Algorithms for the Nonlinear Dynamics of Shells, Computer Methods in Applied Mechanics and Engineering 42, pp.225-251, 1984.

5. Hallquist, J.O., A Procedure for the Solution of Finite-Deformation Contact-Impact Problems by the Finite Element Method, Report UCRL-52066, Lawrence Livermore Laboratory, Livermore, California, April 1976.

6. PAM-CRASH, Theory Manual, Engineering Systems International, 20 Rue Saarinen, Silic 270, 94578 Rungis-Cedex, France, February 1988.

7. PAM-CRASH, User's Manual, Engineering Systems International, 20 Rue Saarinen, Silic 270, 94578 Rungis-Cedex, France, January 1988.

8. Belytschko, T. and Gilbertsen, N., Concurrent and Vectorized Mixed Time Explicit Nonlinear Structural Dynamics Algorithms, Proc. Symp. on Parallel Computations and their Impact on Mechanics, ed. A.K. Noor (ASME Winter Annual Meeting, Boston, Ma, December 1987) ASME, New York, 1987.

9. Aberlenc, F., Development d'un Schema de Sous-Cyclage pour un Logiciel Dynamique Explicite Lagrangien, ESI Report, Engineering Systems International, 20 Rue Saarinen, Silic 270, 94578 Rungis-Cedex, France, June 1987.

10. Scharnhorst, T., Numerical Calculation of the Bending Collapse of two Structural Safety Components, IMech E 1984 C187/84, pp.29-38, London, 1984.

11. Hillmann, J., Grenzlasten und Tragverhalten axial gestauchter Kreiszylinderschalen im Vor- und Nachbeulbereich, Report nb. 85-45, Inst. für Statik, Techn. Hochschule Braunschweig, Braunschweig W.G., 1985.

12. Underwood, P., Dynamic Relaxation, Chapter 5 in Computational Methods for Transient Analysis, Belytschko T. and Hughes T.J.R., Ed., North Holland, 1983.

13. de Rouvray, A. and Haug, E., Failure of Brittle and Composite Materials by Numerical Methods, Chapter 7 in Structural Failure, Ed. T. Wierzbicki, M.I.T., June 6-8, 1988 (Proceedings to appear at Wiley).

14. Belytschko, T. and Bazant, Z.P., Strain Softening Materials and Finite Element Solutions, Department of Civil Engineering, Northwestern University, Evanston, Illinois, 60201, 1984.

196

15. Massoni, E., Bellet, M., Chenot, J.L., Elastic-Plastic Analysis of Thin Plates by the Finite Element Method in Deep Drawing Process". Intl. Conference on Computational Plasticity, Barcelona (April 1987).

16. Haug, E., Engineering Safety Analysis via Destructive Numerical Experiments, Engineering Transactions 29, Polish Academy of Sciences, pp.39- 49, 1981.

17. de Rouvray, A. et al, Numerical Techniques and Experimental Validations for Industrial Applications, Chapter 7 in Structural Impact and Crashworthiness, Vol.1 (Key note lectures), Ed. G.A.O. Davies, Elsevier Applied Science Publishers, pp.193-242, 1984.

18. Haug, E., Scharnhorst, T. and Dubois, P., FEM-Crash, Berechnung eines Fahrzeugfronta-laufpralls, VDI-Reports nb. 613, VDI Verlag, pp.479-505, 1986.

19. Scharnhorst, T. and Schettler-Köhler, FEM-CRASH Experiences at Volkswagen Research, in Supercomputer Applications in Automotive Research and Engineering Development, Editor C. Marino, CRAY Research Corp., Computational Mechanics Publications, pp.75-86, 1986.

20. Chedmail, J.F. et al. and Haug, E. et al., Numerical Techniques, Experimental Validation and Industrial Applications of Structural Impact and Crashworthiness Analysis with Supercomputers for the Automotive Industries, in Supercomputer Applications in Automotive Research and Engineering Development, Editor C. Marino, CRAY Research Corp., Computational Mechanics Publications, pp.127-145, 1986.

21. Nalepa, E.J. and Le-The, H., Crashworthiness Simulation, An Emerging Tool for Vehicle Design Optimization, CRAY Publication, pp.8-11, Winter 1987.

22. Vander Lugt, D.A., Chen, R.J. and Deshpande, A., Passenger Car Frontal Barrier Simulation using Nonlinear Finite Element Methods, SAE Technical Paper Series 871958, Passenger Car Meeting, Dearborn, October 19-22, 1987.

23. Bigi, D., Simulazione numerica dei problemi di crash veicolistico, ATA-Ingegneria automotoristica, Vol.41, nb.5, pp.386-392, 1988.

24. Mens, J. and Bianchini, J.C., Computing Codes for Development of Helicopter Crashworthy Structures and Test Substantiation, paper at the AHS/NAI International Seminar, Nanjing, China, November 6-8, 1985.

25. Bianchini, J.C. and Haug, E., Simulation Numérique d'Ecrasement de Structures Chute d'Helicoptère, Collision de Navires, Ecrasement de Voiture, paper at Colloque Tendances Actuelles en Calcul de Structures, DRET-CCSA, Bastia (Corsica), November 6-8, 1985.

26. Wierzbicki, T. and Abramowicz, W., The Mechanics of Deep Plastic Collapse of Thin Walled Structures, Chapter 9 in Structural Failure, Ed. T. Wierzbicki, M.I.T., June 6-8, 1988 (Proceedings to appear at Wiley).

27. Lasry, D. and Belytschko, T., Gradient-Type Localization Limiters for Strain-Softening Materials, in Advances in Inelastic Analysis, AMD-Vol.88 / PED-Vol.28 (ASME Winter Annual Meeting, Boston, Ma, December 1987), pp.127-144, ASME New York, 1987.

Issues in Vector and Parallel Migration to the IBM 3090/VF of Explicit Finite Element Codes: the PAM-CRASH [TM] Experience

P. Angeleri

IBM/ECSEC (European Center for Scientific and Engineering Computing), Via Giorgione 159, 00147 Rome, Italy

Introduction

In the automotive industry vehicle crashworthiness simulation has become increasingly important. Actually crash design allows to improve the safety of the vehicle occupants and to optimize the design process. Yet, many automotive manufacturing industries still obtain the information on the structural behavior under crash conditions by means of experimental tests on prototypes, which is a costly practise. In the last ten years mathematical models and computer programs have been developed to provide structural crash simulation tools aiding the users in solving, at reduced costs, crashworthiness design. Crash phenomena are some of the most complex problems faced by structural mechanics. Since they are both spatial and temporal phenomena, they involve not only contacts among several surfaces, but also complex non-linear behaviors characterized by large strains. Consequently the solution of these kinds of problems needs such an amount of computer resources as to require the computational power provided by a supercomputer, in order to achieve the results consistent with time constraints of the industrial processes.

PAM-CRASH ^{TM} is a finite element program for the crash analysis of structures. The program is an industrial code designed for automotive crashworthiness analysis. It was developed and is mantained by Engineering Systems International (ESI) S.A. Rungis-France. PAM-CRASH has been enabled in order to exploit the vector processing capabilities of the IBM 3090/VF system providing the automotive manufacturing industries with an engineering system for the crash phenomena simulation. The vector version of PAM-CRASH has been further provided with the capability of exploiting parallel processing on a multi-processor machine.

We discuss the problems encountered and we describe the solutions provided for an efficient migration of the code. Particularly we deal with the problems posed by the class of finite element codes, which use an explicit algorithm to perform the integration of dynamic equations. Emphasis is given to the description of the shell element implementation and to the criteria followed in exploiting vector processing. Runs on real test cases show a vector/scalar speedup between 2.7 and 3.5. Moreover, we show the migration strategy in order to exploit parallel processing using the Multitasking Facility provided by the VS FORTRAN compiler. Performance results, from a two-way to a six-way processor 3090/VF system, are shown.

PAM-CRASH Code Characteristics

PAM-CRASH [1] is a three dimensional lagrangian explicit finite element code analyzing the dynamic response of structures. The code takes both material and geometrical non-linearities into account providing general contact-impact capability. The program is especially designed for automotive crashworthiness analysis.

The aim of this section is to review briefly the formulation of the finite element method in continuum mechanics in order to show the computational problems arising in the implementation of a finite element code, as PAM-CRASH, on a vector multiprocessor machine.

The formulation is strictly Lagrangian: the current position of a point with coordinates $\underline{x}(\underline{X}, t)$ is function of the initial location $\underline{X}$ and of the time t. The dynamic equilibrium equations of a body at time station t^i can be written by means of the Principle of Virtual Work:

$$\int_\Omega \delta\underline{\varepsilon}^{iT}\underline{\sigma}^i d\Omega - \int_\Omega \delta\underline{x}^{iT}(\underline{b}^i - \rho\underline{\ddot{x}}^i)d\Omega - \int_{\Gamma_t} \delta\underline{x}^{iT}\underline{t}^i d\Gamma_t = 0 \qquad (1)$$

where

ρ is the density,

$\underline{x}$ is the coordinate vector,

$\underline{\sigma}$ is the stress vector,

$\underline{b}$ is the vector of applied body forces,

$\underline{t}$ is the vector of surface tractions,

Ω is the body domain,

Γ_t is the section of the boundary Ω that has tractions applied.

The finite element method interpolates the coordinate field by means of the nodal coordinate values $\underline{x}_e$:

$$\underline{x}(\underline{X}, t) = \underline{N}_e(\underline{X})\,\underline{x}_e(t) \qquad (2)$$

where $\underline{N}_e(\underline{X})$ is the matrix of the shape functions. Further, in the finite element representation, the strain field is expressed as:

$$\underline{\varepsilon} = \underline{B}_e(\underline{X})\underline{x}_e \qquad (3)$$

where $\underline{B}_e(\underline{X})$ is the global strain-displacement matrix. The strains are assumed to be linear.

PAM-CRASH is provided with a library of different kinds of elements. Their peculiarity is that of making use of the simplest finite element formulation, based on linear interpolation functions and reduced element quadrature. The experiences of the developers [2], compared with those drawn from the literature [3], show that the utilization of elements based on more complex finite element formulations, e.g. using higher order elements, are not computationally cost-effective for explicit non-linear structural codes. On the other hand both a stabilization procedure, to avoid spurious modes due to the underintegration, and a greater mesh density in areas of severe deformations are required.

The substitution of (2) and (3) into equation (1) leads to the semi-discrete matrix equation of motion:

$$\underline{M}\underline{\ddot{x}}^i + \underline{f}_{int}^i(\underline{x}^i) = \underline{f}_{ext} \tag{4}$$

where

$\underline{M}$ is the mass matrix,

$\underline{f}_{int}$ is the internal force vector,

$\underline{f}_{ext}$ is the external force vector.

In the present formulation the internal forces are functions of the nodal displacements by an inelastic rate material model, and the mass matrix is a diagonal lumped matrix. Therefore the equations set is uncoupled and the solution is trivial. Node velocities and node displacements are evaluated using a central difference approximation scheme:

$$\underline{\dot{x}}^{i+1/2} = \underline{\dot{x}}^{i-1/2} + h^i \underline{\ddot{x}}^i \tag{5}$$

$$\underline{x}^{i+1} = \underline{x}^i + h^{i+1/2} \underline{\dot{x}}^{i+1/2} \tag{6}$$

$$h^{i+1/2} = \frac{1}{2}(h^{i+1} + h^i) \tag{7}$$

where h is the integration step-size.

The conditional stability of an explicit integration scheme forces the integration step-size to be evaluated by a stability criterion. In the case of PAM-CRASH this criterion is the Courant Criterion, limiting the time step-size to be small enough not to allow a sound wave to transverse a mesh element in the single time step. Consequently, steps of the order of microseconds are typical. Alternative implicit schemes are unconditionally stable for defined values of the integration parameters. The time step-size is selected by considerations of accuracy alone, thus allowing larger time steps than those of an explicit scheme. Yet, the main drawback in using explicit schemes consists in solving a non-linear equations set at each time step via an iterative procedure. The developers [2] assert that the advantages offered by unconditionally stable implicit integration schemes, allowing larger time steps, cannot be exploited in crash simulation because of the characteristics of the phenomena under examination. Actually they are characterized by a short duration, a highly non-linear behaviour, and they are dominated by a wide range of frequencies. In order to be correctly described they require a time step-size of the same order as that needed by an explicit code. A factor of up to two orders of magnitude can be observed.

From an implementation point of view an explicit code shows a simpler structure in comparison with the one arising from an implicit formulation. E.g., in the non-linear dynamic analysis an explicit scheme avoids the implementation of an iterative procedure (in order to solve the non-linear equations set arising from the finite element formulation) and consequently of a linear solver for the equations set (generated either at some or at every iteration step). Furthermore, equations set assembling and updating   is avoided, thus allowing a significative reduction of virtual memory requirements. The computational cost per step is much lower for the explicit scheme than for the implicit one .

These considerations show why until some years ago an explicit approach was considered the only possible one to face the most complex problems of structural analysis. Nevertheless now the hardware and system architecture evolution provide new powerful computer resources in terms of CPU, storage, and I/O processing capabilities, thus allowing one to face with an implicit approach the quasi-statics and low-frequencies dominated analyses.

In crashworthiness analysis the analyst   has mainly to model thin structures. Consequently, modelling efficiently the behaviour of that kind of structure is a critical issue from the point of view of result reliability and of computational costs. The shell element, provided by PAM-CRASH, is a bilinear four node quadrilateral element, originally developed by Belytschko [4], based on the Reissner-Mindlin plate theory. This theory differs from the classical Kirchhoff theory in the shear deformation treatment. In the Kirchhoff hypothesis the effect of the transverse shear deformation is neglected and the finite element approximation requires shape functions with C^1 continuity. In the Reissner-Mindlin theory the shear deformation is taken into account and only C^0 continuity is required with great simplification in the element formulation. A reduced integration technique with one point quadrature has been used to evaluate the element forces, allowing to strongly reduce the computational costs. Yet the reduced integration in a bilinear plate element permits either zero-energy or kinematic modes. These modes, named *hourglassing* modes in finite difference literature [5], can destroy the solution because of the introduction of spurious spatial oscillations. An hourglass control scheme has been implemented to avoid that numerical instability phenomenon. Some comparison results in terms of performance and reliability with other formulations can be found in literature [6].

A wide variety of material laws are available to model elastic, non-linear and failure conditions. Rigid walls may be defined to provide external impact surfaces, while a slide line algorithm can prevent penetration of internal structure surfaces that may collide during the crash simulation. Various slide line interface conditions are available, including sliding, separation, and friction. The user has to specify which region of the structure must be considered for crash purposes, by listing the elements which are part of it. Either partially or totally automatic 'boxing strategy' is under development, according to which the user has no longer to list the elements belonging to the sliding surfaces. Two basic types of contact phenomena can be represented. The first type involves two surfaces coming into contact, the second one involves a single surface buckling and coming into contact with itself. Both situations include friction interface behavior and are based on the penalty method.

The IBM 3090/VF System

The PAM-CRASH program has been enabled to exploit the vector/parallel processing capabilities of the IBM 3090 vector multi-processor system in providing the automotive users with an engineering system for crashworthiness simulation. The IBM 3090/VF system [7] is a general-purpose data processing system, including several models mainly differing both for the number of processors and for the storage system configuration. The 3090/VF system structure is given by:

- one or more central processors,
- the vector facility for each processor,
- the storage system,
- one or two channel subsystems.

Each central processor is microcode controlled and consists of an *instruction element* controlling the sequence of all instructions and of an *execution element* executing the instructions set up by the instruction element. The central processor provided with the vector facility allows significantly increased levels of performance for many numerically-intensive engineering and scientific applications. The vector facility feature [8] is an extension of the *instruction* and *execution elements* of the central processor. The vector facility provides:

- the vector facility registers:
  - 16 vector registers
  - a vector-mask register
  - a vector-status register
  - a vector-activity count
- 171 vector instructions.

The length of the vector registers is model dependent, and in the following discussion we assume that the vector registers are 256 elements (of 32 bits each) long, as on the IBM 3090/VF S model systems. The sixteen vector registers can also be coupled to form eight vectors of 256 elements for 64-bit arithmetic computation.

The vector facility provides floating-point and integer arithmetic vector instructions, logical vector instructions, load and store vector instructions, and allows the conditional execution of arithmetic and logical instructions depending on the vector mask register. The vector facility provides two floating-point compound instructions, multiply and accumulate and multiply-add, performing an addition and a multiplication for each element.

The floating-point arithmetic instructions (addition, multiplication, subtraction, division, multiply-add, and multiply and accumulate) exist in several formats, depending on whether the input operands are:

- vector registers (e.g. addition of two vector registers);
- vector registers and memory (e.g. sum a vector register and data in memory);
- vector register, memory, and a scalar floating-point register (e.g. multiply a vector by a constant which is in a floating-point register).

In all cases the result of these vector instructions is placed in a vector register. After a start-up time of approximately 20 machine cycles, these vector instructions deliver one result per machine cycle (15 *ns*). After the first 256 elements of a vector have been processed the operations must be re-initialized on the next section. The two compound instructions, multiply and accumulate and multiply-add, perform a multiplication and an addition in a single machine cycle.

Arithmetic vectors in storage may be loaded or stored either by sequential addressing (contiguously or with stride), or by indirect element selection. The load or store by sequential addressing requires one machine cycle per element after a start-up time. Indirect element selection, gather-scatter, permits vector elements to be loaded or stored in an arbitrary sequence, requiring two machine cycles to process each element.

In the IBM 3090/VF data from the main memory is available to the CPU only after it has been brought into an intermediate high-speed buffer memory, the cache. When a storage location, which is not already in the cache, is accessed by the CPU a cache miss occurs. The operation that caused the cache miss will be completed only after one or more cache-lines of 128 bytes, which contains the data, are transferred to the cache, where they displace the least recently referenced cache lines. Then, if the next instruction references another word in the same cache line, it will already be available. In particular a vector operation with an operand from memory delivers one result per machine cycle provided the data is in cache. Otherwise, it may require approximately two machine cycles per vector element before the entire operation, data transfer to cache and vector operation, is completed.

The storage system of a 3090/VF system consists of three different levels of storage, being different for the data access speed: namely a cache for each central processor, the central storage, the expanded storage feature. Further, the storage system is provided with a *system control element* providing access to the central storage for the central processors and for the channel subsystem. The cache is 128 Kb high-speed buffer with pipelined two cycle access. It is used to hold the portions of central storage that have been referenced most recently and cannot be directly programmed. The expanded storage feature is a new separate storage designed for block transfers of 4 Kb pages between itself and central storage, under the control of the operating system. The expanded storage storage feature runs synchronously with respect to the central processor. A complete page transfer operation has been measured in about 75 µs. Up to 512 Mb of real storage and 2 Gb of expanded storage are available on the largest 3090/VF S model systems.

The channel subsystem handles all I/O operations controlling communication between channels, control units and devices.

Vector Feasibility Analysis

The vector feasibility analysis is a basic step to collect those elements necessary to carry out a cost evaluation. That analysis allows one to determine if the price/performance requirements are met. The vector feasibility analysis has been carried out in these steps:

1. definition of the objectives and requirements;

2. analysis of the overall program structure, of the logic organization, and of the coding implementation to build the execution flows and to identify the main code sections;

3. evaluation of the computer resource distribution among the code sections identified in the previous step;

4. analysis of the algorithms of those code sections which are responsible for a comparatively large amount of computer resources utilization, and definition of the algorithm characteristics for a possible vector implementation;

5. evaluation of the vector content in the present implementation;

6. comparison of the results of steps 4 and 5 to identify the potential vector content, to locate the inhibitors preventing vectorization, to evaluate the level effort required to promote vectorization, and finally to estimate the algorithm performance improvement;

7. evaluation of the overall effort and estimation of the global performance improvement according to the objectives defined at step 1;

8. definition of time schedule and planning of human resources.

The primary objective of a vector migration process consists in the processing time reduction by exploiting the processing capability of the vector facility feature. The secondary objective, which can still yield benefits, consists in the system resource exploitation such as the virtual storage and the I/O processing.

PAM-CRASH carries out the integration of the dynamic equation using a finite difference explicit algorithm. Thanks to the characteristics of this algorithm, widely used in structural mechanics, the virtual memory request is highly reduced in comparison with that of alternative implicit schemes. According to that, the virtual memory provided by S/370 architecture is enough to use efficiently the code in standard applications. Yet, to meet customers growing future requirements in terms of model sizes, the exploitation of virtual memory over the 16 Mb line, allowed by 370/XA architecture [9], whose present extension is 2 Gb, has been set among the objectives.

The PAM-CRASH program produces a large amount of data, hundreds of megabytes, as result of the analysis. These data can be used as input for the graphic post-processing program aiding the user in analyzing the results of the simulation, in checking if the design requirements are met, and eventually in proceeding to further modifications. The I/O content can be considerable and the time the program spends to perform the I/O activity can become a considerable fraction of the total time. An I/O algorithm, either not well tuned or designed for a different system architecture, can produce a degradation in performances. We have decided to include, among our objectives, the I/O activity tuning if performance tests would ascertain any degradation in elapsed time performances due to the I/O processing.

A great part of the widely used engineering codes had been written during the 1970(s). The system architecture, they were designed for, has extensively evolved during the last twenty years. Yet, the software houses, that develop and maintain those codes, have adopted a conservative policy regarding software upgrading. Consequently our experience in the vector migration is that we must expect to face preliminary problems to map the code architecture on the hardware characteristics in order to optimize the system resource utilization.

The software houses generally deliver different versions of the same programs in order to be employed in different environments. Yet, they reasonably prefer to limit, as much as they can, the differences among the versions. Therefore we had to define a policy to be adopted regarding the introduction of code changes. That has been conceived taking into account portability, maintenance, and readability requirements.

We have then faced the problem of identifying the program structure by analyzing the physical formulation, the algorithms involved, and their relation with the code implementation. This implies:

- to build the logic program organization and locate the main code routines;

- to draw the tree of the execution flows and trace the routine calling sequence;

- to identify the data organization and the process of data transmission.

Logic organization and data structure of a finite element code making use of an explicit algorithm for the time integration are well known in literature [10]. Yet, either requirements or programming styles can make the code implementation of two programs very different although they carry out the same tasks using the same algorithms. E.g. modularity requirement can affect the mapping of the logical organization of the code sections and hardware requirement can affect the overall data organization. Flow-chart, shown in Figure 1, describes the program organization and outlines the main sections.

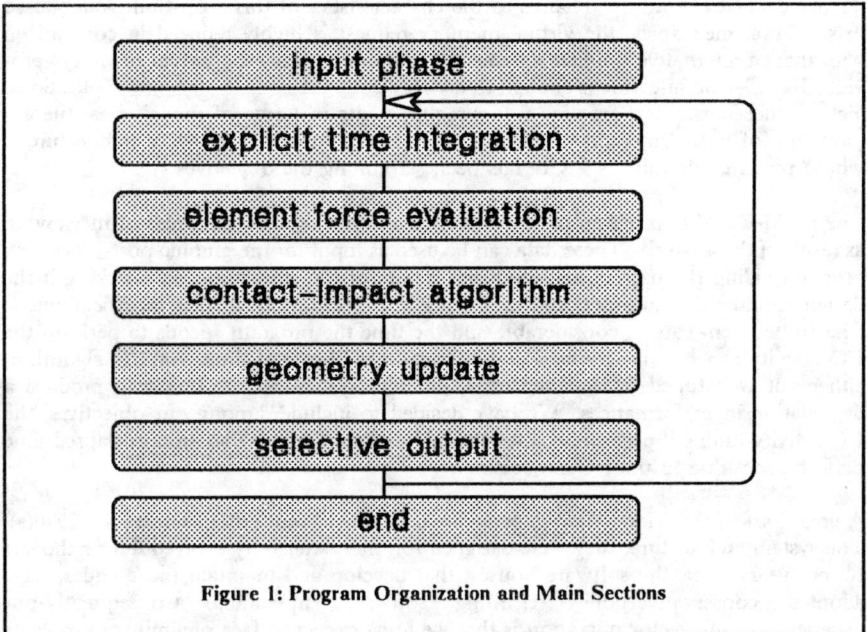

```
input phase
explicit time integration
element force evaluation
contact-impact algorithm
geometry update
selective output
end
```

Figure 1: Program Organization and Main Sections

From a preliminary analysis this information has been collected for each code section:

- *element force evaluation*: over 90% of the elements used is the non-linear shell element ;

- *selective output*: the I/O activity is not negligible and it affects the elapsed time;

- *contact algorithm*: the incidence of the execution time for the impact phonemena treatment is dependent on the modelling choices since the user can select the region to be analyzed for crash purpose.

Performing a hot-spot analysis we have obtained the percentage distribution of the processing time among the sections of the code. We have carried out this task using the VS FORTRAN Execution Analyzer, that allows the user to obtain detailed timing information at subroutine and statement levels. The time distribution depends on the type of the analysis. The user has to select the region of the structure to be analyzed for crash purposes.

The following table shows the variation range of the processing time distribution among the different code sections for the test cases provided by ESI. Since those cases are a complete set of the applications presently employed in the design process of a manufacturing industry which builds automotive vehicles, that range can be reasonably considered as representative of the general class of problems in automotive industry.

Code Section	CPU Percentage
1 - element force evaluation	**70-82**
2 - geometry update	**7-9**
3 - explicit time integration	**3-4**
4 - contact algorithm	**17-3**
1 + 2 + 3 + 4	**97-98**

The remaing 2-3 % of the CPU time is distributed among other code sections whose single incidence is less than 1%. The incidence of the *input phase* section is negligible.

We have analyzed the algorithms of those code sections which were responsible for computer resources, for a comparatively large amount utilization, and we have defined the characteristics for a possible vector implementation. We found that:

- the code sections of the *geometry update* and of the *explicit time integration* are completely vectorizable;

- the *element force evaluation* algorithm is vectorizable, except for the section assembling the element forces into the global force vector. We have estimated its percentage incidence being about 10 % of the force execution time of the element force section;

- the *contact algorithm* is vectorizable, but we have ascertained from the analysis of the present implementation that vector enabling was feasible just carrying out a complete reorganization of the code.

We can assert that a good estimation of the vectorizable percentage f, defined as the percentage of the scalar execution time spent by the vectorizable code sections on the total execution time, varies in a range between 73 and 87 %. The table shows the vectorizable percentage estimated for each section and compares it to the total CPU percentage.

Code Section	CPU Percentage	Vector Percentage
1 - element force evaluation	**70-82**	**63-74**
2 - geometry update	**7-9**	**7-9**
3 - explicit time integration	**3-4**	**3-4**
4 - contact algorithm	**17-3**	**0-0**
1 + 2 + 3 + 4	**97-98**	**73-87**

The theoretical model, known as the *Amdahl's Law* [11], allows to estimate the performance improvement of a vectorizable application:

$$S = \frac{1}{(1 - f) + f / sr} \tag{8}$$

where:

- S is the overall vector/scalar speedup factor;

- f is the vectorizable percentage;

- sr is the vector/scalar speedup obtained by performing the computation of the vectorizable code sections with the vector facility.

The vector/scalar speedup design range of the vector facility is between 3 and 5. This value is optimized for the mid-range of vectorizable percentage. Figure 2 shows the improvement ratio of a vector application's execution time as a function of the vectorization percentage. Several vector/scalar speedup ratios are given.

The parameter sr depends both on the hardware characteristics of the vector facility feature and on the characteristics of the computation performed inside the code. We mention the most important factors affecting the estimation of sr.

- Kind of operations involved in the computation. Different vector operations have different startup times and different speeds in comparison with the scalar instructions.

- Vector length. A vector length of 128 elements is close to the optimum speedup. The threshold above which vector code performances are better than scalar ones is approximately 12.

- Stride of the vector element in storage. The best vector performances are achieved with the lowest stride: i.e. stride 1. Actually it is important to reduce the stride in order to minimize cache misses and paging. For particular stride values the scalar execution is preferred because it usually results in faster execution than the vector execution.

- Presence of conditional processing. The excessive presence of IF statements can cause poor performance. When an IF statement is vectorized, the time needed to process the IF statement and the operations governed by it, is the same both when

Speedup Factor vs Vector/Scalar Speed Ratio

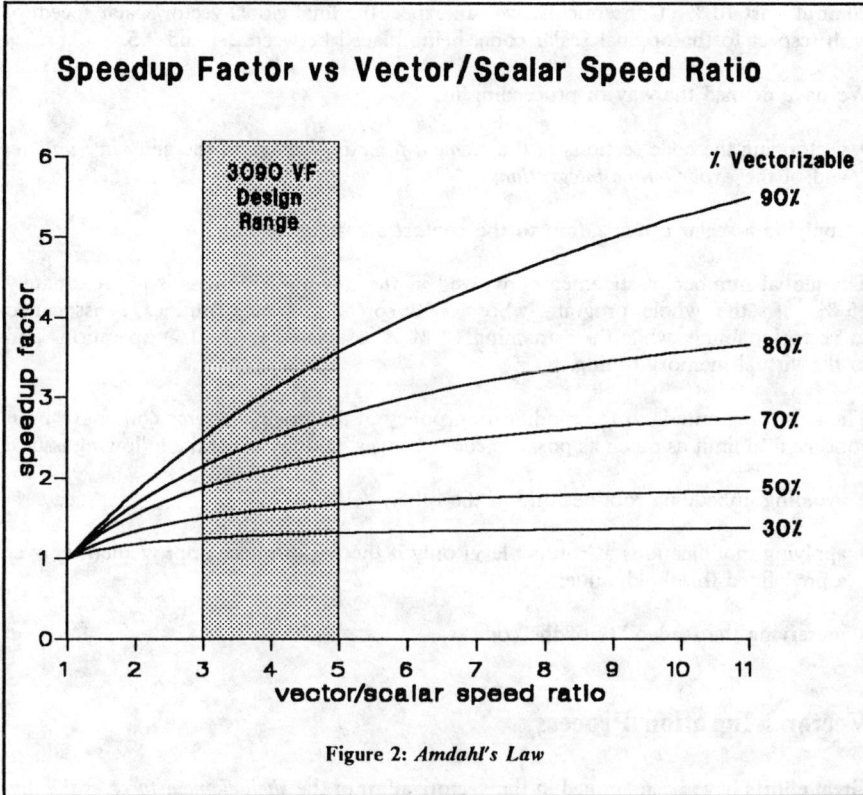

Figure 2: *Amdahl's Law*

the condition is verified and when it is not; i.e. the speedup is the more adversely affected the more the condition is not verified.

The operations of the vectorizable code sections are essentially arithmetic in nature. The only functions are the trigonometric functions, sine and cosine, and the square root. The presence of conditional processing is minimal. The data organization is complex and requires a great number of arrays forming the database of the analysis. We have ascertained a large presence of operations performed using the indirect element selection.

On the basis of the previous considerations and of our practical experience in vector migration, the value 4 has been selected as a reasonable estimation of *sr* parameter.

By means of the *Amdahl's Law*, we have estimated between 2.2 and 2.9 the global speedup *S* of the final enhanced version running in vector mode with respect to the same one running in scalar mode. Yet, we have to consider that this speedup takes into account only the performance improvement obtained exploiting the vector feature. But a migration process involves a general code optimization, with a consequent improvement in scalar performances. A good percentage of the efforts required to exploit the vector facility involve good programming practises rather than specific vector techniques, yielding a reasonable further improvement in performance of about

20 or at least 10 %. Consequently we can expect the final global vector/scalar speedup, with respect to the original scalar code, being placed between 2.4 and 3.5.

We have defined the way of proceeding in:

- vectorizing the code sections of the *element force evaluation*, of the *geometry update*, and of the *explicit time integration*;

- applying a scalar optimization to the *contact algorithm* section.

The global number of statements involved in the migration process is approximately 15 % of the whole program, whose 5 %, spread in twelve routines, is associated to vector enabling, while the remaining 10 % is associated to the I/O operations and to the virtual memory tuning.

The informing criteria in the modification policy of the original source code have been conceived to limit as much as possible code changes, taking care of the following issues:

- avoiding imbedding routines one in the other;

- applying modifications at routine level only if the performance improvement is over a predefined threshold value;

- preserving the readability of the code.

Vector Migration Process

Great efforts have been applied in the vectorization of the *shell element force evaluation* section, because of its large influence on the processing time. We show the restructuring process we have applied to promote vectorization taking advantage of the IBM VS Fortran Compiler Version 2.1 [12].

From a first analysis we have immediately realized that the original coding form has been conceived having in mind an implementation on a vector machine with a Cray-like architecture. The code was well structured and organized in a DO-loop pattern with a count value of 128, which is one of the optimum values associated to the section size of the vector register. With the aid of the compiler output relative to the vectorization analysis we have ascertained that about 50 % of the DO-loops have been correctly vectorized. The remaining 30 % have presented some inhibitory reasons preventing the vectorization. The inhibitors can be classified according to the following different situations:

- the use of intrinsic functions not available in vector mode: MAX/MIN functions;

- the presence of statements inhibiting the vectorization: CALL and GO TO statements;

- the detection by the compiler of recurrence conditions.

In the last 20 % of the possible situations the compiler has recognized the DO-loops eligible to be vectorized but vector code has not been generated because of performance considerations.

Before approaching the problem of the mere elimination of the inhibitors we have thought a reformulation of the algorithm implementation necessary to obtain the best results from the migration process. Particularly we have thought that the data organization has to be redefined since the current structure did not present a suitable form for getting the best results from vectorization.

The data of each shell element, once pre-processed in the input phase, is stored in a database made of several arrays. For the purpose of computing element forces, the program makes use of pointer arrays to locate in the database, anywhere necessary, the data associated to each element. Consequently the program has to perform sparse matrix computation with a large utilization of the indirect element selection. The original developers were not deeply aware of the influence this kind of data organization could produce on the performance of systems consisting of central processor units provided with caches.

The first step of the migration process has consisted in restructuring the data organization in order to localize data references and to promote an efficient utilization of the cache. Our efforts have been directed to achieve the following goals:

- collecting the gather operations in a localized code section at the beginning, before the force computation starts;

- collecting the scatter operations in a localized code section at the end, after the force computation has been completed;

- performing the force computation on contiguous data;

- reducing and optimizing the usage of those temporary arrays used to store either gathered data or intermediate results.

We have then approached the problem of eliminating the inhibitory reasons in order to promote the vectorization of those DO-loops for which the compiler does not generate a vector code .

The MAX function did not vectorize at the time of the enabling because the VS FORTRAN Compiler Version 2.1 does not support maximum/minimum functions in vector mode. The solution has consisted in replacing the function by an IF construct. The problem has been solved since the most recent VS FORTRAN Compiler Version 2.3 now supports min/max functions.

A main routine calls another one evaluating the element stresses by means of a material relation defined by the user. This routine is of small dimensions and we have imbedded it inside the calling routine. In this way we have promoted the vectorization of the DO-loop whose range the CALL statement was included in.

The presence in a DO-loop range of a GO TO statement inhibits the vectorization. We have replaced it by an IF construction to promote vectorization. Yet, we have not got satisfactory results in performances. So then we have splitted up the original DO-loop into two blocks. In the first block the program now computes the number of times the condition for conditional processing is met, and a pointer vector is initialized in order to identify the array elements on which the computation has to be performed. In the second block the operations executed earlier under the control of the IF statement are processed selecting the array elements by means of the pointer vector previously evaluated. This second block is executed either in vector or in scalar mode depending on the count value. The directive, allowing the user to provide information to the compiler on the DO-loop count, forces the compiler to generate scalar code, when the count value is smaller than a threshold.

The recurrence conditions are detected by the compiler when the program performs scattering operations in order:

- to save element information to be used in the next time step, and

- to assemble the global force vector.

Those operations are performed by means of pointer vectors, indirectly selecting the array elements on which the computation has to be performed. The compiler is not able to recognize if the recurrence condition involves a backward dependence. Since we know the algorithm, we are sure that no backward recurrences are associated with the former situation. Vectorization has been promoted by means of the specific directive. In the latter situation the recurrence condition is detected when the global force vector is assembled. The vectorization can not be promoted because the assembling algorithm is serial since it involves an unbreakable recurrence condition.

20% of the DO-loops are recognized by the compiler as eligible for vector analysis, but the compiler prefers, for performance reasons, to generate a scalar code. Vectorization has been promoted by means of the directive providing information on the DO-loop count.

We describe now some fine tuning techniques to optimize the utilization of the vector registers and of the compound instructions. They have yielded remarkable improvements in performance.
Vector register usage can be optimized keeping in mind that the vector facility feature makes available sixteen vector registers in single precision and eight registers in double precision. The compiler uses the registers to store the intermediate results of the operations in the DO-loop range, and it does that trying to optimize the load and store operations. The programmer can explicitly assist the compiler in the optimization process using a particular programming style. This intervention consists in minimizing, inside the DO-loop range, the work vectors used to store intermediate results and in replacing them with fictitious scalar variables. This method forces the compiler to hold in the vector registers (when a sufficient number of them are available) the intermediate results of the computation. The programmer is assisted in the vector register optimization techniques by the listing of the object module in pseudo-assembler language. In general, successful optimization will be enhanced by minimizing load and store operations.

The vector facility provides two compound instructions, multiply-add and multiply-accumulate, performing two operations in a single machine cycle. We have tried to increase the number of the compound instructions automatically generated by the compiler. Once the statements with compound instructions had been identified, we have verified by means of the object listing if compound instructions had been generated too. Otherwise, we have tried to reorder the statement operation keeping in mind that the compiler follows arithmetic rules in performing operations and that three operands can not reside all together in the vector registers.

We present some performance results showing the improvement the migration work has yielded. In the following tables we report the performance results of two test cases in terms of CPU time for the first table and in terms of speedups for the second table. The two selected test cases have been chosen since they can be considered to represent the lower and the upper bound in performance results, due to the different incidence of vector code exploitation. In the first table the first column reports the CPU time needed by the original ESI scalar version run in scalar mode, the second one reports the CPU time of the final enhanced version, result of the migration work, compiled and run in scalar mode and the third one reports the time of the same enhanced version run in vector mode.

Test Case (s)	Original Version Scalar Run	Enhanced Version Scalar Run	Enhanced Version Vector Run
S1	1234.	971.	449.
S2	6273.	5303.	1774.

In the second table the two columns show the speedups of the final enhanced version run in vector mode with respect to the original version and to the enhanced one both run in scalar mode.

Speedup Enhanced Version	Speedup vs Original Version Scalar Run	Speedup vs Enhanced Version Scalar Run
S1	2.7	2.2
S2	3.5	3.0

The benchmark has been carried out on a 3090/VF 200 E model system.

Parallel Implementation

Once the vectorization has been completed, we have faced the problem of parallelizing the program by dividing up the computation among the processors in a multi-processor configuration. The IBM 3090/VF system is a tightly-coupled multiprocessor system and it can be classified as a 'course-grain machine'[13]. The parallelization efforts have been driven in order to take advantage of the 3090 multi-processor architecture, whose main characteristics, from the parallel exploitation point of view, can be summarized in being provided with a limited number of powerful processors and with a relatively big amount of central and expanded storage accessible by all processors. In particular, the main goal of our efforts has consisted in developing a standard version of the program, provided with the capability of exploiting parallel processing, and delivered without any limitations to IBM customers. The design requirements have been given by:

• turnaround time improvement in a shared environment;

- elapsed time reduction in a dedicated environment.

The software selected to support the development of the PAM-CRASH parallel version, has been the Multitasking Facility (MTF), a standard feature provided to the customers with the IBM VS FORTRAN compiler. The MTF feature makes use of the system macros, provided by the MVS operating system, to execute in parallel selected FORTRAN routines and to synchronize their executions. The programmer can easily introduce in the program suitable calls to simple ('fork' and 'join') MTF primitives, in order to specify the sections of the code to be executed in parallel and to synchronize their executions.

The overhead associated with execution scheduling of a parallel subroutine using MTF is relatively high. The value of 75 μs is the selected reference value of the MTF overhead, obtained by dispatching and synchronizing an empty routine [14]. Consequently the programmer must take care in verifying the overhead costs to be negligible in comparison with the execution time of the code sections running in parallel. The Multitasking Facility is an efficient tool in exploiting parallelism at quite high level ('coarse grain parallelism'). Presently some other tools are available on IBM machines to exploit parallelism from fine to coarse grain level. Nevertheless, all of them are at prototype level. In February '88, the IBM Corporation announced its own Parallel Fortran Compiler in order to make available a standard new powerful tool to face parallelism in the engineering-scientific environment.

As we expected, through the vector migration process, the percentage incidence (with respect to the total processing time) of the vectorized code sections, eligible to be parallelized, considerably went down. Actually that occurs in all those analyses involving a scarse utilization of the contact algorithm, for which no vector content has been identified. On the other hand that percentage has nearly remained constant in the analyses characterized by a large utilization of serial code. In the following table, the first column reports the processing time percentages associated to the code sections executed in scalar mode; the second one reports those associated to the same sections executed in vector mode. The two values of each column define the range of variation depending on the type of analysis.

Code Section	CPU Percentage Scalar Mode	CPU Percentage Vector Mode
1 - element force evaluation	70-82	52-83
2 - geometry update	7-9	4-6
3 - explicit time integration	3-4	4-6
4 - contact algorithm	17-3	36-2
1 + 2 + 3 + 4	97-98	96-97

On the basis of the previous considerations and of the results of the vector migration process, a parallel feasibility analysis has been performed in order to define the conditions allowing the parallelism to be efficiently exploited and to estimate costs, times and resources. Since the variation range of the time execution, shown in the previous table, is not negligible for a wide range set of analyses, we have selected as reference test cases for parallelism, those needing such an amount of computer resources to require, in terms of design requirements, the exploitation of the parallel processing. The crash analysis of a full vehicle (about 10 CPU hours with the vector version) has been chosen.

The preliminary approach to the parallel exploitation has been the simplest one. It has consisted in verifying the possibility of restructuring the serial algorithms (presently implemented in PAM-CRASH) into concurrent algorithms. The approach has consisted in applying the following key issues: restructuring the code in order to allow concurrent computation, and breaking a task up into smaller subtasks to be treated independently. The criteria, followed in identifying the code sections to be parallelized, have consisted in selecting the sections which have taken benefit from vector processing. Yet, those sections have to comply with the requirement that their vector processing time be greater than a threshold, above which the overhead due to dispatching and synchronization can be considered negligible. The requirement has been met for those code sections performing the *shell element force evaluation* and the *geometry update*. The following table shows the processing time percentage of each single section and the parallel percentage we have estimated for the reference test case, consisting in the crash analysis of a full vehicle.

Code Section	CPU Percentage Vector Mode	Parallel Percentage
1 - element force evaluation	82	74
2 - geometry update	5	5
3 - explicit time integration	6	0
4 - contact algorithm	3	0
1 + 2 + 3 + 4	96	79

We have estimated the parallel percentage, defined as the percentage of the vector execution time relative to those code sections which can benefit by parallelization, being about 79 %. We have used a relation, based on the *Amdahl's law*, to estimate the improvement in performance to be expected by exploiting parallel processing. That relation takes into account the degradation in performance due both to the task unbalance and to the parallelization overhead, by means of specific normalized factors, respectively o_p and lb_p.

$$S = \frac{1}{(1 - f_p) + (f_p / n_p) * (1 + o_p + lb_p)} \tag{9}$$

where

S is the overall parallel speedup factor;

f_p is the percentage of parallelization,

n_p is the number of processors,

o_p is the overhead factor associated to task scheduling and syncronization,

lb_p is the task unbalancing factor.

The overhead factor o_p takes into account the overhead associated to task dispatching and synchronization and it is normalized with respect to the average execution time of each task:

$$o_p = \frac{t_o}{(f_p / n_p)} \tag{10}$$

where t_o is the time needed to perform the task scheduling and synchronization. The task load unbalancing factor lb_p takes into account the possibility that the workload among the parallel tasks is not equally distributed. This condition is verified when more computational paths can be followed inside the parallel code.

The parameter lb_p is defined as follows:

$$lb_p = \frac{t_{max}}{(f_p / n_p)} - 1 \tag{11}$$

where t_{max} is the maximum execution time among the parallel tasks.

Assuming the load balancing factor lb_p being equal to 0, i.e. the workload is equally distributed among the parallel tasks, and imposing the overhead o_p not exceeding 5 %, we have estimated the performance improvement to be 1.5 for dyadic processor system and 2.4 for four-way processor system.

The algorithm of the *shell element force evaluation* section shows an inherent parallelism, except for the section assembling the element forces into the global forces vector, whose present implementation is serial. Indeed, this section can be executed in parallel only if the elements assigned to different processors, at any one time, have no common nodes, but it does not generally occur. In order to exploit concurrent processing for this section too, a global element numbering, e.g. through a coloring technique, must be applied. Yet, once we have evaluated the restructuring efforts associated with these modifications, as well as the overhead in terms of increase of execution time, we have drawn the conclusion this optimization process is not effective from cost/performance aspects. The solution selected has consisted in assigning to each processor the forces computation relative to 128 elements. The synchronization is scheduled, before the forces assembly is performed. The number of dispatching operations is equal to the number of 128 element groups, and the number of synchronizations is equal to the number of dispatching operations divided by the number of available processors. The numer of elements dispatched in the last parallel set can be less than 128.

From the point of view of the coding structure restructuring, an interface routine has been created to re-organize the data communication among the main task and the parallel ones. Before, this operation was perfomed using the storage areas defined by the COMMON statement, while it is now carried out by means of the subroutine arguments. The section of the element forces assembly has been moved into a new routine whose execution is scheduled after the synchronization. This parallel migration phase has been quite laborious because we have been forced to apply heavy changes in the code structure and many efforts have been spent in order to preserve the code readability.

The task unbalance lb_p has been discovered during the program test to negatively affect performances. A load unbalance among the tasks rises because the shell element is a non-linear element. Different computational paths can be followed depending on the deformation and stress state of the structure, generating a task load unbalance, that does not remain constant but differs step by step. Nevertheless, performance measurements have further shown that the overhead factor does not exceed the values of 2%.

The *geometry update* section has been parallelized by means of the MTF capabilities following a 'divide and conquer' approach. Each parallel task performs the update in the finite element model of a defined region whose dimension depends on the number of available processors.

Benchmarks have been performed on the test case consisting of a full vehicle model and in other selected cases suitable to take benefits from parallelization. The speedup figures, shown below, provide an average estimation of the parallel PAM-CRASH version performance.

3090 System	Parallel Speedup
Triadic	2.0
Four-Way	2.5
Six-Way	3.0

Benchmark has been carried out on 3090/VF E model systems.

Conclusions

The result of our work has consisted in delivering a new IBM version of PAM-CRASH exploiting vector and parallel capabilities of the 3090/VF system. The quality of the results, in terms of PAM-CRASH performance improvement, has been far beyond the initial expectation. The new PAM-CRASH version presents a performance improvement, with respect to the old scalar one, of 300% on a uni-processor 3090/VF, of 600% on a triadic 3090/VF, and of 900% on a six-way 3090/VF system. At present PAM-CRASH is installed and successfully runs on the 3090/VF system of a Japanese automotive manufacturing industry.

We have discussed and commented on problems encountered in the PAM-CRASH explicit finite element code. We have shown the advantages that an explicit approach in finite element computation has in comparison with an implicit approach, if vector/ parallel computation must be exploited. Nevertheless, we can finally assert that the migration process will be successful only if the code is well structured and it has been designed taking into account some key issues concerning vector/parallel computation.

PAM-CRASH is a registered trademark of the Engineering Systems International (ESI) S.A. Rungis-France

Acknowledgments

We are grateful to Jan Clinckemaillie (ESI), Shunichi Katoh (IBM Japan) and Susumu Suda (IBM Japan) for their active contributions in developing the vector version/parallel version of PAM-CRASH code.
We are also grateful to G. Radicati (IBM/ECSEC) and M. Vitaletti (IBM/ECSEC) for having provided the authorization to publish excerpts, concerning the IBM 3090/VF

vector architecture, from their paper 'Sparse matrix-vector product and storage representations on the IBM 3090 with Vector Facility' [15].

References

1. PAM-CRASH User Manual, Engineering Systems International S.A., Rungis, France.

2. J.F. Chedmail, P. Du Bois, A.K. Pickett, E. Haug, B. Dagba, and G. Winkelmuller (1986), Numerical Techniques, Experimental Validation and Industrial Applications of Structural Impact and Crashworthiness Analysis with Supercomputers for the Automotive Industries, Proceedings of the International Conference on Supercomputer Applications in the Automotive Industry, Zurich, Switzerland.

3. G.L. Goudreau and J.O. Hallquist (1982), Recent Development in Large-Scale Finite Element Lagrangian Hydrocode Technology Problems, Computer Methods in Applied Mechanics and Engineering, Vol. 33, pp = 725-757.

4. T. Belytschko and J.I. Lin (1984), Explicit Algorithms for the Nonlinear Dynamics of Shells, Computer Methods in Applied Mechanics and Engineering, Vol. 42, pp = 225-251.

5. T. Belytschko, J.S. Ong, W.K. Liu, and J.M. Kennedy (1984), Hourglass Control in Linear and Nonlinear Problems, Computer Methods in Applied Mechanics and Engineering, Vol. 43, pp = 251-276.

6. G.L. Goudreau, D.J. Benson, J.O. Hallquist, G.J. Kay, R.W. Rosinsky, and S.J. Sackett (1986), Supercomputers for Engineering Analysis, Proceeding of the Winter Annual Meeting of the American Society of the Mechanical Engineers, Anaheim, California, U.S.

7. 3090 Processor Complex: Functional Characteristics, SA22-7121, IBM Corporation; available through IBM branch offices.

8. W. Buchholz (1986), The IBM System/370 Vector Architecture, IBM Systems Journal, Vol. 25 No. 1, pp = 51-62.

9. IBM System/370 Extended Architecture Principles of Operation, SA22-7085, IBM Corporation; available through IBM branch offices.

10. K.J. Bathe (1982), Finite Element Procedures in Engineering Analysis, Prentice-Hall, Englewood Cliffs, N.J.

11. G.M. Amdahl (1967), Validity of the Single Processor Approach to Achieving Large Scale Computing Capabilities, AFIPS Conference Proceedings, Vol. 30, pp = 483-485.

12. VS Fortran Version 2 Programming Guide, SC26-4222, IBM Corporation; available through IBM branch offices.

13. A.K. Noor (1988), Parallel Processing in Finite Element Structural Analysis, Engineering with Computers, Vol. 3, pp = 225-241.

14. P. Carnevali, P. Sguazzero, and V. Zecca (1986), Microtasking on IBM multiprocessors, IBM Journal of Research and Development, Vol. 30 No. 6, pp = 574-582.

15. G. Radicati di Brozolo, and M. Vitaletti (1986), Sparse matrix-vector product storage on the IBM 3090 with Vector Facility, G513-4098, IBM Corporation; available through IBM branch offices.

The Process of Finite Element Validation

A.J. Morris

College of Aeronautics, Cranfield Institute of Technology, Cranfield, Bedford, MK43 OAL, UK

1   Introduction

All ideas have a natural rhythm to their development and, eventually, application. Initially, they are the province of a few research workers who originate the basic concepts and then develop them to a state where they can, hopefully, be used in the solution of practical problems. At this early stage, those involved with the theory have a deep understanding of the underlying principles and often have a very acute awareness of the limitations. Because of the high level of knowledge, the theory is rarely abused at this point in the development process. Misuse, if it is to occur, must await the application of the methods by the less experienced general user. However, in the case of classical closed form methods, there are usually safeguards against blatant misuse since any application requires some understanding of the equations before use. The advent of computer based methods, particularly commercially supported software packages, has changed the scene. Whilst the research workers developing the ideas still have the same profound understanding of the basis theories employed and their limitations, the applications engineer need have no knowledge whatsoever of the underlying principles.

The freedom of action being afforded the inexperienced user has been compounded in the case of the Finite Element Method by commercial pressures to develop the range of application at too rapid a rate. Elements are employed for the solution of problems which lie outside the valid range of application or are extended by ad-hoc heuristic modification which often has no theoretical justification. In this situation, even the most prudent user finds there is no sensible method for validating the solution. Unfortunately, most users do not realise the need for prudence and the relative ease of use provided by modern FE systems gives many possibilities for error. Thus, the way the method is currently being developed and the relative inexperience of the user implies that some process for error avoidance should be devised.

In addition to the natural desire of design engineers to circumvent erroneous results, there is a growing legal requirement to assure quality control in software and other products. This legal aspect to the requirement for quality assurance confidence has recently been augmented by the implications of product liability legislation and, in certain industries, because of statutory and regulatory requirements which are imposed., This latter is particularly important in the nuclear, chemical, aviation and pharmaceutical industries.

Outside of these specific industries there is an increasing awareness of the importance of quality systems in the design process which includes the associated software. Although there are no direct moves to incorporate FE software into the international ISO 9000 series the UK is seeking to include this in its own British Standard quality system BS 5750.

Against this general background, there is a growing international recognition of the need for the creation of accreditation and validation procedures for commercially available FE systems. Within the UK, a start has been made by the British Government in founding the National Agency for Finite Element Standards and Methods (NAFEMS). This has been given additional emphasis by the move within the European Aircraft Industry, supported by the certification bodies, to progessively replace expensive larger scale testing by finite element analyses. Although NAFEMS has been running for 4 years, the criteria and underlying philosophy required to set up the necessary and sufficient conditions for a valid verification process are far from being established. Extensive research is still needed in this area of increasing importance. But some very worthwhile work has been done and a steady flow of validation procedures and tests are being made available, which reflect a variety of viewpoints. The remaining part of this paper reports some of the work which has come from NAFEMS activity. The background philosophical approach is outlined first and then aspects relating to benchmark tests are illustrated for static, dynamic and temperature problems.

2 Sources of Potential Error in the Finite Element Solution of Design Problems

The first task to be performed in any validation procedure is to decide what is meant by error. Here we shall adopt a commonsense approach and mean by an error the difference between the results predicted by the FE method for some parameter of interest, stress (say), and that occurring in the real structure. Of course, by this definition we bypass entirely the important question of how we know what the actual stress values in a real structure are! It is tempting to say that these are found by experiment but the notorious difficulties associated with achieving reliable and accurate experimental results are well known. These relate to the problems in applying representative loads, providing both accurate and adequate measuring devices, and the very complex problems of results interpretation.

By ignoring experimental aspects, we can concentrate on the ability of the FE method to represent abstract reality and within this framework we can see three broad types of error. First, there may be something wrong with the FE code or the way it works; the elements may have faults, the solution procedure may be inadequate etc and for convenience these are labelled simply SYSTEM errors. Secondly, there are possible errors due to the fact that the real structure is not adequately represented by the FE model and these may be called SIMULATION errors. Finally, there is the growing difficulty with modern software that the users may introduce problems due to inexperience, lack of training etc and these are usually called USER errors. We can now examine these in a little more detail to give an overview of how various specific sources of error within these three areas can be categorised. Once categorised, it is possible to devise methods for validating systems so that the effect of such errors can be identified and, thus, avoided. This process of categorisation is clearly an important part of the validation and verification philosophy of the NAFEMS organisation. It should be emphasised that the current paper is not attempting to provide a comprehensive list of all possible error sources associated with the FE analysis of structural behaviour.

2.1 System Errors

Within the broad area of System errors there are several sub-divisions of categorisation relating to the properties of the individual elements, the solution method used, bugs in the code etc. Turning first to the individual element property there are, at least, two error categories,; Inherent and Imposed. By Inherent is meant errors which occur because of the very nature of the FE method; Imposed implies additional errors which may occur because of additional modifications made by the analyst to 'improve' element performance.

In order to discuss Inherent error, let us consider how a specific finite element stiffness matrix is devised. In order to perform this task, various approaches may be adopted but all are essentially equivalent to the classical variational method. In this method a functional is sought which characterises the problem in terms of an energy formulation with specified boundary conditions. An examination of this functional allows the analyst to correctly select the appropriate connectors, continuity conditions and the order of the interpolating polynomials, which play the role of shape functions. If this process is followed and one has 'correctly' devised a stiffness matrix for a specific application there are still potential sources of error - these are the Inherent errors due to the fact that we are approximating the original problem. For example, these would include:-

- Shape sensitivity errors
- Inability to model certain behavioural modes (even when complete polynomials are used)
- Rank deficiencies
- Discretisation errors for certain materials (particularly for non-linear behaviour)
- Integration errors
- etc

In many cases, these errors will be overcome as further research provides new elements with enhanced performance for specific problems eg. the use of 'drilling' freedoms.

Imposed errors are different in kind from Inherent errors since they are something which is not, necessarily, coming from inadequacies in the basic theoretical approach but are imposed on the method by the analyst. Ironically, some errors in this category have come about in an attempt to overcome certain Inherent errors as, for example, in the case of thin shell analysis. Although we may describe these errors their presence, in many cases, cannot be proved but may be suspected after ad-hoc modifications have been made to an element stiffness matrix. For example many elements are modified by the preferential adjustment of the numerical integration points to overcome a defect. Whilst such a heuristic approach may cure the specified defect it may also give rise to a new fault which often only becomes apparent at a later date. Unless the analyst has a very clear understanding of the full effect of such a heuristic modification it is usually impossible to say if, or where, a new problem might arise. It is, therefore, prudent to suspect Imposed error whenever an element is modified in this manner or indeed, fully developed by heuristic methods in preference to the standard variational approach. However, in this latter case, it is more convenient to adopt the concept of Formulation error.

The underlying concept of Formulation error is relatively straightforward. It implies the potential generation of error when the basic variational approach is consciously abandoned before the stiffness matrix has been generated. Included in this category are errors which might arise from:-

- the use of a non variational approach
- the use of incomplete polynomials
- the use of non-conforming elements etc

It is worth noting that some of these can be incorporated into an extended classical derivation. For example, non-conforming elements can be included in a rigorous manner by penalty methods but not all such elements have necessarily, been treated in this way.

The next level of errors are those related to the techniques employed to achieve a solution to the series of equations generated by the method. These Solution errors may not give rise to concern in the case of linear statics problems but do give rise to potential difficulties in other regimes. Thus, we might expect to encounter this type of error in problems related to:-

- dynamically loaded structures
- geometric and material non-linearities
- transient problems requiring time stepping
- reduction methods

The last item could be included under other headings but is also happily contained within a Solution error description.

Finally, within this overall category, are the usual bugs which are inadvertently written into all programs and included here under the heading of Program error. It is also convenient to cover faults which may occur in moving programs from computer to computer where differences in word length or operating procedures upset the working of the program.

2.2   Simulation Error

All the above sources of error are related to the internal working of the FE programs or faults in the underlying theory. Outside of this are the more subtle errors which are associated with the problem of relating the FE model within the computer to the outside real-world structure. Often such errors are called modelling errors but are, more properly, Simulation errors. This is a very broad category because it is both structure and problem dependent. So that, for example, the problem associated with creating an appropriate FE model for a specific structure will be different when considered from a static or dynamic viewpoint.

The course of Simulation error lies in the nature of the Finite Element method. Although the method is a discrete representation of the original structure and the applied loads, the underlying assumption is that the basic problem is concerned with a smooth, continuous structure with clearly defined properties. In reality, we are dealing with something quite different. First of all, many structures are built-up and fabricated from a variety of components which are joined together by fasteners, adhesives and other connections. Secondly, no structure is smooth except in very special circumstances and will have changes in section, abrupt changes in material properties, holes etc. A full list of potential error sources in creating a satisfactory simulation model for a complex structure would be long, yet the type would be similar in nature to the two listed above.

Any competent FE analyst has a range of 'tricks' to help in modelling a given structure. These would include the use of special elements i.e. for riveted joints, or modifying element material or other properties to simulate the specific structural parameter being modelled. However, these are usually generated for a given type of structure with specific fabrication or design aspects and may well be inappropriate for a different structure or even the same structure when the fabrication methods are changed. The great danger is that even a competent analyst can fail to recognise the importance of features which are unfamiliar. In fact, there are many examples of competent FE teams making serious Simulation errors when new (to them) structural concepts are encountered.

Simulation error is, therefore, potentially a serious source of error in the Finite Element analysis of complex structures. It is insidious and many FE specialists regard it as the most serious source of error in the whole process of analysing a structure by FE methods. Unfortunately, it has, to date, received little attention possibly because it is one of the most difficult areas to quantify.

2.3 User Error

Many of the developers of Commercial Finite Element codes claim that a scientific survey of FE related analysis errors would show that the majority were traceable to User error. The user may be an individual or a team who are perhaps new to a specific program, or inexperienced in the use of FE methods or are new to the problem type. In these situations, the potential errors are limitless but some typical ones may be indentified:-

- use of elements outside of legitimate range
- incorrect boundary conditions
- incorrect loads
- mis-modelled structure
- inability to interpret results and thereby trap false information

3 Validation Procedures

In setting up validation procedures NAFEMS has become the leading organisation in endeavouring to establish appropriate methods and getting them accepted by FE users and developers alike. So this section concentrates on the NAFEMS approach, though the policy adopted does not differ significantly from that followed by the major developers. Effort has been focused on System error though a limited activity has been directed at User error by giving some attention to Education Accreditation matters. At the present time, Simulation error has not been considered primarily due to the difficulty of systematically identifying and quantifying specific errors. We shall, therefore, concentrate on System error in the sequel.

In dealing with System error, it is clear that a validation procedure must be founded on some form of testing philosophy. Because the end point of these activities is to support the design engineer the procedure must be relatively easy to follow and not require a high level of expertise. Whilst it would be beneficial to ask an engineer who wishes to validate a specific FE system to distinguish between elements which might exhibit Formulation, or perhaps, Inherent errors only this is not a practical question. Thus a validation philosophy must assume that any element in any code might exhibit any or all of the three error types; Formulation, Inherent and Imposed. The tests should, then, be able to stand in place of the usual demand that the element stiffness matrix should be generated from a legitimate variational principle in addition to sifting for the other error types. Clearly Solution error can be discriminated for separately.

Although most of the error categories are to be covered by a broad set of tests, these must change as one moves from problem type to problem type. Thus, the tests specific to FE analyses for statically loaded structures must be augmented to cover dynamics problems and further augmented for other areas. Thus far, NAFEMS has started Working Groups which look at tests for Statics, Dynamics, Nonlinear, Fracture mechanics and results are emerging from all these areas. In order to focus attention we concentrate on the test procedures currently being advanced for Statics and Dynamics.

The principles which have been discussed above indicate the basis of a verification tests sequence which is practical in nature but sufficient to discriminate between 'good' and 'bad' elements. The general requirements may be interpreted as follows:-

1. The tests must examine the ability of the element to accurately model the basic stress/strain states which make up the general states the element purports to cover. For example, in the case of thin shells, a general solution to a complex loading and geometric configuration is a mixture of inextensional bending stresses, membrane stresses and edge effects. Thus, a set of validation tests should demonstrate that a specific element is capable of separately modelling each of these states in an adequate fashion for all Gaussian curvatures. In addition to actual stress states, a demonstration that rigid body modes can be accommodated is required.

2. As part of the process of checking against Imposed error elements should be checked against known inadequacies ie locking phenomena, false energy modes, geometric non-invariance etc.

3. In the case of Inherent error, tests to evaluate shape sensitivity, sensitivity to integration rule etc are required.

4. Tests should be organised in an hierarchy starting with basic tests (possibly single elements) to more complex (benchmark assemblies) and taking account of increasing complexity of the problem types: Static, Dynamic, Nonlinear ...

5. Target values for assessment should, if possible, be obtained from known analytic results. If this is not possible, results from convergence studies by Finite Element or other numerical analyses are acceptable providing the converged values can be justified.

6. From a practical viewpoint tests should not need excessive data preparation and be unambiguous in geometry and restraining conditions.

7. In the case of assemblages (benchmark tests) the Finite Element meshes should be sufficiently well defined such that minor changes in the mesh do not produce

significant changes in the results.

8. Benchmarks should not be problems which cannot be posed properly and should be such that Finite Element systems can be expected to produce acceptable results.

Although most of the NAFEMS activity has interpreted these eight points in terms of benchmark tests involving an assemblage of elements, it is possible to evaluate certain element properties by single element tests. This is the approach adopted by Robinson [1] which was the precursor of the present NAFEMS interest in the subject.

As already emphasised, the above list represents the current thinking of NAFEMS on what constitute the basic rules to validation testing for System error. It is a staging post on the way to a more comprehensive and sophisticated philosophy which will require more detailed research than that devoted to the subject in the past.

4 Static Benchmark Tests

Having discussed the NAFEMS approach to validating FE elements and systems we can now see how this has been applied in practice. To this end, we use the Selected Benchmarks [2] compiled from the full NAFEMS set devised to evaluate Finite Elements against a limited number of validation tests. These are used because results are available of how various FE systems perform with respect to these tests [3]. Most of the systems used are well known and consist of the British codes, PAFEC, ASAS, LUSAS, BERSAFE, MELINA and MELISSA and US codes; NASTRAN, SUPERTAB, MARC, GIFTS. The tests are specified in figures 1-8 and the results in tables 1-8. The first test, LE1, is a membrane benchmark which, as indicated by Davies, has three possible errors which are essentially Inherent. First, the target stress at D is a stress concentration sensitive to curvature which focuses on discretisation problems. Second, low order elements may not accurately represent the varying field. Finally, if the element uses the Gauss points for stress evaluation then incorrect values are likely to appear. The results indicate that the test is reasonably discriminating and shows up triangular elements in a relatively unfavourable light.

The tests LE2 and LE3 again look at Inherent error and focus on two of the basic stress states for thin shells mentioned earlier; membrane and inextensional bending (nearly) respectively. The results, tables 2 and 3, indicate how difficult the thin shell problem still is for the FE method. The results also examine, to a limited extent, the effect of integration rules. The total effect given by these two tests does not give rise to great confidence in the current element libraries ability to handle thin shell problems. Yet the negative Gaussian Curvature test LE4 appears very benign! The final purely stress-based test in the static category is provided by the cantilever subject to pure torsion LE5. The target stress required is indicated in figure 4 with results given in table 4.

Turning to the thermal benchmarks, we see that T1 targets stresses induced by a prescribed temperature distribution whilst T3 and T4 target point values of temperature. The purpose of T1 is to assess the ability of elements to, essentially, handle discontinuities in the thermal region. The discontinuity at D poses modelling and results difficulties which, on occasion, required certain codes to employ heuristic devices to achieve reasonable answers. The full set of results shown in Table 6 show some very curious trends which give rise to the suspicion that errors may be self-cancelling.

T3 is an example of a NAFEMS mistake in that the test is not realistic. When the test had been proposed, discussion between NAFEMS and the Contractor led to the selection of material properties relevant to concrete. Once this selection had been made it became clear that the analytic solution used to generate the target was in error. Correcting this produced a target solution where the thermal boundary layer is restricted to a region of length about one cm at the point where the temperature step is imposed. Obviously, a poor test for an FE system but, perhaps, a good try at isolating Inherent error. For completeness, the results are shown at table 7.

The static benchmarks conclude with T4 which tests errors in modelling heat conduction problems with discontinuous boundary conditions. The geometry is kept simple to reduce geometry modelling errors so that the results shown in table 8 essentially test the temperature features. As with the first thermal test, the results from the various systems show significant variations which are difficult to interpret in a coherent manner.

Overall the results of these tests are interesting and shed some light on the behaviour of elements which purport to solve a range of static problems. However, comparing these few tests with the philosophy described in earlier sections, we see there is still a considerable gap between theory and practice.

5 Free Vibrations Benchmarks

In the case of Dynamics the Validation process has started with providing benchmarks for finite element assemblies in free vibration. Only isotropic stress free assemblies are considered and attention is focused on the most popular types of finite elements in commercial use. The point of the tests is twofold: to attempt to generate some tests which target Inherent error and, to look at Solution error by dealing with various aspects of the eigenvalue solution methods. A new aspect of this phase is the desire to introduce an educational component to the tests. In this way benchmarks can be employed in the accreditation process which is a feature of the NAFEMS role not discussed in this paper.

The full range of free vibration benchmarks constructed to date amount to some 72 tests [4]. However, this represents too extensive a range and it would be unreasonable to expect a developer or user to cover all these in addition to static, nonlinear and other benchmarks. A reduced set of tests is recommended and shown in figures where they retain the number ascribed in reference [4] but even these may be further reduced in time. In certain cases, the tests represent comparison with a reference solution which is analytic and, in other cases, the reference target solution is obtained by a "converged" finite element solution. These numerical targets are obtained using the British Systems ASAS and confirmed using the American Systems ANSYS. A further analysis has been performed, where possible, at the College of Aeronautics using another British System LUSAS. A glance at figures shows that the proposed tests fall into specific structural categories.

The first three represent a sample of the beam benchmarks designed to test an element's ability to cover longitudinal torsional and flexural properties. Specific features which can be tested are:

- the ability to represent rigid body modes
- the correct coupling between flexural, torsional, shear and extensional modes
- the ability to predict multiple repeated modes

- the ability to handle mass offset from the flexural axis

The three tests 2, 4 and 5 pick up most of these attributes as evidenced by the 'attributes' section in the single page test definition. The results for the ASAS runs are contained with the test definitions. These are, of course, relatively simple tests and other finite element runs confirm the results. Though LUSAS was unable to tackle test 4 not having an offset lumped mass capability.

The next set of tests 12, 15 and 16 are a selection of the thin plate tests. Once again, a series of attributes were identified as targets which would imply Inherent or Imposed or Solution error:-

- again the ability to represent rigid-body modes
- the ability to predict repeated roots
- distorted elements

Also by introducing plate tests, we have moved to another structural class in the attempt to produce a comprehensive set. The thick case is covered, inadequately, in this reduced set by test 22 which endeavours to cover the next level of structural category and to include non-square elements.

The sole membrane test number 32 assumes that for the most part, membrane elements in free vibration are likely to prove able to adequately represent the basic situation modes. Thus, this single test again intends to cover a wide range of aspects which target Inherent error. However, the test is rather peculiar in that elements are rather mis-shaped at the tip and one might expect that errors would occur unless the integration rule is altered to cater for special problems.

Having examined two dimensional structures, the free-vibrations tests move onto solid objects, initially for axisymmetric structures tests 41, 42 and subsequently, for 3-D solids test 52. As before, the main target is Inherent error though two of the tests (41, 52) evaluate the robustness of systems solution algorithms. The specific attributes tested are indicated in the 'attributes' box in the accompanying figures.

The final test, number 73, selected for presentation here is taken from a group specifically constructed to test the effect of accuracy of the systems. In this test, the effect of master degrees of freedom selection on frequencies is examined. This clearly focuses on Solution error as do the others in the same class of four presented in reference [4].

6 Conclusions

The earlier sections have indicated that problems do exist in the use of Finite Element and, indeed, other methods for the solution of continuum problems. The line of attack advanced for resolving the situation is limited but does have some merit in aiding the user in addressing the main issues raised. Although the approach has been focused on just the one numerical solution method it is, in its generality, applicable to other areas. This is evidenced by the fact that the NAFEMS organisation is seriously considering the implications of its standard forming process for non FE methods and their application.

In the specific case of Finite Element Methods the approach, though limited, is seen to be effective in discriminating between good and poor elements and in aiding the rational application of specific elements. This is evidenced by the results presented above and by unpublished discussions and documents emanating from the NAFEMS work.

For the super-computing field, the validation procedure is even more vital than in conventional computer implementations. Large and fast computers are used, in many cases, as a last resort because the application is complex with possibly non-linear behaviour patterns. In these situations, past experience and intuition on the part of the user will have little relevance and an even greater reliance will be placed on the numerical output from the computer. Confidence in the system employed will be vital to the creation of confidence in the eventual construction made on the basis of stress and deformation calculations using numerical methods. The application of the philosophy developed here and its continued development will form a necessary link in the overall QA process. Not only will this pressure come from the engineering community but it will also occur because of an informed public demanding the highest levels of safety in aircraft structures, nuclear containment vessels etc.

References

1. Robinson, J., Element Evaluation - A set of Assessment Points and Standard Tests, Finite Element Methods in the Commercial Environment, pp. 218-247, published by Robinson & Associates, 1978.

2. Davies, G., et al, Selected FE Benchmarks in Structural and Thermal Analysis, published by NAFEMS August, 1986.

3. Davies, G., Results for Selected Benchmarks, in Benchmark, pp. 8-12. Oct 1987.

4. Abbassian, F., Dawswell, D.J. and Knowles, N.C., Free Vibrations Benchmarks, NAFEMS, November 1987.

ΠAFEMS ELLIPTIC MEMBRANE	TEST No LEI	DATE/ISSUE 1-7-86/1

ORIGIN	NAFEMS report C1

ANALYSIS TYPE	Linear elastic

GEOMETRY

$$\left(\frac{x}{3.25}\right)^2 + \left(\frac{y}{2.75}\right)^2 = 1$$

$$\left(\frac{x}{2}\right)^2 + y^2 = 1$$

All dimensions in metres
Thickness = 0.1

LOADING	Uniform outward pressure of 10MPa at outer edge BC Inner curved edge AD unloaded

BOUNDARY CONDITIONS	Edge AB, symmetry about y axis, e.g. zero x displacement Edge CD, symmetry about x axis, e.g. zero y displacement

MATERIAL PROPERTIES	Isotropic, $E = 210 \times 10^3$ MPa, $\nu = 0.3$

ELEMENT TYPES	Plane stress quadrilaterals or triangles

MESHES	(Corner nodes only given)

Coarse

Fine – Approx. halving of coarse mesh

OUTPUT	Tangential edge stress (σ_{yy}) at D	TARGET	92.7 MPa

	Table 1							
	NAFEMS BENCHMARK TEST No. LE1 **For membranes**							
	QUADS				**TRIANGLES**			
SYSTEM	**COARSE** **4** Nodes	**8**	**FINE** **4** Nodes	**8**	**COARSE** **3** Nodes	**6**	**FINE** **3** Nodes	**6**
NASTRAN	-30.7	-10.3	-7.8	-0.7	-11.6	+3.2	14.7	3.2
ANSYS	-35.7	-7.4	-12.7	-1.7	-61.6	-8.6	-12.7	4.7
GIFTS	-21.8	-9.2	+2.9	+1.0	-43.9	+2.8	-	-
MELINA	-23.9	-8.8	-8.1	-1.2	-62.5	-17.4	-39.9	-6.8
MELISSA		-13.6		+3.3		-14.7		+9.2
MARC	-	-12.5	-	-3.4	-	-	-	-
PAFEC	-	-11.4	-	-0.8	-	-9.5	-	-7.4
ASAS	-18.4	-4.2	-7.8	-1.9	-42.5	-3.7	-	-
LUSAS	-23.8	-8.4	-7.8	-1.5	-42.9	-5.7	-	-
FINEL	-18.0	-12.0	-5.0	-2.0	-40.0	-13.0	-19.0	-3.0
SUPERTAB	-28.0	-9.6	-5.9	-0.8	-42.9	+11.1	-23.5	+1.3
BERSAFE	-23.3	-11.0	11.5	-1.6	-32.0	-12.6	-21.1	

ΠAFEMS CYLINDRICAL SHELL PATCH TEST	TEST No LE2	DATE/ISSUE 21-11-86/2

ORIGIN   NAFEMS report TSBM

ANALYSIS TYPE   Linear elastic

GEOMETRY

Thickness = 0.01m

LOADING

Case 1

Uniform normal edge moment, on DC, of 1.0 kNm/m

Case 2

Uniform outward normal pressure, at mid-surface ABCD, of 0.6 MPa
Tangential outward normal pressure, on edge DC, of 60.0 MPa

BOUNDARY CONDITIONS   Edge AB, all translations and rotations zero
Edge AD and edge BC, symmetry about r θ plane
e.g. z translations and normal rotations all zero

MATERIAL PROPERTIES   Isotropic E = 210 x 10³ MPa, ν = 0.3

ELEMENT TYPES   Quadrilateral shells

MESH

(Corner nodes only given)

OUTPUT Outer (convex) surface tangential (θ - θ) stress at point E	**TARGET** 60.0 MPa for both cases

Table 2				
NAFEMS BENCHMARK TEST No. LE2 **For shells**				
	CASE 1		**CASE 2**	
SYSTEM	**Nodes** **4**	**8**	**Nodes** **4**	**8**	
NASTRAN	-15.3	-10.5	-17.0	-6.6	
ANSYS	-18.0	+32.0	-55.0	+5.0	
GIFTS	*		+2.9		* Membrane locking
MELINA	-98.2	-9.8	-2.3	-5.0	
MELINA	-2.2		+0.8		9-Node element
MELISSA		-7.0		-6.2	
MARC	-	-	-	-	
PAFEC		+5.7		-1.6	
ASAS		-55.0		+4.5	
LUSAS		-		-	
FINEL		-10.0		+2.0	
SUPERTAB	-18.8	-25.8	-40.7	-15.8	
BERSAFE		-2.2		-8.2	Reduced integration
		-58.7		-2.1	Full integration

NAFEMS HEMISPHERE-POINT LOADS	TEST No LE3	DATE/ISSUE 21-11-86 / 2

ORIGIN	NAFEMS report CI

ANALYSIS TYPE	Linear elastic

GEOMETRY

$x^2 + y^2 + z^2 = 100$

r = 10m

2KN

2KN

Thickness = 0.04m

LOADING	Concentrated radial loads of 2KN outwards at A, inwards at C

BOUNDARY CONDITIONS	Point E, zero z displacement
	Edge AE, symmetry about zx plane
	    e.g. zero y displacement, zero normal rotation
	Edge CE, symmetry about yz plane,
	    e.g. zero x displacement, zero normal rotation
	All other displacements on edge AC are free

MATERIAL PROPERTIES	Isotropic. E = 68.25 x 10^3 MPa, ν = 0.3

ELEMENT TYPES	Quadrilateral shells

MESHES

Coarse

Fine

Equally spaced nodes on AC, CE, EA
Point G at $x = y = z = \frac{10}{\sqrt{3}}$
Equally spaced nodes on BG, DG, FG, (all great circles)

OUTPUT	x displacement at point A	TARGET	0.185m

Table 3					
NAFEMS BENCHMARK No. LE3 **Hemisphere with pinch loading**					
	COARSE		**FINE**		
SYSTEM	**Nodes** **4**	**8**	**Nodes** **4**	**8**	
NASTRAN ANSYS GIFTS MELINA	+3.2 * -60.0 -100.0	- -19.4 -63.2	+1.1 * -6.5 -100.0	+3.2 1.1 -40.5	* Warp flag
MELINA	-5.4		-8.1		9 Node Shell
MELISSA MARC PAFEC ASAS LUSAS FINEL SUPERTAB BERSAFE	 +1.6 	-23.8 - -11.3 - -9.8 -95.2 -31.0 -31.9 -14.7 -94.9	 -0.5 	-0.5 - 0.0 - - - -2.2 -4.3 -0.1 -59.2	 2x2 Integration 3x3 Integration 2x2 Integration 3x3 Integration

NAFEMS AXISYMMETRIC HYPERBOLIC SHELL, EDGE LOADING	**TEST No** LE4	**DATE/ISSUE** 21 - 11 - 86 /3

ORIGIN	NAFEMS report TSBM

ANALYSIS TYPE	Linear elastic

GEOMETRY

$$\tan \phi = \tfrac{1}{\sqrt{2}}$$

thickness = 0.01m

LOADING	Distributed outward edge load at $y = 0$ meridional direction, magnitude $F = \cos 2\theta$ MN/radian

BOUNDARY CONDITIONS	Antisymmetric about $r\theta$ plane. at $y = 1.0$ ie. around waist circle through B :- zero displacement in θ direction zero displacement in r direction

MATERIAL PROPERTIES	Isotropic, E = 210 x 10^3 MPa, ν = 0.3

ELEMENT TYPES	Axisymmetric shells

MESH

On half model, 10 elements of equal projection on y axis

OUTPUT	Shear stress at $y = 0$, $\theta = 45°$ outer surface	TARGET	81.65MPa

Table 4		
NAFEMS BENCHMARK LE4 **Hyperbolic shell**		
SYSTEM		
NASTRAN ANSYS GIFTS	+5.2	
MELINA MELISSA MARC	-2.0	Axisymmetric 8-Node solids
PAFEC ASAS LUSAS	+0.1	
FINEL SUPERTAB	+4.0	
BERSAFE	-1.5 +0.7	2-Node elements 3-Node elements

ΠᴀFΣᴍS	Z-SECTION CANTILEVER	**TEST No** LE5	**DATE/ISSUE** 1-7-86/1

ORIGIN	NAFEMS report CI

ANALYSIS TYPE	Linear Elastic

GEOMETRY

Thickness = 0.1m

LOADING	Torque of 1.2MNm applied at end $x = 10$ by two uniformly distributed edge shears, S = 0.6MN at each flange

BOUNDARY CONDITIONS	At edge $x = 0$, all displacements zero

MATERIAL PROPERTIES	Isotropic, E = 210 x 10^3 MPa, v = 0.3

ELEMENT TYPES	Quadrilateral plates

MESH

Uniform mesh of eight elements along length, one element across width of flange or web

OUTPUT	Axial $(x-x)$ stress at mid-surface, point A	TARGET	−108MPa (compression)

SYSTEM	Nodes 4	8	
Table 5			
NAFEMS BENCHMARK TEST No. LE5 Z-Section cantilever			
NASTRAN	-46.0	+2.0	
ANSYS	+2.0	+2.3	
GIFTS	-22.0		
MELINA	-87.3	-30.1	
MELINA	-29.8		9-node shell
MELISSA		+1.9	
MARC	-23.0		
PAFEC		+1.9	
ASAS		+1.1	
LUSAS	-15.0	+1.9	
FINEL		+2.0	
SUPERTAB	+1.8	-3.7	
BERSAFE	-	+0.5	

ΠΛFΣΠS	MEMBRANE WITH HOT-SPOT	TEST No TI	DATE/ISSUE 1-7-86/1

ORIGIN	NAFEMS report C3

ANALYSIS TYPE	Linear thermo-elastic

GEOMETRY

Thickness = 1.0mm

LOADING	Within hot-spot $(0 \le r \le 1.0mm)$ thermal strain $(\alpha\,T) = 1.0 \times 10^{-3}$ Outboard of hot-spot $(r > 1.0mm)$ thermal strain $(\alpha\,T) = 0$

BOUNDARY CONDITIONS	On quarter model, at $y = 0$, symmetric about x, e.g. y displacement zero at $x = 0$, symmetric about y, e.g. x displacement zero

MATERIAL PROPERTIES	Isotropic, $E = 100 \times 10^3$ MPa, $\nu = 0.3$

ELEMENT TYPES	Plane stress quadrilaterals

MESH

All dimensions in mm

OBJECTIVE	Direct stress, y direction, at point D, outside hot-spot	TARGET	50.0MPa

Table 6			
NAFEMS BENCHMARK TEST No. T1			
Membrane with Hot-Spot			

SYSTEM	Nodes		
	4	8	
NASTRAN	–	–	
ANSYS	-7.2	-13.2	
GIFTS	-43.5	1.0	Using dummy rods at interface
MELINA	1.8	-11.8	
MELISSA	–	3.0	
MARC	–	-1.7	
PAFEC	–	7.8	
ASAS	–	–	
LUSAS	1.9	3.1	3x3 Integration
FINEL	–	7.0	
SUPERTAB	10.8	2.7	
BERSAFE	–	2.6	

ΠAFEMS	TWO DIMENSIONAL HEAT TRANSFER WITH CONVECTION	TEST No T4	DATE/ISSUE 1-7-86/1

ORIGIN YARD (preliminary) report 3087 (Test 10)

ANALYSIS TYPE Heat conduction

GEOMETRY

Convection to ambient temperature of 0°C

Zero flux

Insulator

Prescribed temperature of 100°C

1.0m

0.2m

0.6m

Uniform thickness

LOADING Zero internal heat generation

BOUNDARY CONDITIONS Edge AB, temperature = 100°C
Edge DA, zero heat flux
Edges BC, CD, convection to ambient temperature of 0°C

MATERIAL PROPERTIES Conductivity = 52.0W/m°C
Surface convective heat transfer coefficient
(edges BC, CD) = 750.0W/m² °C

ELEMENT TYPES Two-dimensional quadrilateral or triangular heat transfer elements

MESHES

Uniform mesh spacing

OUTPUT Temperature at point E	**TARGET** 18.3°C (Ref 2 p168)

Table 8

NAFEMS BENCHMARK TEST No. T4

Two Dimensional Heat Transfer with Convection

SYSTEM	QUADS		TRIANGLES	
	Nodes		Nodes	
	4	8	3	6
NASTRAN	0.9	–	–	–
ANSYS	5.5	-10.4	–	–
GIFTS	5.7	–	–	–
MELINA	-53.6	-0.0	-24.6	-21.3
MELISSA	–	-2.2	–	16.9
MARC	–	–	–	–
PAFEC	–	-2.2	–	-1.6
ASAS	–	–	–	–
FINEL	–	–	–	–
SUPERTAB	–	–	–	–
BERSAFE	–	-2.2	–	-1.6

TEST 2	

PIN–ENDED DOUBLE CROSS : IN–PLANE VIBRATION

GEOMETRY & MESH

Exact beam:
    4 elements per arm

Iso–parametric beam:
    4 elements per arm

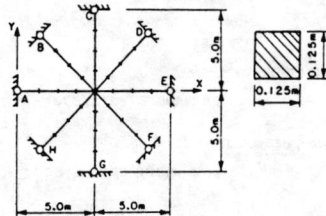

All arms equal length

ATTRIBUTES:
(a) Coupling between flexural and extensional behaviour
(b) Repeated and close eigenvalues
(c) Educational value

BOUNDARY CONDITIONS: x = y = 0 at A, B, C, D, E, F, G and H

MATERIAL PROPERTIES: $E = 200 \times 10^9 \text{ N/m}^2$, $\rho = 8000 \text{ kg/m}^3$

ELEMENT TYPES:
Exact: Exact 2–D beam element
Iso: 3–noded iso–parametric 2–D beam element

FREQUENCIES & MODE SHAPES

$f_r^* = 11.336$ Hz
$f_t = 11.336$ Hz [Exact]
   $= 11.332$ Hz [Iso]

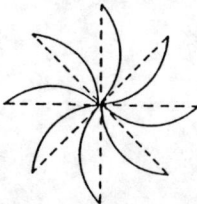

Mode 1

$f_r^* = 17.709$ Hz
$f_t = 17.687$ Hz [Exact]
   $= 17.670$ Hz [Iso]

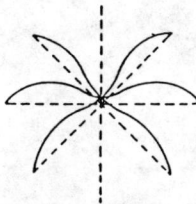

Modes 2 & 3

$f_r^* = 17.709$ Hz
$f_t = 17.715$ Hz [Exact]
   $= 17.698$ Hz [Iso]

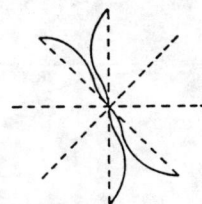

Modes 4, 5, 6, 7 & 8

$f_r^* = 45.345$ Hz
$f_t = 45.477$ Hz [Exact]
   $= 45.667$ Hz [Iso]

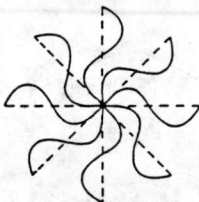

Mode 9

$f_r^* = 57.390$ Hz
$f_t = 57.364$ Hz [Exact]
   $= 57.719$ Hz [Iso]

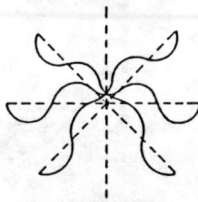

Modes 10 & 11

$f_r^* = 57.390$ Hz
$f_t = 57.683$ Hz [Exact]
   $= 58.052$ Hz [Iso]

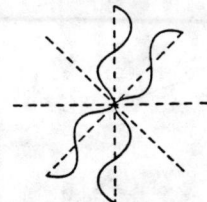

Modes 12, 13, 14, 15 & 16

TEST 4

CANTILEVER WITH OFF–CENTRE POINT MASSES

GEOMETRY & MESH

$M_1 = 10000$ kg (along X, Y, Z)
$M_2 = 1000$ kg (along X, Y, Z)

Exact beam :
    5 elements along cantilever

Iso–parametric beam :
    3 elements along cantilever

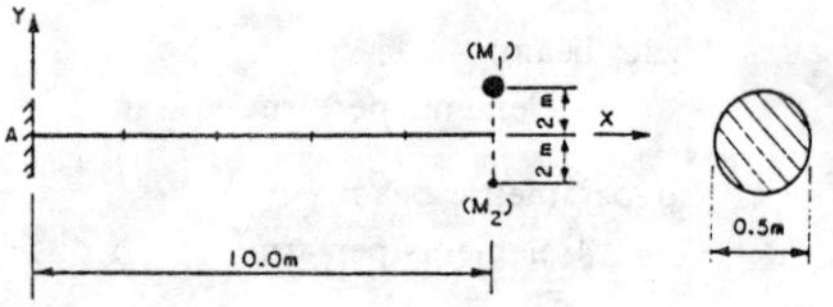

ATTRIBUTES:

(a) Coupling between torsional and flexural behaviour
(b) Inertial axis non–coincident with flexibility axis
(c) Close eigenvalues
(d) Discrete lumped mass, rigid links

BOUNDARY CONDITIONS: $x = y = z = Rx = Ry = Rz = 0$ at A

MATERIAL PROPERTIES: $E = 200 \times 10^9$ N/m^2 , $\nu = 0.3$, $\rho = 8000$ kg/m^3

ELEMENT TYPES:

Exact: Exact 3–D beam element
Iso: 3–noded iso–parametric 3–D beam element

FREQUENCIES & MODE SHAPES

$f_r = 1.723$ Hz
$f_t = 1.723$ Hz [Exact]
   $= 1.722$ Hz [Iso]

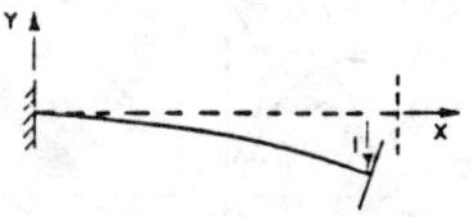

Mode 1

$f_r = 1.727$ Hz
$f_t = 1.727$ Hz [Exact]
   $= 1.726$ Hz [Iso]

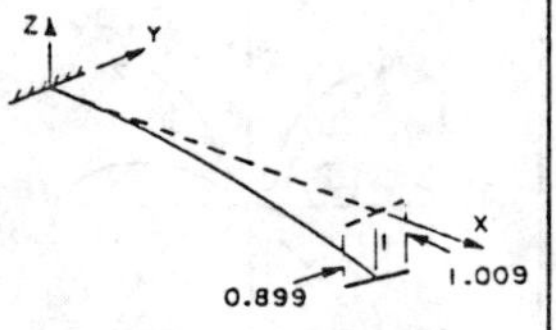

Mode 2

$f_r = 7.413$ Hz
$f_t = 7.413$ Hz [Exact]
   $= 7.412$ Hz [Iso]

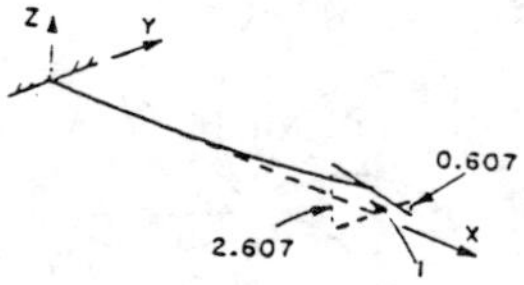

Mode 3

$f_r = 9.972$ Hz
$f_t = 9.972$ Hz [Exact]
   $= 9.953$ Hz [Iso]

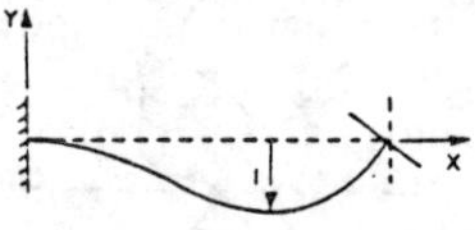

Mode 4

$f_r = 18.155$ Hz
$f_t = 18.160$ Hz [Exact]
   $= 18.166$ Hz [Iso]

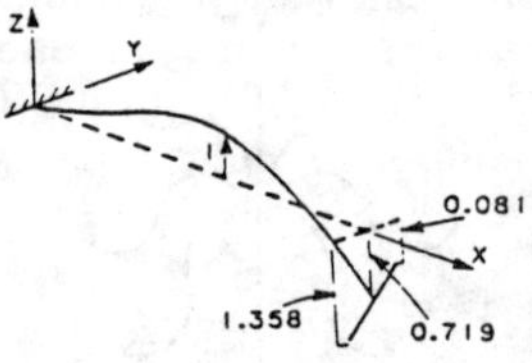

Mode 5

$f_r = 26.957$ Hz
$f_t = 26.972$ Hz [Exact]
   $= 27.034$ Hz [Iso]

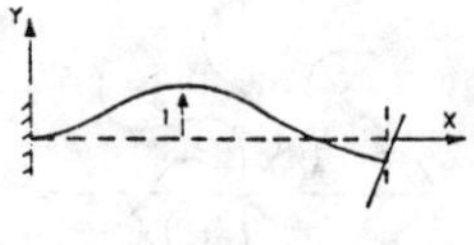

Mode 6

TEST 5

DEEP SIMPLY–SUPPORTED BEAM

GEOMETRY & MESH

Exact beam :   5 elements

Iso–parametric beam :   5 elements

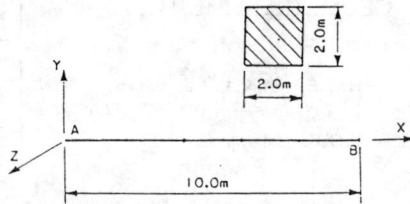

2.0m

2.0m

10.0m

ATTRIBUTES: (a) Repeated eigenvalues

(b) Shear deformation and rotary inertia (Timoshenko Beam)

(c) Possibility of missing extensional modes when using iteration solution methods

BOUNDARY CONDITIONS: $x = y = z = Rx = 0$ at A, $y = z = 0$ at B

MATERIAL PROPERTIES: $E = 200 \times 10^9$ N/m^2 , $\nu = 0.3$, $\rho = 8000$ kg/m^3

ELEMENT TYPES: Exact: Exact 3–D beam element

Iso: 3–noded iso–parametric 3–D beam element

FREQUENCIES & MODE SHAPES

$f_r^* = 42.649$ Hz
$f_t = 42.568$ Hz [Exact]
$\quad = 42.657$ Hz [Iso]

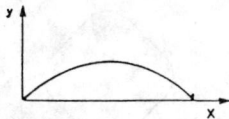

FLEXURAL
Modes 1 & 2

$f_r^* = 77.542$ Hz
$f_t = 77.841$ Hz [Exact]
$\quad = 77.522$ Hz [Iso]

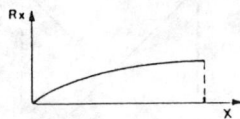

TORSIONAL
Mode 3

$f_r^* = 125.00$ Hz
$f_t = 125.51$ Hz [Exact]
$\quad = 125.00$ Hz [Iso]

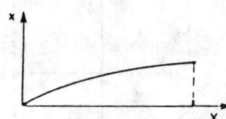

EXTENSIONAL
Mode 4

$f_r^* = 148.31$ Hz
$f_t = 145.46$ Hz [Exact]
$\quad = 148.71$ Hz [Iso]

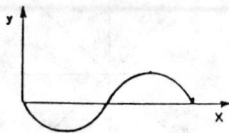

FLEXURAL
Modes 5 & 6

$f_r^* = 233.10$ Hz
$f_t = 241.24$ Hz [Exact]
$\quad = 232.69$ Hz [Iso]

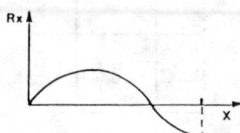

TORSIONAL
Mode 7

$f_r^* = 284.55$ Hz
$f_t = 267.01$ Hz [Exact]
$\quad = 287.81$ Hz [Iso]

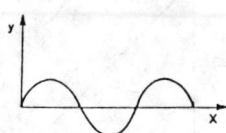

FLEXURAL
Modes 8 & 9

TEST 12

FREE THIN SQUARE PLATE

GEOMETRY & MESH

H.O.E.   4 x 4 (as shown)

L.O.E.   8 x 8

t = 0.05 m

10.0m × 10.0m

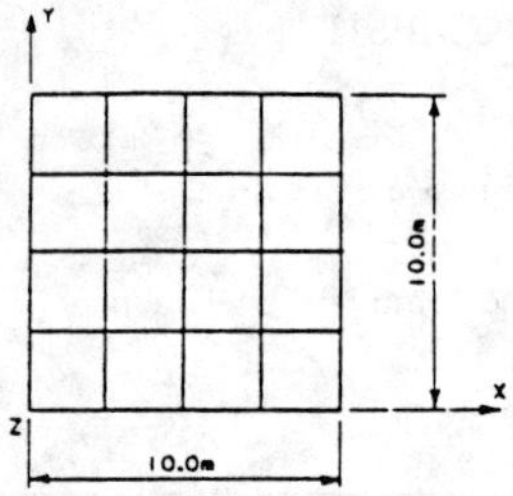

ATTRIBUTES:
(a) Rigid body modes (3 modes)
(b) Repeated eigenvalues
(c) Kinematically incomplete suppressions

BOUNDARY CONDITIONS: x = y = Rz = 0 all nodes

MATERIAL PROPERTIES: $E = 200 \times 10^9$ N/m^2 , $\nu = 0.3$, $\rho = 8000$ kg/m^3

ELEMENT TYPES:
H.O.E.:   8-noded semi-loof thin shell element
L.O.E.:   4-noded iso-parametric shell element

FREQUENCIES & MODE SHAPES

$f_r^* = 1.622$ Hz
$f_t = 1.532$ Hz [HOE]
  $= 1.632$ Hz [LOE] 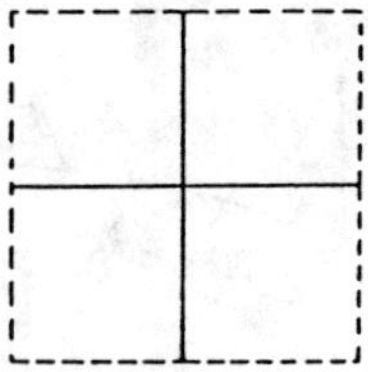
Mode 4

$f_r^* = 2.360$ Hz
$f_t = 2.356$ Hz [HOE]
  $= 2.402$ Hz [LOE] 
Mode 5

$f_r^* = 2.922$ Hz
$f_t = 2.861$ Hz [HOE]
  $= 3.006$ Hz [LOE] 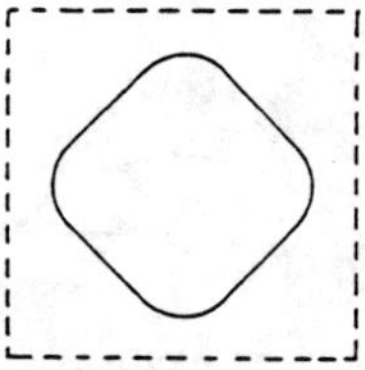
Mode 6

$f_r^* = 4.233$ Hz
$f_t = 4.122$ Hz [HOE]
  $= 4.251$ Hz [LOE] 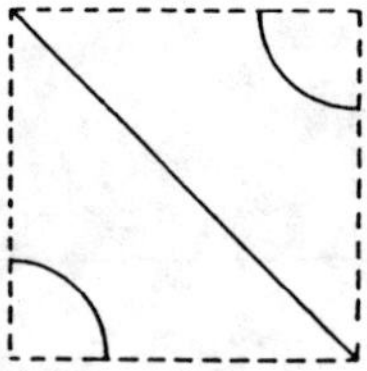
Modes 7 & 8

$f_r^* = 7.416$ Hz
$f_t = 7.363$ Hz [HOE]
  $= 7.859$ Hz [LOE]
Mode 9

$f_r^* = $ (Not available)
$f_t = 7.392$ Hz [HOE]
  $= 8.027$ Hz [LOE] 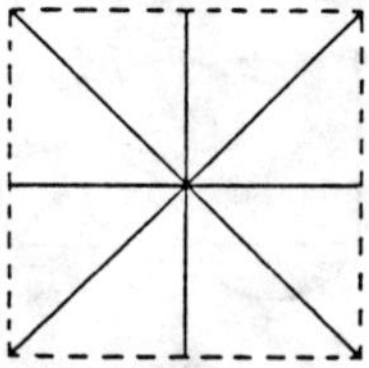
Mode 10

TEST 15

CLAMPED THIN RHOMBIC PLATE

GEOMETRY & MESH

H.O.E. : 6 x 6 (as shown)

L.O.E. : 12 x 12

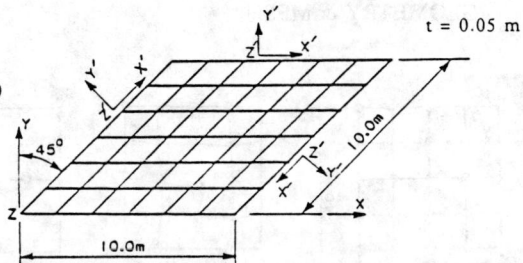

t = 0.05 m

10.0m

ATTRIBUTES: (a) Distorted elements

BOUNDARY CONDITIONS: x = y = Rz = 0 at all nodes,
z′ = Rx′ = Ry′ = 0 along all 4 edges

MATERIAL PROPERTIES: $E = 200 \times 10^9 \text{ N/m}^2$, $\nu = 0.3$, $\rho = 8000 \text{ kg/m}^3$

ELEMENT TYPES: H.O.E.:   8-noded semi-loof thin shell element

L.O.E.:   4-noded iso-parametric shell element

FREQUENCIES & MODE SHAPES

$f_r^{\bullet}$ =   7.938 Hz
f_t =   7.873 Hz [HOE]
  =   8.142 Hz [LOE]

$f_r^{\bullet}$ =   12.835 Hz
f_t =   12.480 Hz [HOE]
  =   13.891 Hz [LOE]

$f_r^{\bullet}$ =   17.941 Hz
f_t =   17.312 Hz [HOE]
  =   20.036 Hz [LOE]

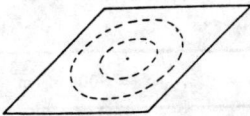

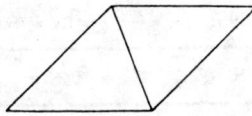

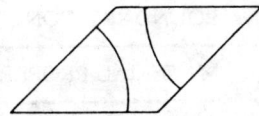

Mode 1	Mode 2	Mode 3

$f_r^{\bullet}$ =   19.133 Hz
f_t =   18.738 Hz [HOE]
  =   20.165 Hz [LOE]

$f_r^{\bullet}$ =   24.009 Hz
f_t =   22.704 Hz [HOE]
  =   27.950 Hz [LOE]

$f_r^{\bullet}$ =   27.922 Hz
f_t =   25.883 Hz [HOE]
  =   32.046 Hz [LOE]

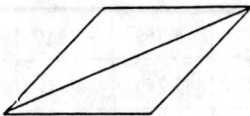

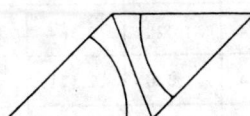

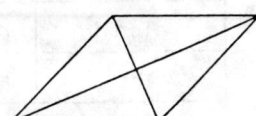

Mode 4	Mode 5	Mode 6

TEST 16

CANTILEVERED THIN SQUARE PLATE

GEOMETRY & MESH

t = 0.05 m

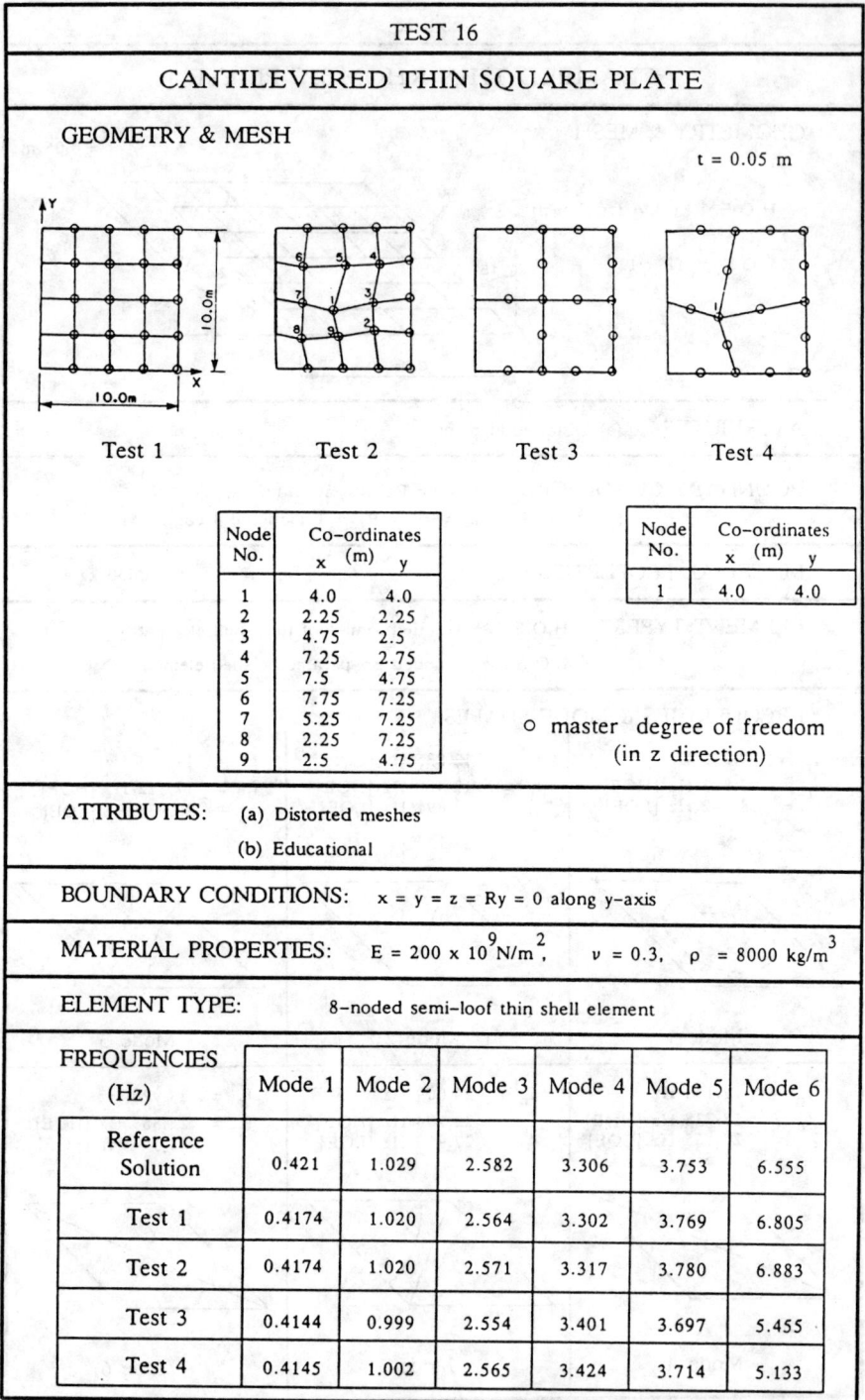

Test 1   Test 2   Test 3   Test 4

Node No.	Co-ordinates x (m) y	
1	4.0	4.0
2	2.25	2.25
3	4.75	2.5
4	7.25	2.75
5	7.5	4.75
6	7.75	7.25
7	5.25	7.25
8	2.25	7.25
9	2.5	4.75

Node No.	Co-ordinates x (m) y	
1	4.0	4.0

o master degree of freedom (in z direction)

ATTRIBUTES: (a) Distorted meshes

(b) Educational

BOUNDARY CONDITIONS: x = y = z = Ry = 0 along y-axis

MATERIAL PROPERTIES: E = 200 x 10^9 N/m^2, ν = 0.3, ρ = 8000 kg/m^3

ELEMENT TYPE: 8-noded semi-loof thin shell element

FREQUENCIES (Hz)	Mode 1	Mode 2	Mode 3	Mode 4	Mode 5	Mode 6
Reference Solution	0.421	1.029	2.582	3.306	3.753	6.555
Test 1	0.4174	1.020	2.564	3.302	3.769	6.805
Test 2	0.4174	1.020	2.571	3.317	3.780	6.883
Test 3	0.4144	0.999	2.554	3.401	3.697	5.455
Test 4	0.4145	1.002	2.565	3.424	3.714	5.133

TEST 22

CLAMPED THICK RHOMBIC PLATE

GEOMETRY & MESH

H.O.E. :   6 x 6 (as shown)

L.O.E. :   10 x 10

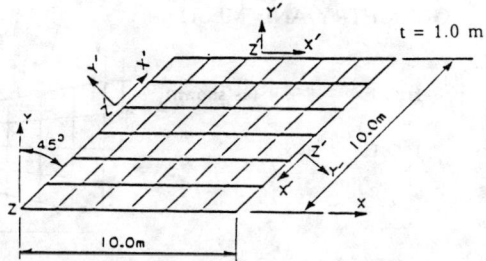

t = 1.0 m

45°

10.0m

10.0m

ATTRIBUTES: (a) Distorted elements

BOUNDARY CONDITIONS: $x = y = Rz = 0$ at all nodes,

$z' = Rx' = Ry' = 0$ along all 4 edges

MATERIAL PROPERTIES: $E = 200 \times 10^9$ N/m^2 , $\nu = 0.3$, $\rho = 8000$ kg/m^3

ELEMENT TYPES: H.O.E.: 8-noded iso-parametric thick shell element

L.O.E.: 4-noded iso-parametric shell element

FREQUENCIES & MODE SHAPES

$f_r^* = $ 133.95 Hz
$f_t = $ 133.86 Hz [HOE]
$= $ 137.80 Hz [LOE]

$f_r^* = $ 201.41 Hz
$f_t = $ 203.34 Hz [HOE]
$= $ 218.48 Hz [LOE]

$f_r^* = $ 265.81 Hz
$f_t = $ 271.38 Hz [HOE]
$= $ 295.42 Hz [LOE]

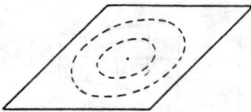

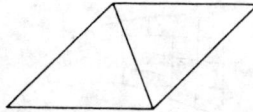

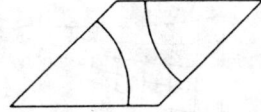

Mode 1

Mode 2

Mode 3

$f_r^* = $ 282.74 Hz
$f_t = $ 283.68 Hz [HOE]
$= $ 296.83 Hz [LOE]

$f_r^* = $ 334.45 Hz
$f_t = $ 346.41 Hz [HOE]
$= $ 383.56 Hz [LOE]

$f_r^* = $ (Not available)
$f_t = $ 386.62 Hz [HOE]
$= $ 426.59 Hz [LOE]

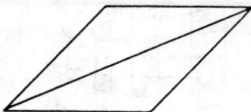

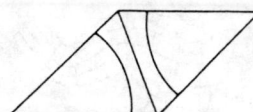

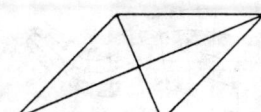

Mode 4

Mode 5

Mode 6

TEST 32

CANTILEVERED TAPERED MEMBRANE

GEOMETRY AND MESH

H.O.E. : 8 x 4 (as shown)

L.O.E. : 16 x 8

ATTRIBUTES:
(a) Shear behaviour
(b) Irregular mesh
(c) Symmetry

BOUNDARY CONDITIONS: z = 0 at all nodes,   x = y = 0 along y-axis

MATERIAL PROPERTIES: $E = 200 \times 10^9$ N/m^2 , $\nu = 0.3$, $\rho = 8000$ kg/m^3

ELEMENT TYPES:
H.O.E.: 8-noded iso-parametric membrane element

L.O.E.: 4-noded iso-parametric membrane element

FREQUENCIES & MODE SHAPES

f_r = 44.623 Hz	f_r = 130.03 Hz	f_r = 162.70 Hz
f_t = 44.636 Hz [HOE]	f_t = 130.14 Hz [HOE]	f_t = 162.72 Hz [HOE]
= 44.905 Hz [LOE]	= 132.12 Hz [LOE]	= 162.83 Hz [LOE]

Mode 1	Mode 2	Mode 3

f_r = 246.05 Hz	f_r = 379.90 Hz	f_r = 391.44 Hz
f_t = 246.63 Hz [HOE]	f_t = 382.02 Hz [HOE]	f_t = 391.55 Hz [HOE]
= 252.99 Hz [LOE]	= 393.31 Hz [LOE]	= 396.26 Hz [LOE]

Mode 4	Mode 5	Mode 6

TEST 41

FREE CYLINDER : AXI-SYMMETRIC VIBRATION

GEOMETRY & MESH

H.O.E. :   8 x 1   (as shown)

L.O.E. :   16 x 3

ATTRIBUTES:
(a) Rigid body mode (one mode)
(b) Close eigenvalues
(c) Coupling between axial, radial and circumferential behaviour

BOUNDARY CONDITIONS: Unsupported

MATERIAL PROPERTIES: $E = 200 \times 10^9 \ N/m^2$, $\nu = 0.3$, $\rho = 8000 \ kg/m^3$

ELEMENT TYPES:
H.O.E.:   8-noded iso-parametric axi-symmetric element
L.O.E.:   4-noded iso-parametric axi-symmetric element

FREQUENCIES & MODE SHAPES

$f_r^* = $ 243.53 Hz
$f_t = $ 243.50 Hz [HOE]
$ = $ 244.01 Hz [LOE]

$f_r^* = $ 377.41 Hz
$f_t = $ 377.46 Hz [HOE]
$ = $ 379.41 Hz [LOE]

$f_r = $ 394.11 Hz
$f_t = $ 394.28 Hz [HOE]
$ = $ 395.41 Hz [LOE]

Mode 2

Mode 3

Mode 4

$f_r = $ 397.72 Hz
$f_t = $ 397.94 Hz [HOE]
$ = $ 401.35 Hz [LOE]

$f_r = $ 405.28 Hz
$f_t = $ 406.41 Hz [HOE]
$ = $ 421.87 Hz [LOE]

Mode 5

Mode 6

TEST 42

THICK HOLLOW SPHERE :   UNIFORM RADIAL VIBRATION

GEOMETRY & MESH

H.O.E. :   5 x 1 (as shown),
α = 10°

L.O.E. :   10 x 1,
α = 5°

1.8m   4.2m

ATTRIBUTES: (a) Curved boundary   (skew coordinate system)
(b) Constraint equations

BOUNDARY CONDITIONS: z' displacement = 0 at all nodes,
nodes at same R' are constrained to have same
r' displacement

MATERIAL PROPERTIES: $E = 200 \times 10^9$ N/m^2 , $\nu = 0.3$, $\rho = 8000$ kg/m^3

ELEMENT TYPES: H.O.E.:   8-noded iso-parametric axi-symmetric element
L.O.E.:   4-noded iso-parametric axi-symmetric element

FREQUENCIES & MODE SHAPES

f_r^* =   369.91 Hz
f_t =   370.01 Hz [HOE]
  =   370.64 Hz [LOE]
	f_r^* = 838.03 Hz
	f_t = 838.08 Hz [HOE]
	  = 841.20 Hz [LOE]
		f_r^* = 1451.2 Hz
		f_t = 1453.0 Hz [HOE]
		  = 1473.1 Hz [LOE]

Mode 1	Mode 2	Mode 3

f_r^* =   2117.0 Hz
f_t =   2131.7 Hz [HOE]
  =   2192.2 Hz [LOE]
	f_r^* = 2795.8 Hz
	f_t = 2852.8 Hz [HOE]
	  = 2975.7 Hz [LOE]	

Mode 4	Mode 5	

TEST 52

SIMPLY–SUPPORTED 'SOLID' SQUARE PLATE

GEOMETRY & MESH

H.O.E. :

    4 x 4 x 1 (as shown)

L.O.E. :

    8 x 8 x 3

ATTRIBUTES: (a) Well established

                   (b) Rigid body modes (3 modes)

                   (c) Kinematically incomplete suppressions

BOUNDARY CONDITIONS: z = 0 along the 4 edges on the plane Z = –0.5m

MATERIAL PROPERTIES: $E = 200 \times 10^9$ N/m^2 , $\nu = 0.3$, $\rho = 8000$ kg/m^3

ELEMENT TYPES: H.O.E.:   20–noded iso–parametric brick element

                           L.O.E.:   8–noded iso–parametric brick element

FREQUENCIES & MODE SHAPES

$f_r^* = 45.897$ Hz $f_t = 44.762$ Hz [HOE] $= 51.654$ Hz [LOE]	$f_r^* = 109.44$ Hz $f_t = 110.52$ Hz [HOE] $= 132.73$ Hz [LOE]	$f_r^* = 167.89$ Hz $f_t = \begin{array}{l}169.08 \text{ Hz [HOE]}\\194.37 \text{ Hz [LOE]}\end{array}$
OUT OF PLANE Mode 4	OUT OF PLANE Modes 5 & 6	OUT OF PLANE Mode 7
$f_r = 193.59$ Hz $f_t = 193.93$ Hz [HOE] $= 197.18$ Hz [LOE]	$f_r = 206.19$ Hz $f_t = 206.64$ Hz [HOE] $= 210.55$ Hz [LOE]	
IN PLANE Mode 8	IN PLANE Modes 9 & 10	

TEST 73
CANTILEVERED THIN SQUARE PLATE

GEOMETRY & MESH

t = 0.05m

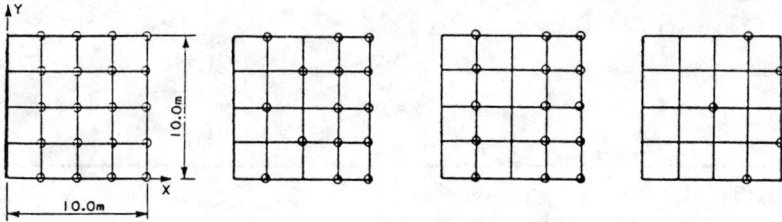

Test 1   Test 2   Test 3   Test 4

○ Master degree of freedom (in Z direction)

ATTRIBUTES: (a) Effect of master degree of freedom selection on frequencies

BOUNDARY CONDITIONS: x = y = z = Ry = 0 along y−axis

MATERIAL PROPERTIES: $E = 200 \times 10^9 \, N/m^2$, $\nu = 0.3$, $\rho = 8000 \, kg/m^3$

ELEMENT TYPE: 8−noded semi−loof thin shell element

FREQUENCIES

	Frequencies (Hz)					
	Mode 1	Mode 2	Mode 3	Mode 4	Mode 5	Mode 6
Reference Solution	0.421	1.029	2.582	3.306	3.753	6.555
Test 1	0.4174	1.020	2.564	3.302	3.769	6.805
Test 2	0.4174	1.020	2.597	3.345	3.888	7.517
Test 3	0.4175	1.021	2.677	3.365	4.035	7.495
Test 4	0.4184	1.032	2.850	3.571	5.466	–

A Knowledge Based Approach For Boundary Element Mesh Design

Jerome J. Connor

Dept. of Civil Engineering, Massachusetts Institute of Technology, Cambridge, MA 02139 USA

SUMMARY

Key tasks in the problem solving process for engineering analysis are concerned with the creation of an idealized model, the selection of an appropriate discrete element analysis method, and the specification of the spatial discretization. The two most frequently applied discrete element methods are the finite element and boundary element techniques. Both methods have evolved into powerful computational tools: boundary element methods are most suitable when the quantities of interest are on the boundary; finite element methods are applicable for all situations but may not be the optimal strategy for certain situations. Much experience has been generated through the application of these methods, but this knowledge has not been formalized. Rather, it resides with the individual practitioners.

The initial mesh design is the key step in the problem solving process. Considerable research has been directed at establishing analytical procedures for modifying a mesh after the solution has been obtained, but it is more desirable to avoid having to solve the problem more than once, especially for large problems. With the current trend towards more automation, the experienced user tends to be bypassed, and poorly designed meshes are now more common. This situation has generated interest in the potential of using a knowledge based system to design the initial mesh.

The approach followed here combines a knowledge based system with a boundary element analysis program. The scope is restricted to the BEM since the implementation is simpler, i.e., one has only to discretize the boundary. Also, the problem domain is limited to two dimensions. An object orientated framework is employed to establish a set of generic objects and rules which allow one to identify characteristic geometric features of the boundary, such as corners, notches, cut-outs, etc. from an examination of the geometric boundary data. These features have local solution patterns associated with them, and this deep knowledge about behaviour, together with heuristic knowledge on mesh graduation for different accuracy levels generated through numerical experimentation, is utilized to design a boundary discretization. The solution is reviewed using a different set of rules, and revisions are suggested. The emphasis here is on the representation and processing of the knowledge to simulate an experienced analyst, and the feasibility of using an object oriented approach for mesh design.

1 Introduction

Key tasks in the problem solving process for engineering analysis are concerned with the following issues: i) creation of an idealized model, ii) selection of an appropriate analysis method and iii) interpretation of the results generated with the idealized model. Creation of an idealized model for a physical system requires a comprehensive understanding of the behaviour of the system. This background is established through formal training and problem solving experience. One needs to define the "idealized" system geometry, specify the material properties, identify the loading and corresponding boundary conditions, and decide on an appropriate behavioural model. The last task issue is the most dependent on "accumulated " experience. Questions such as: i) is the behaviour linear? ii) is the response quasi-static? iii) is the spatial distribution of the response localized in certain regions? iv) can certain behavioural modes be neglected, such as bending versus stretching? and v) what are reasonable values for the material properties? need to be answered in this phase.

Once the model and problem data are defined, the next step concerns the selection of an analysis procedure. For simple cases, one may be able to apply an analytical method. Usually, one has to resort to a computer based technique. The three most common computational methods are: i) finite difference, ii) finite element and iii) boundary element. These strategies treat the spatial discretization aspect. In addition, one needs to specify the temporal discretization strategy. The critical decisions here are: i) time vs. frequency domain and ii) choice of numerical integration scheme for a time domain approach, i.e., implicit vs. explicit; order of the integration operator. If the behaviour is nonlinear, one also has to decide upon a strategy for treating the nonlinearity. One option is to formulate the problem as pseudo-linear, and iterate on the nonlinear terms using as the "basic" system either the initial linear system or an updated "tangent" system.

Selection of the spatial discretization method is influenced by the following issues:

- Geometrical considerations
  How irregular is the geometry?

- Desired output
  Are the quantities of interest on the boundary?
  Is the distribution over the entire domain desired?

- Linear versus nonlinear behaviour
  If the problem is nonlinear, what is the extent of the nonlinear region?

- Static vs. dynamic behaviour
  Are the time derivative terms important?

Finite difference methods are not convenient when the geometry is irregular. Boundary element methods are most suitable in these cases as the quantities of interest are on the boundary [1], such as the tangential stresses and, in addition, the behaviour is linear and static. Finite element methods are applicable for all situations, but may not be the optimal strategy for certain situations, especially when the ratio of surface area to volume is small.

Once the "method" is selected, the problem reduces to defining an appropriate spatial discretization. Here, one draws upon past experience with the method and knowledge about the behavioural modes for simple models which allows one to generate insight on the behaviour of the actual system. Accumulated experience provides the correlation between element size and "accuracy". Simple analytical models can be applied to establish behavioural patterns associated with geometric and load discontinuities. This information is needed in order to modify the spatial discretization in the neighbourhood of local disturbances.

Mesh design is a key step in the problem solving process. A poorly designed mesh can result in erroneous results. Considering the current trend toward complete automation of the process, there is a need for improved capabilities for spatial and temporal discretization. These tasks cannot be handled by geometric modellers, i.e., technicians. They require skilled engineering professionals.

Recent advances in Artificial Intelligence have resulted in methods for building computer programs that are capable of solving problems at a level equivalent to that of a skilled practitioner. The expertise resides in a knowledge base which can be customized for a particular problem. They are generally referred to as knowledge based assistants. The special attractiveness of knowledge based methods is the flexibility they offer in representing knowledge. Since mesh design involves a combination of heuristic information based on observations and analytic information generated with scientific principles, flexibility in knowledge representation is a critical factor.

In what follows, the development and implementation of a knowledge based system approach to the design of a boundary element mesh is described. The first section is concerned with identifying the cognitive process and criteria that an "expert" boundary element practitioner follows in selecting a mesh. Representation of this knowledge using an object orientated programming environment is discussed next. The implementation is restricted to two-dimensional geometries since our primary objective was to evaluate the feasibility of the approach. Extension of these ideas to three-dimensional surfaces is presently being examined.

2 Mesh Design Strategy

The boundary for the two-dimensional case is defined by a set of line segments connected at their ends to form one or more closed loops, as illustrated in Figure 1. The line segments are continuous "smooth" curves between the boundary points and are usually represented by polynomials such as quadratic, cubic, etc. Slope and curvature changes are restricted to occur only at boundary points. Geometric discontinuities, such as slope changes, may introduce a singular type behaviour which requires a finer mesh in the neighbourhood of the discontinuity. Therefore, one aspect of mesh design is to develop a strategy for identifying those line patterns that result in high stress gradients.

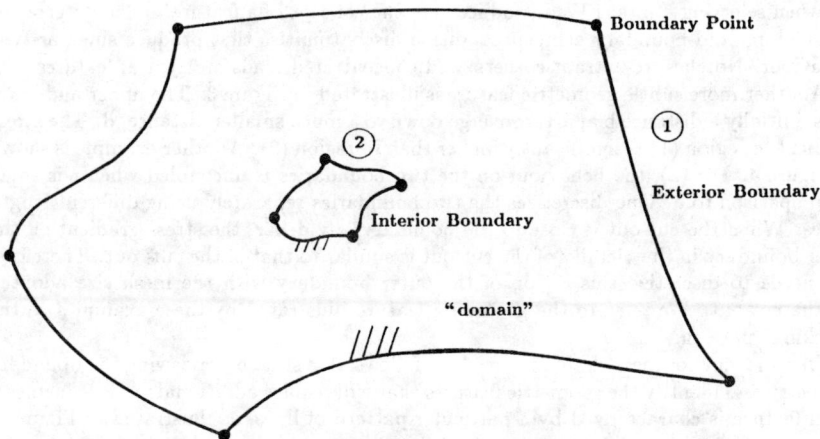

Figure 1: A Typical 2D Boundary

In addition to geometrical considerations, one also has to consider the loading and boundary conditions specified for the problem. Figure 2 illustrates these other aspects. In region (1), there is a V notch with a circular fillet. The behaviour here depends on geometric measures, primarily the magnitude of the fillet radius and the angle of opening of the notch. In region (2) the boundary conditions are discontinuous: on "a" the displacements are prescribed; on "b" the surface tractions are specified. One needs to treat the boundary conditions at the corner point in a special way, either by using discontinuous elements or introducing an additional node at the corner. The shearing stress and transverse normal stress distributions require more mesh refinement than one would normally assign to this area. A similar situation exists for region (3). Loading discontinuities can produce high stress gradients. If one is interested mainly in the global behaviour, loading discontinuities can be ignored. Actually, one tends to "lump" loads in this case. However, if the stress distribution in the loaded region is the goal, then the loading discontinuity must be accounted for by refining the mesh locally.

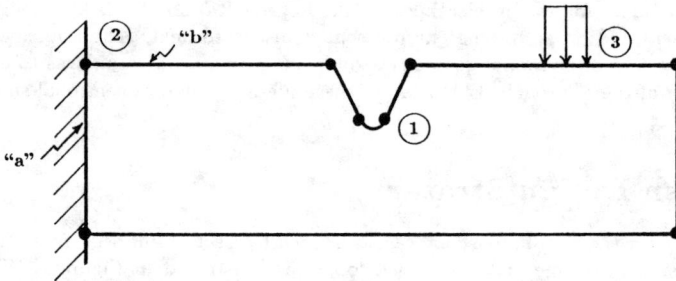

Figure 2: Loading and Boundary Condition Features

The examples described above illustrate different perspectives that the analyst considers when selecting a mesh. We introduce here the concept of a feature to denote geometric, loading, and boundary element condition discontinuities that produce singular-type behaviour. Notches, re-entrant corners, and concentrated loads are typical features.

Another more subtle geometric feature is illustrated in Figure 3. The upper and lower lines, initially a distance b apart, converge down to a much smaller distance, d. The stress gradient in region (1) is significantly higher than in region (2). Another example is shown in Figure 4. For (a), the behaviour on the two boundaries is uncoupled when r is small in comparison to b. One discretizes the two boundaries separately using different length scales. When the cut-out is close to the boundary, say d < r, the stress gradient on the outer boundary in the vicinity of the cut-out is similar to that of the cut-out. Therefore, one needs to discretize this region of the outer boundary with the mesh size adopted for the cut-out. We refer to the geometric feature illustrated by these examples as the "necking" pattern.

Our strategy for mesh design proceeds as follows. We start by reviewing the boundary geometry and identify the geometric features that will require additional mesh refinement. Each feature is characterized by a particular pattern of lines, as illustrated in Figure 5. Also, each feature has a characteristic length measure which defines the zone of influence of the stress perturbation. Outside of the influence zones, one selects a mesh size based on

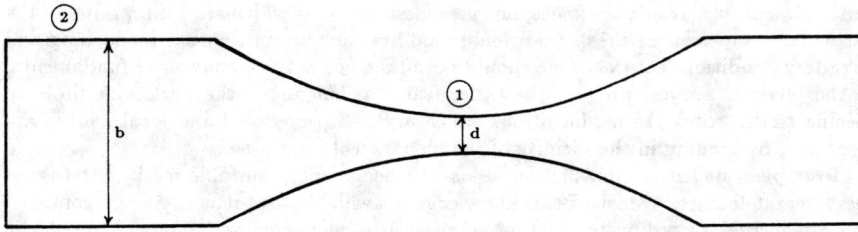

Figure 3: The Necking Pattern

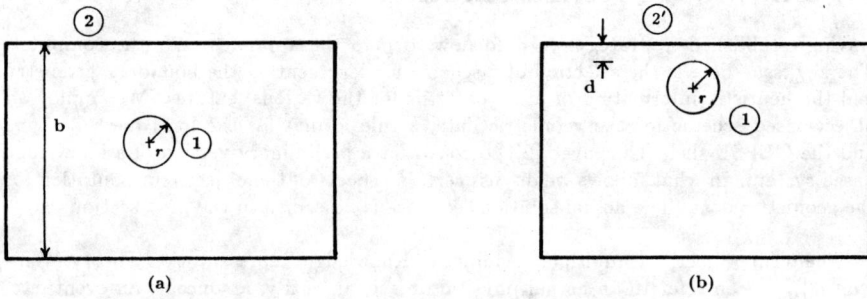

Figure 4: Boundary - Boundary Interaction

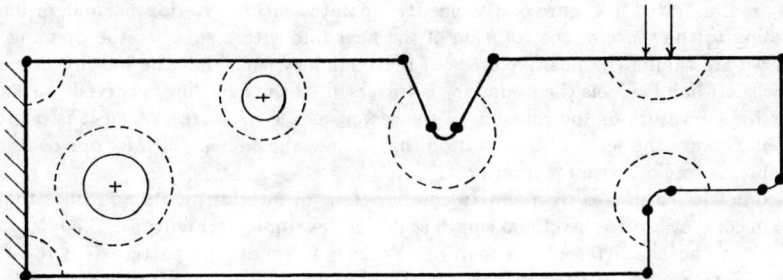

Figure 5: Illustration of Local Zones

global dimensions. Inside the zones, one uses mesh patterns which are appropriate for the particular geometric features. Additional modifications are introduced for loading and boundary condition features. One should note that geometric features are fundamental to the physical system whereas the other features depend on the particular problem specifications. After the modifications are introduced, the global and local meshes are reconciled by grading in the vicinity of the global-local interfaces.

To implement this approach, one needs knowledge about suitable mesh patterns for the different features. Most of this knowledge is available, but it needs to be compiled and categorized according to the type of analysis model, such as BEM with quadratic expansion, and the degree of accuracy obtained. Quadratic elements are generally regarded as the most efficient, and therefore are adopted here as the standard element. Correlating accuracy with mesh size is an ambitious undertaking, and will require a systematic long term study. Since our focus is more on how to implement the strategy we use some "reasonable" mesh patterns in the initial prototype version.

3 Knowledge Representation

Having established a strategy we need now to map the approach onto the computer. The key issue here is the selection of a scheme for representing the boundary geometry and the heuristic information of mesh patterns for the various features. We employ an object-based scheme for the geometric data, a rule format for the local mesh designs, and the GEPSE shell (Chehayeb [2]) to construct a preliminary version of a knowledge based system. In what follows we discuss certain aspects of the object representation for the geometric data. The actual solution sequence is described in the next section.

The term, object, has multiple meanings in Knowledge Representation Theory (Winston [3]). It can refer to an actual physical object, an entity, a concept, an event, etc. We use it to represent a point, a line, and a set of lines, comprising a geometric feature. An object is defined by specifying its attributes (properties) and their corresponding values. Knowledge base system building tools such as GEPSE provide a highly interactive environment for developing the system. For example, one can create objects and add attributes dynamically during the execution of the program. Conventional programming languages such as FORTRAN are more rigid and consequently, much more difficult to work with for these operations.

By definition, a boundary point corresponds to a transition between line segments. The angle change at the point is a key bit of information that provides an indication of a potential stress concentration (high stress gradient) in the region of the point. Various cases are illustrated in Figure 6. We classify a point as either exterior, normal, or interior according to the sense of the rotation of the next line with respect to the previous line. Here, we are taking the positive sense of the boundary such that the interior domain is on one's left hand side as the boundary is traversed. If the next line is curved, we can use either the curvature or the rotation of the secant as the measure. When it is a circular segment, we use the sense of the rotation that defines the circle, which for our convention corresponds to clockwise for interior.

Geometric features correspond to combinations of interior points and lines that are small in comparison to the global length scale. For example, in Figure 7a, the fillet radius determines the character of the feature. When r is small, the pattern is a re-entrant corner; when r is large, the behaviour is global and the segment is meshed according to a global scale. Figure 7b illustrates V notches and keys. The magnitude and variation of the stresses at the interior points depend on the angle α. When d is large, the stress perturbation is confined to a segment of the line and the feature is a re-entrant corner. In this case, one can mesh the E end of the line with a global size, and grade down to the scale required at the I end.

The key aspect of a line is its relative length with respect to a global line measure. We adopt the largest length as a standard, and consider a line to be of "local" size when its length is less than 20 per cent of the global standard. The "local" attribute of line and "interior" attribute of boundary point are necessary to have features such as re-entrant corners and notches.

We have been discussing various properties of points and lines that play a role in mesh design. These properties are represented as attributes for the various objects. We use generic point and line objects to form a list of the individual objects and other global information such as the largest length. Typical object representations are:

```
OBJECT:        POINT
   NAME             (a dummy variable)
   MEMORY LIST      (a list of the names of the individual points)
OBJECT:        LINE
   NAME             (a dummy variable)
   LARGEST LENGTH
   MEMORY LIST      (a list of the names of the individual lines)
```

The individual objects with the attributes identified to this point are:

```
OBJECT:        POINT_N
   NAME:        POINT_N
   NUMBER:          N
     X_COOR
     Y_COOR
     PREVIOUS_LINE
     NEXT_LINE
     ANGLE_CHANGE
     TYPE             (interior, normal, exterior)
OBJECT:        LINE_M
   NAME:        LINE_M
   NUMBER:          M
   START_POINT
   END_POINT
   LENGTH
   ORIENTATION   (angle)
   TYPE          (local vs. global)
```

If a circular line, we add

```
   CENTRE_POINT
   DIRECTION      (clockwise or counter-clockwise)
   RADIUS
```

Our objective is to define the number of elements and mesh sizes for each line. We achieve this goal by executing a number of loopings over the points and lines, and applying heuristic rules for mesh size. In the process, we create additional attributes and, depending on the problem, may also create new point and line objects. We expand on these aspects in the following section.

4 Mesh Specification Procedure

We start by looping over the list of lines contained in the MEMORY_LIST slot of the LINE object. We compute the length, orientation, and update if necessary the LARGEST_LENGTH attribute of LINE. At the end of this pass, we shift to a looping over the list of points contained in the MEMORY_LIST slot of the POINT object.

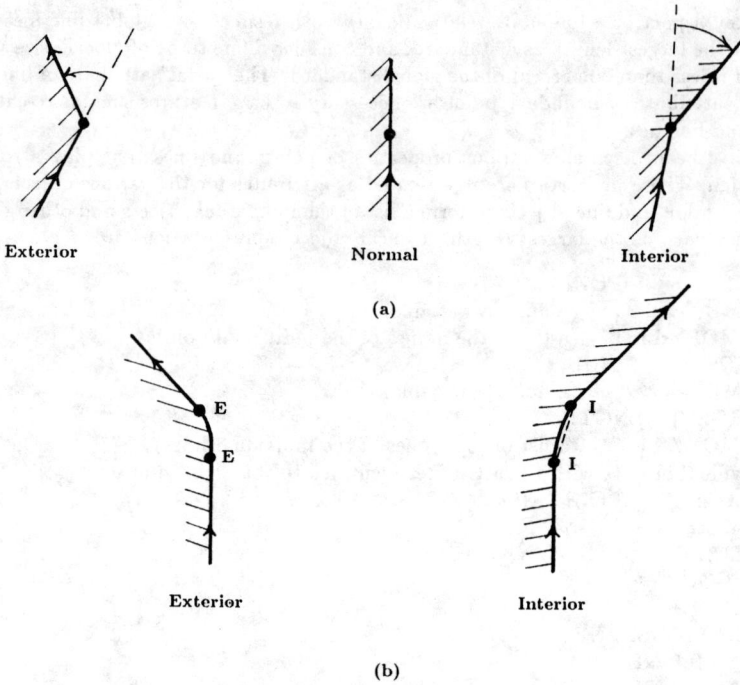

Exterior   Normal   Interior

(a)

Exterior   Interior

(b)

Figure 6: Corner Point Classification

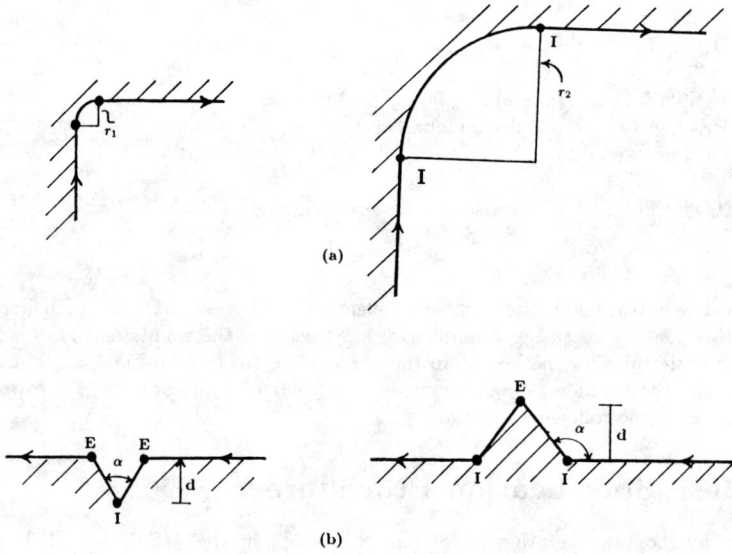

(a)

(b)

Figure 7: Local Versus Global Length Scales

The orientation information for the lines incident on a point (PREVIOUS_LINE and NEXT_LINE) provides the value of the ANGLE_CHANGE and establishes the type of the point, i.e. interior, normal, or exterior. Then, one more pass over the line list establishes the type of each line, i.e., local or global size. With this information, we can now proceed with the mesh design.

Our approach is based on the concept of continuity of mesh size at a boundary point. By continuity, we mean the size of the two elements adjacent to a boundary point are equal (see Figure 8). We enforce this condition by introducing a POINT_MESH_SIZE attribute for the POINT_N object, and requiring the mesh for an individual line to start and end with the appropriate point mesh size. We found this approach to be sufficiently flexible to handle a variety of geometric features.

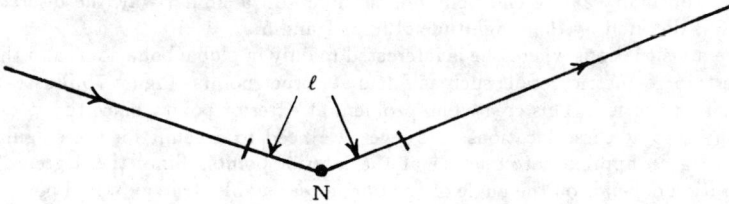

Figure 8: Mesh Continuity at a Boundary Point

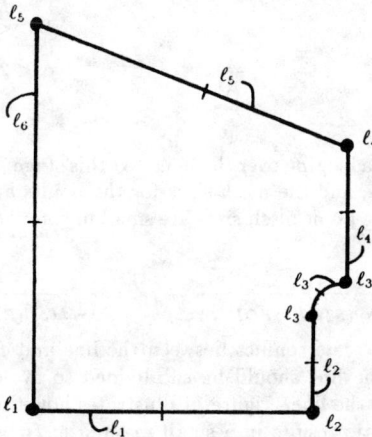

Figure 9: Boundary Point Mesh Sizes

The mesh size for the boundary points is determined in the following manner. We loop over the lines, and establish a representative mesh size measure for each line using:

$$\ell_R = \frac{L}{N} \tag{1}$$

where L is the total length of the line, ℓ_R is the representative mesh size attribute, and N is an integer which depends on the element expansion. We assume quadratic elements and take:

$$N = 2 \quad \text{for a straight line} \tag{2}$$

$$N = 2\left\{\frac{\alpha}{90}\right\} \quad \text{for a circular segment}$$

where α is the subtended angle in degrees. Equation (2) gives $N = 2$ for a $90°$ sector, which translates to a node every $22.5°$. Experience suggests this is a reasonable approximation. With ℓ_R known, we examine the points at each end and compare ℓ_R with ℓ_{MIN}, the minimum point mesh size attribute for the boundary point. When the looping is finished, we have generated a set of boundary point mesh sizes which reflect the constraints introduced by local lengths on the global discretization. Figure 9 illustrates this process. The small mesh size for the fillet radius is propagated into the adjacent lines by enforcing mesh size continuity at the end points of the fillet. In a similar way, the discretization for line 6 is dictated by the magnitude of lines 1 and 5.

There are situations where one is interested mainly in global behaviour, and therefore omits certain geometric details such as fillets at corner points. Figure 7b illustrates this type of simplification. This creates no problem at exterior points, since the stresses are identically zero at these locations. However, we need to account for the singular type behaviour in an approximate manner at the interior points. Since the degree of stress concentration depends on the angle of opening, a reasonable strategy would be to modify the point mesh size as a function of this angle. Referring to Figure 10, we replace ℓ_{MIN} for a sharp interior corner with ℓ_{MIN}^* where

$$\ell_{MIN}^* = \frac{\ell_{MIN}}{f(\alpha)} \tag{3}$$

and $f(\alpha)$ is based on heuristic information. We use the following simple relation:

$$f(\alpha) = \frac{180}{\alpha} \qquad\qquad 20° < \alpha \leq 180° \tag{4}$$

$$f(\alpha) = 9 \qquad\qquad \alpha < 20°$$

The last step involves a looping over the lines. At this stage, we have the representative mesh size for the line, ℓ_R, and the mesh sizes for the points at the ends of the line. We need to check whether the point mesh sizes are small in comparison to the line mesh size. The ratios

$$r_s = \frac{\ell_R}{(\ell_{MIN})_{(START\_POINT)}} \qquad r_e = \frac{\ell_R}{(\ell_{MIN})_{(END\_POINT)}} \tag{5}$$

provide a measure of the scale conflict between the line and end points. If both r_s and r_e are large, say > 3, the line should be subdivided to avoid an excessive number of elements being placed on the line. Figure 11 illustrates how this situation can arise. The small depth to length ratio results in a small mesh size, ℓ_1, at the end points whereas the line mesh is on the order of ℓ_2. We introduce an additional boundary point located according to a weighting of the end ratios (see Figure 12):

$$L' = L\frac{r_s}{r_s + r_e} \qquad L'' = L - L' \tag{6}$$

and update the representative line mesh size and corresponding point mesh sizes to account for the change in the length using (1):

$$\ell_R^{()} = \frac{L^{()}}{N} \tag{7}$$

$$(\ell\_MIN)_{NEWPOINT} = \text{smaller of } \ell_R' \text{ and } \ell_R''$$

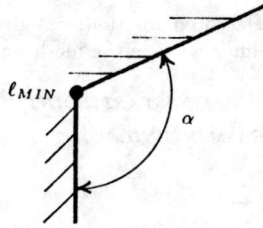

Figure 10: A Sharp Interior Point

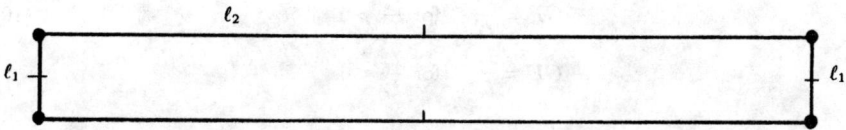

Figure 11: Local Versus Global Length Scales

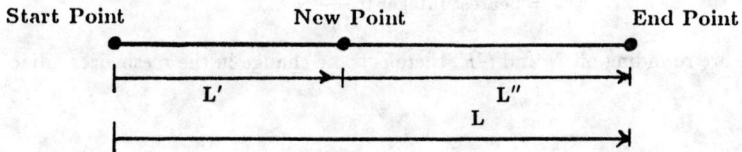

Figure 12: Notation For Line Subdivision

After modifying the lines to account for the local-global scale conflict, we can now define the final mesh. For each line, we have the mesh sizes for the points at the ends,

$$\ell_s = (\ell_{MIN})_{START\_POINT} \tag{8}$$
$$\ell_e = (\ell_{MIN})_{END\_POINT}$$

We need to select a graded pattern of line segments which starts with ℓ_s and ends with ℓ_e. We use a geometric progression for grading and adjust the parameters such that the grading factor and number of elements are integers. One starts with the relations (see Figure 13):

$$\ell_1 = r\ell_o \quad \ell_2 = r^2\ell_o \quad \ell_{N-1} = r^{N-1}\ell_o \tag{9}$$
$$L = \ell_o \left[\frac{1 - r^N}{1 - r} \right]$$

Equation (9) is based on N segments. We define the grading factor as the ratio of the end mesh sizes:

$$GF = \frac{\ell_e}{\ell_s} \quad \text{for } \frac{\ell_e}{\ell_s} > 1 \tag{10}$$
$$GF = \frac{\ell_s}{\ell_e} \quad \text{for } \frac{\ell_e}{\ell_s} < 1$$

From (9),

$$GF = \frac{\ell_{N-1}}{\ell_o} = r^{N-1} \tag{11}$$

We compute ℓ_e/ℓ_s, round down to the nearest integer, and then solve N with

$$N = 1 + \frac{\ln GF}{\ln r} \tag{12}$$

taking r = 1.5. Another approach would be to use the "geometric average" value,

$$\ell_{av} = \ell_o \, r^{(N-1)/2} \approx \{\ell_s * \ell_e\}^{\frac{1}{2}} \tag{13}$$
$$N = \text{nearest integer to} \frac{L}{\ell_{av}}$$

Since we are rounding off N and GF, there is some change in the mesh sizes adjacent to the points.

5 Illustrative Examples

We present here some examples which demonstrate the ability of the approach to detect geometric features and adjust the mesh in local regions. The first example is a well known test case that has been analysed in considerable detail by both Finite Element and Boundary Element proponents (Floyd [4]; Brebbia et al. [5]; Sussman et al. [6]). Using the heuristics described earlier, e.g. 2 elements per $90°$ circular arc, produces the mesh shown in Figure 14. A subsequent analysis with this grid gave results for the maximum tangential boundary stress which are in close agreement with the predictions of more detailed analyses (Brebbia et al). [5].

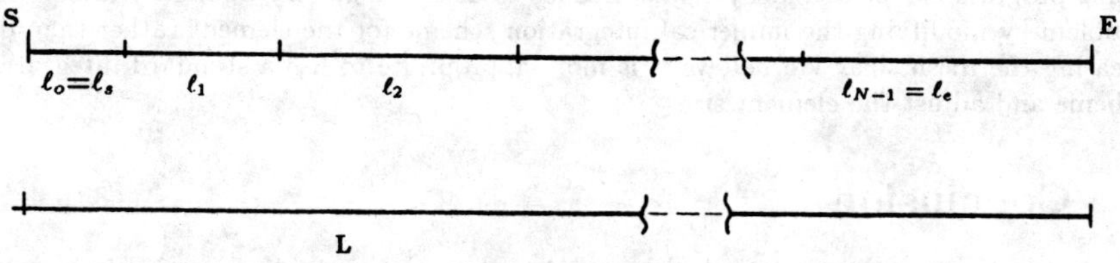

Figure 13: Geometric Progression

Figure 14: The Floyd Problem

The next example demonstrates boundary-boundary interaction. In Figure 15a, the circular cut-out is essentially independent of the outer boundary, and both boundaries are meshed separately. In Figure 15b, we move the cut-out closer. We need to introduce a fictional cut in order to link the two curves. This is one of the limitations of the existing program. Most boundary element codes treat the boundary-boundary interaction problem by modifying the numerical integration scheme for the element rather than decreasing the mesh size. We believe it is more appropriate to use a standard integration scheme and adjust the element size.

6 Conclusion

The mesh design strategy described here is based on the concept of continuity of mesh size at boundary points. With this concept, one takes as control values the mesh sizes at the boundary points and adjusts the interior mesh sizes to produce a suitable gradation. Heuristics are used to decrease the "control" mesh size for boundary points that correspond to interior points. The examples show that the proof of concept version yields quite reasonable results for 2-D problems, even though it is based on limited heuristics.

Our future efforts will be directed at improving the quality of the heuristics and extending the methodology to three-dimensional bodies. We believe the mesh continuity concept also applied for the 3-D case. Here, one will need a multilevel strategy; i) mesh the intersection lines using the boundary point mesh constraints, and ii) starting with the boundary discretization, mesh the area of each surface component. To handle the boundary-boundary interaction problem, we believe it is more advantagous to employ an object representation for the geometric shapes, using the concepts suggested by Pentland [7].

References

1. Brebbia, C.A. and Dominguez, J. Boundary Elements An Introductory Course, Computational Mechanics Publications, Southampton; McGraw Hill, New York 1989.

2. Chehayeb, F. A Framework for Engineering Problem Solving, PhD. Thesis, MIT 1987.

3. Winston, P.H. Artificial Intelligence, Addison-Wesley, 1984.

4. Floyd, C.G. The Determination of Stress Using a Combined Theoretical and Experimental Analysis Approach. Computational Methods and Experimental Measurements, Proc. of 2nd Int. Conf. June/July 1984, (Ed.Brebbia,C.A.) Springer-Verlag Berlin and NY and Computational Mechanics Publications, Southampton, 1984.

5. Brebbia, C.A. and Trevelyan, J. On the Accuracy and Convergence of Boundary Element Results for the Floyd Pressure Vessel Problem, Technical Note, Computers and Structures, 24, pp. 513-516, 1986.

6. Sussman, T. and Bathe, K-J. Studies in Finite Element Procedures - On Mesh Selection, Computers and Structures, 21, pp. 257-264, 1985.

7. Pentland, A. Perceptual Organization and the Representation of Natural Form, Artificial Intelligence Journal, Vol.2 pp.1-38,1986.

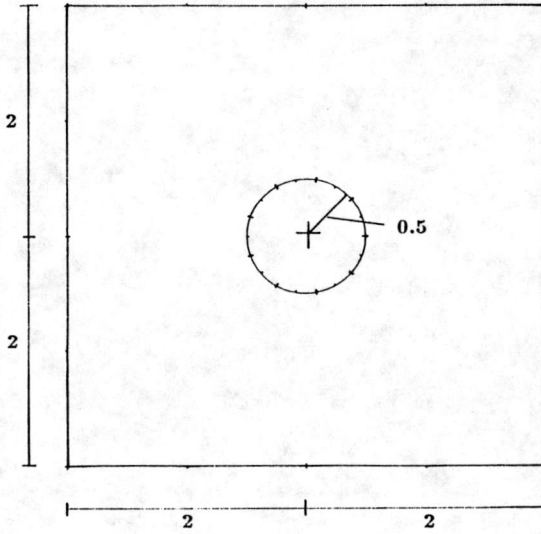

Figure 15 (a)

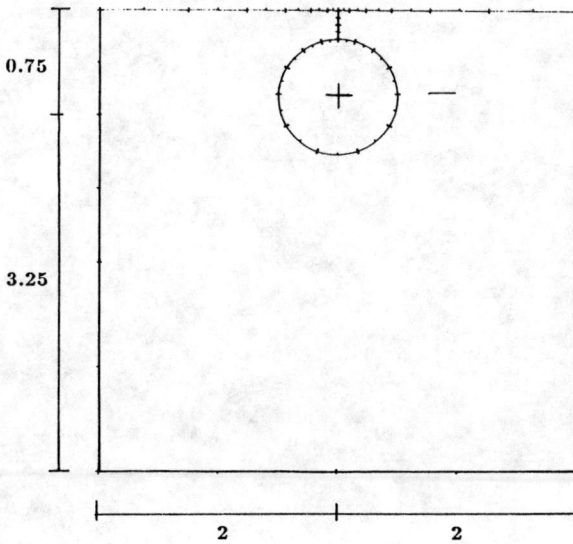

Figure 15 (b)

Figure 15: Illustration of Boundary - Boundary Interaction

Optimization and Supercomputing

Garret N. Vanderplaats

Engineering Design Optimization, Inc., 1275 Camino Rio Verde, Santa Barbara, California 93111 USA

ABSTRACT

Modern numerical optimization methods are discussed with particular reference to structural optimization. The basic concepts are first described, noting that this is a computationally intensive process that is strongly dependent on the power of today's supercomputers. Recent research results in structural optimization are next described, followed by a discussion of large scale engineering optimization. Examples are offered to demonstrate the present state of the art.

INTRODUCTION

The concept of optimization is not new. Indeed, it is as old as mankind. From the development of the first crude wheel, we have continually refined, and enhanced our designs. This is the essence of optimization. This is not to say that the entire design process is one of optimization. On the contrary, the basic invention or concept development is a creative process that is difficult to formalize or quantify. However, the process of developing the design into a useful product is what we define here as optimization.

The purpose here is to briefly outline modern, computer based, optimization methods and to identify the state of the art in the specific field of structural optimization. In discussing structural optimization, it is assumed that the underlying analysis will be based on the finite element method, although optimization is in no way restricted to this. Also, we will limit the discussion to numerical optimization methods, as opposed to the classical variational methods. Examples will be offered to identify the present capabilities of this technology.

While modern linear programming was developed as early as the 1940's, numerical methods for the solution of the general constrained nonlinear problem began their development in ear-

nest in the late 1950's. Serious research in structural optimization began with the classical paper of Schmit in 1960
[1]. He solved a simple three-bar truss design problem, using
linearly elastic analysis (two independent displacement degrees
of freedom) and stress limits only, by combining finite element
analysis with mathematical programming (numerical optimization)
methods. He showed that the design obtained in this fashion
was better than that obtained by the traditional fully stressed
design method. The computer time used to solve this simple two
design variable problem was one-half hour on an IBM 653 computer. In his review of this technology in 1981 [2], Schmit
commented that "only a congenial optimist could have been so
enthusiastic about future prospects." Today, this same problem
is solved in about one second on an IBM PC-AT and one mil-
lisecond on a supercomputer. Much of the improvement in computer time comes from the advances in the optimization technology itself which has led to about a factor of 500 improvement
in basic computational efficiency. The remaining factor of
about 10,000 in efficiency improvements has come from the computer technology. Extending this simple problem to a practical
structural design task by today's standards, we can easily en-
vision a 500 design variable problem for a structure modeled
with 50,000 displacement degrees of freedom. The reader is
left to estimate the computational cost using yesterday's computers! Yet this larger problem is computationally simple compared to the design problems, including dynamic response, and
perhaps with geometric and material nonlinearities, we cannot
solve for lack of computational power. Also, in addition to
member sizing, we wish to treat the structural shape as variable in the design process. Thus, it is clear that more computational power is needed, and this will come from advances in
supercomputers, much more than from advances in design
methodology.

Having agreed that super-supercomputers are needed, we now
turn to the general methodology of optimization.

NUMERICAL OPTIMIZATION CONCEPTS

Mathematical programming (numerical optimization) solves
the general nonlinear, constrained problem, Find the set of
design variables, X_i, i=1,n, contained in vector $\underline{X}$, that will

$$\text{Minimize} \quad F(\underline{X}) \tag{1}$$

Subject to;

$$g_j(\underline{X}) \leq 0 \qquad j=1,m \tag{2}$$

$$\underline{X}^L \leq \underline{X} \leq \underline{X}^U \tag{3}$$

Typically the objective function to be minimized is the

weight of the structure, although any response can be minimized or maximized. For example, maximization of a fundamental vibration frequency is also a common objective function. The design variables contained in $\underline{X}$ consist of member dimensions and joint coordinates, although here too the set can be expanded to include non-structural parameters such as control variables. Common constraints include limits on stress, strain, displacement, buckling load factors, vibration frequencies and dynamic response, aeroelastic response, etc. For example, a typical stress constraint would be written in normalized form as

$$g(\underline{X}) = S_{ijk}/S_{allow} - 1 \leq 0 \tag{4}$$

i = element number, j = stress component, k = load case

Noting that in a typical structural analysis there are several stress components to be constrained per element, several hundred elements and ten or more load cases, it is clear that the total number of such constraints, m in Equation 2 can be in the hundreds of thousands. Consequently, without the aid of a rational design method, there is little hope of obtaining even a near "optimum" design.

A multitude of numerical search algorithms are available to solve this general optimization problem [3]. Most of the more powerful methods update the design by the following relationship;

$$\underline{X}^{q+1} = \underline{X}^q + a^*\underline{S} \tag{5}$$

where q is the iteration number, $\underline{S}$ is a vector search direction and a* is a scalar move parameter determining the amount of change in the design. In other words, the product $a^*\underline{S}$ is the design modification at the current step. An initial design $\underline{X}^0$ must be provided, but need not satisfy all of the constraints (indeed, one of the most powerful uses of optimization is to find a feasible solution to a complicated problem). In order to determine a search direction, $\underline{S}$, that will improve the design, gradients of the objective and critical constraints must be supplied. Ideally, these are computed analytically as described below, but if they must be computed by finite difference, this may be effectively done in a parallel processing environment. Finally, a "one-dimensional" search is performed by trying several values of a* and interpolating for the one that gives a minimum objective while satisfying the constraints. During this process, the objective and constraints must be repeatedly evaluated. Here again, parallel processing can be valuable by taking several trial values for a* and simultaneously calculating the needed functions.

STRUCTURAL OPTIMIZATION

Now consider the general methodology of structural optimization (often called structural synthesis).

Virtually all modern structural optimization is based on the finite element displacement method, where for linearly elastic analysis we have the familiar set of simultaneous equations to be solved;

$$K\underline{u} = \underline{p} \tag{6}$$

where K is the master stiffness matrix, $\underline{u}$ is the vector of joint displacements and $\underline{p}$ is the vector of joint loads. The load and displacement vectors usually contain multiple columns, representing separate loading conditions. Equation 6 is of course only the most restricted case and, in general, may be expanded to include mass, dynamic and aeroelastic terms. Also geometric and material nonlinearities as well as time dependent loads may be included. Thus, it is clear why we chose to work with such a standard. While the discussion here is limited, it should be remembered that the state of the art is well beyond this in terms of the classes of problems that can be solved.

Effective structural optimization requires that the gradients (sensitivities) of the structural responses in terms of the design variables be calculated. Since member stresses are based on the displacements calculated in Equation 6, once displacement sensitivities are available, sensitivity of stresses is calculated at the element level. It is interesting to note that, prior to 1965, all gradients were calculated by finite difference. This is simply because no one had observed that a simple chain rule differentiation of Equation 6 yields the required information;

$$[\partial K/ \partial X_i]\underline{u} + K[\partial \underline{u}/ \partial X_i] = \partial \underline{p}/ \partial X_i \tag{7}$$

from which

$$\partial \underline{u}/ \partial X_i = K^{-1}[\partial \underline{p}/ \partial X_i - (\partial K/ \partial X_i)\underline{u}] \tag{8}$$

Noting that K^{-1} is already available in decomposed form, the solution of Equation 8 requires only the information in the brackets and then solution of the original equations for a new set of "loads." Since this first observation, analytic or semi-analytic (where $\partial K/ \partial X_i$ is calculated as a finite difference subproblem) gradients have been calculated for system buckling, frequency, dynamic response, aeroelastic response, and nonlinear time dependent response as examples [4,5].

Efficient structural optimization does not require that the finite element analysis and gradient computations be done whenever the optimization program requests them. What we actually do is perform a detailed analysis at the beginning, together with an evaluation of all of the constraint functions. The constraints are then sorted and only those that are critical or near critical are retained for the current design step. Typically, this requires only retaining about 5*n constraints, where n is the number of design variables. Gradients are then calculated for the objective and this retained set of constraints. This information is then used to create a "high quality approximation" to the original problem.

For example, the simplest approximation would be a Taylor series expansion of the form;

$$F(\underline{X}) = F(\underline{X}^0) + \underline{\nabla} F(\underline{X}^0)*(\underline{X} - \underline{X}^0) \tag{9}$$

where $F(\underline{X})$ is any objective or constraint function we wish to approximate and $\underline{\nabla}$ is the gradient operator. If we create such an approximation for the objective function and all retained constraints, we could then solve this with optimization without continually calling the finite element analysis code for function and gradient information. After an approximate optimization is complete, we would repeat the process until it has converged to the final nonlinear optimum. Such an approach is referred to as Sequential Linear Programming and experience has shown it to be a reasonably efficient method for engineering applications.

However, in structural optimization, we have additional insights that allow us to make a better approximation to the original problem. Consider for example a simple truss element where we wish to calculate the stress. Then

$$S = P/A \tag{10}$$

were S is the stress, P is the force in the element, and A is the cross-sectional area, which is to be minimized.

This leads to the simple optimization problem;

Minimize A   (11)

Subject to;

$$P/A \leq S_{allow} \tag{12}$$

While the objective function is linear, the constraint is clearly nonlinear in the design variable, A, so a direct linearization of this would not be accurate. However, we are free to choose another variable for design, so long as we link it to the cross-sectional area. Therefore, let the design variable be

$$X = 1/A \tag{13}$$

from which the stress is now linear in X;

$$S = P*X \tag{14}$$

and the optimization problem becomes;
Minimize 1/X   (15)

Subject to;

$$P*X \leq S_{allow} \tag{16}$$

Here, we have made a trade where the objective is now non-linear, but explicit, and the constraint is linear. Creating a linear approximation in X to this constraint is precise for statically determinate structures and is a significantly improved approximation for indeterminate structures. Therefore, relatively large changes may be made to the design during the approximate optimization stage before a new finite element analysis is needed. Also, if the objective function is the weight of the structure, it can be easily evaluated in its non-linear form during the approximate optimization.

This basic concept is referred to as an approximation technique and was first introduced in the mid 1970's [6]. The methods are applicable to basic member sizing where the element stiffness matrix is the product of the original design variable (area or thickness) to some power, and a constant matrix. For structures where this is not true, such as beam elements or when geometric variables are considered, these methods are not directly applicable. One approach to the more general problem is to create a conservative, convex, approximation to the original problem [7]. The basic concept is that, by observing the algebraic sign on each component of the function gradient, an automatic switching may be performed between direct variables (Equation 11) and reciprocal variables (Equation 15). This creates an approximation that is conservative relative to a direct linearization and usually provides improved efficiencies.

Whatever approximation is used, the overall design process is summarized as follows:

1. Input user defined data.

2. Analyze the initial design.

3. Evaluate the objective function and constraints.

4. Sort the constraints and retain those that are critical or near critical.

5. Calculate the sensitivity of the member forces with respect to some intermediate variables (e.g. Section Properties).

6. Create and solve the approximate optimization problem based on these sensitivities.

7. Analyze the proposed design.

8. Evaluate the objective function and constraints.

9. Check for convergence to the optimum. If satisfied, exit. Otherwise go to step 4.

The effect of this simple transformation to a high quality

approximation is approximately two orders of magnitude improvement in design efficiency over previous methods.

RECENT RESEARCH IN STRESS APPROXIMATIONS

Research in efficient approximation methods has continued, and recent results relative to stress constraints (the most common constraints considered in structural design) have demonstrated significant improvements over those obtainable by the approach of Equations 13 and 14 [8].

Consider again a simple bar element, but now instead of creating an approximation to the stress, approximate the force in the member;

$$P(\underline{X}) = P^0 + \underline{\nabla} P(\underline{X}) * (\underline{X} - \underline{X}^0) \tag{17}$$

where here the design variable, X, is the original member size, A. Now proceed with the approximate optimization as before, except when stresses are needed they are calculated from

$$S = P(\underline{X})/A \tag{18}$$

Using this approach, the approximate stress is nonlinear in the design variable, A, but is still explicit and easily evaluated. However, noting that displacements are proportional to load and stresses are related to the rate of change of displacements, it follows that Equation 18 is a higher order approximation than Equation 14. Also, it is no more difficult to calculate the sensitivity of element forces than to calculate stress sensitivities.

For truss structures, this modification is almost trivial, but dramatically improves the design efficiency. For frame structures, this allows us to calculate member forces in terms of section properties, but treat the individual dimensions of the beam as design variables during the approximate optimization [8,9]. Thus for beam structures, the approximate optimization uses the following steps;

1. When the optimizer requires stresses, the member dimensions are provided to the approximate analysis program.

2. The section properties are calculated exactly from the member sizes.

3. The member end forces are approximated in a form like Equation 17, except that the design variables are replaced by section properties since these provide a better approximation.

4. Finally the stresses are recovered in the usual FEM manner and returned to the optimizer.

Note that this process requires the solution of numerous identical subproblems that are ideally suited for parallel processing.

For more complicated elements such as three dimensional continuum elements, an additional step is required, which is again well suited for parallel computation. Having approximated the element nodal forces, it is necessary to determine the internal element displacement state, from which stresses are recovered. This entails first detecting and removing the rigid body degrees of freedom from the element and then solving a reduced set of equations at the element level. In the case of a 20 node brick element, this would require the solution of a 54 DOF set of equations. Following this the needed stresses are recovered in the usual manner [10]. Again, since this must be done for all elements for which stresses are required, this is a process that is ideally suited for parallel computation.

To date, this method has been tested for truss, beam, and three dimensional continuum problems. It is interesting that this approach works especially well for shape optimization of trusses [11,12] and three dimensional continuum shape optimization [10,13].

RECENT RESEARCH IN LARGE SCALE OPTIMIZATION

In recent years, research has begun in the development of multilevel methods for design optimization. Two principal motives are offered for this effort. The first is that engineering organizations operate in a multilevel and multidiscipline organizational structure. Thus, in order to include all engineering disciplines in the optimization process, it is desirable to provide techniques encompassing all disciplines while allowing each group to work with reasonable independence. The second motivation is to effectively break the overall optimization task into smaller parts that are more amenable to solution using numerical optimization methods. In structural optimization, this may include separating an aircraft into wing, body and vertical and horizontal tail structures as one set of subsystems. Further subsystems may be the individual panels, spars, ribs, etc. of the substructures.

Much of the work in this area has followed the concepts presented by Sobieszczanski-Sobieski and co-workers [14,15]. The basic approach defined in these works is to decompose the overall design task into two or more levels, while maintaining the essential interdependence among them. Building on this basic concept, recent work [16,17] has used a multilevel linearization method to solve this problem. In discussing multilevel optimization, it is important to remember that this is fundamentally different than substructuring which is often performed in structural analysis. Here we are solving the same general nonlinear constrained minimization problem defined above, so the interactions between the subsystems (even if the analysis is linear) is nonlinear and often highly coupled.

The basic concept of multi-level design may be understood by considering the simple cantilevered beam shown in Figure 1. The objective is to minimize the material volume subject to limits on the deflection at the beam junction and at the tip, and on the maximum bending stresses in the members. The design variables of interest are the width, B_i, and height, H_i, of each beam, and the length, L_1 ($L_2 = L - L_1$). Clearly, for such a simple problem, this would be solved directly. However, for demonstration purposes, it is possible to formulate it as a multilevel problem with a system level and two subsystems.

The system level problem may be stated as, find the beam length, L_1, and dimensions B_1, H_1, B_2 and H_2 to

Minimize $\quad V = B_1 H_1 L_1 + B_2 H_2 (L-L_1)$ $\qquad$ (19)

Subject to:

$$\delta_1 \leq \bar{\delta}_1 \qquad \qquad \delta_2 \leq \bar{\delta}_2 \qquad\qquad (20)$$

Here, the system level design includes all variables which control the system objective and all constraint functions. Thus,

$$\underline{Y} = \begin{Bmatrix} B_1 \\ H_1 \\ B_2 \\ H_2 \\ L_1 \end{Bmatrix} = \begin{Bmatrix} \underline{x}_1 \\ \underline{x}_2 \\ \underline{X} \end{Bmatrix} \qquad\qquad (21)$$

The subsystem design problem for subsystem i is to find the member dimensions, B_i and H_i that will

Minimize $\quad V_i = B_i H_i L_i$ $\qquad\qquad$ (22)

Subject to:

$$\sigma_i \leq \bar{\sigma} \qquad\qquad (23)$$

$$H_i \leq 10*B_i \qquad\qquad (24)$$

where

$$\underline{x}_i = \begin{Bmatrix} B_i \\ H_i \end{Bmatrix} \qquad\qquad (25)$$

Equation 23 is a stress constraint and Equation 24 limits the depth to ten times the width.

Note that, at the system level, all design variables are included. At the subsystem, the design variables that are important to that subsystem are considered, but the strictly system level variable, L_1, is held fixed.

Because of the interdependence between the system and subsystem variables, each level will effect the other. The key issue is how to account for these interactions and, in the general case, how to account for competition between subsystems.

In previous methods, for this example, the system level variables would be taken as the cross-sectional areas and moments of inertia, since these directly effect the system objective and constraints. Then, at each subsystem optimization, equality constraints would be imposed to insure that the system variables and functions do not change. This has the advantage that the integrity of the system optimization problem is maintained, and so the interaction among subsystems is directly controlled. In the present approach a much heavier burden will be imposed at the system level, in terms of the number of design variables and constraints, in order to reduce the overall nonlinearity of the problem.

For brevity, only two levels of the optimization problem are presented here. It is clear that multiple levels can be considered simply by treating each sub-level as a system, with further subsystems. The general program organization is shown in Figure 2, with a single system and multiple subsystems.

The complete set of design variables are contained in vector $\underline{Y}$ as;

$$\underline{Y}^T = \{(X_1, X_2, X_3, X_4), (x_5, x_6), (\ldots), (x_{N-2}, x_{N-1}, x_N)\} \qquad (26)$$

where the number of components in parentheses, ( ), relating to the system and subsystems is arbitrary and is given here as an example. Capital letters, X, refer to the system and lower case letters, x, refer to the subsystems. The vector, $\underline{Y}$, contains the complete set of design variables, including those changed only by the system as well as all subsystems.

Now, the system level design task can be stated in mathematical terms as, find the set of system variables, $\underline{X}$ and subsystem variables, $\underline{x}_i$, that will

Minimize $F(\underline{Y})$ \qquad\qquad\qquad (27)

Subject to;

$$G_j(\underline{Y}) \leq 0 \qquad\qquad j=1,MS \qquad\qquad (28)$$

$$g_{ji}(\underline{X}, \underline{x}_i) \leq 0 \qquad j=1,mss \quad i=1,nss \qquad (29)$$

$$\underline{X}_i \leq X_i \leq \overline{X}_i \qquad i=1,NS \qquad\qquad (30)$$

$$\underline{x}_{ij} \leq x_{ij} \leq \overline{x}_{ij} \qquad i=1,ns, \; j=1,nss \qquad (31)$$

where nss and mss refer to the number of subsystems and their constraints, respectively, and NS and MS refer to the system design variables and system constraints, respectively. Here, it is understood that the subsystem constraints, $g_{ji}(\underline{x}_i)$ are linearized to give the form;

$$g_{ji}(\underline{X},\underline{x}_i) = g_{ji} + \nabla_X g_{ji}(\underline{X},\underline{x}_i) \cdot (\underline{X} - \underline{X}^0) +$$

$$\nabla_{xi} g_{ji}(\underline{X},\underline{x}_i) \cdot (\underline{x}_i - \underline{x}_i) \qquad (32)$$

Note again that upper case letters refer to the system and lower case letters refer to the subsystems.

At each subsystem optimization, the problem solved is: find the set of subsystem variables, $\underline{x}_i$, that will

$$\text{Minimize} \quad f(x_i) \qquad (33)$$

Subject to;

$$G_j(\underline{x}_i) \leq 0 \qquad j=1,MS \qquad (34)$$

$$g_j(\underline{X},\underline{x}_i) \leq 0 \qquad j=1,mss \qquad (35)$$

$$x_i \leq x_i \leq x_i \qquad i=1,ns \qquad (36)$$

Now, it is understood that the system constraints, $G_j(\underline{x}_i)$ are the linearized form of the actual constraints in terms of the subsystem variables;

$$G_j(\underline{x}_i) = G_j + \nabla_x G_j(\underline{X},\underline{x}_i) \cdot (\underline{x}_i - \underline{x}^0) \qquad (37)$$

Furthermore, there is an additional set of inputs to the subsystem that must be accounted for. They may be considered to be the boundary conditions to the sub-optimization problem. For example, in the case of a beam, the end forces on the elements are functions of the design variables. This is a departure from the method of reference 14, where the section properties were held constant at the subsystem and so the end forces were constant as well. Here, these forces must be approximated in terms of the design variables. Now, when the subsystem variables are changed, the boundary conditions must first be updated as;

$$BC = BC^0 + \nabla_x BC \cdot (\underline{x} - \underline{x}^0) \qquad (38)$$

where BC is taken to be any input to the subsystem that is dependent on the values of the subsystem variables themselves. Now when the subsystem variables are changed, the boundary conditions are first updated by Equation 38, and then the necessary functions are calculated. Note also that, if the system constraints are functions of the boundary conditions, the appropriate additional terms must be added to Equation 37 as well.

The key idea here is that system level functions are sent to the subsystems in linearized form and the subsystem functions are sent to the system in linearized form. The net effect is that there is two-way communication relative to the functions involved. Furthermore, if the subsystems wish to send constraints relative to their own objective functions, this is possible. For example, a subsystem may send an upper bound on its own objective function (such as weight) to the system, and this must be respected in the system level optimization.

In this approach, more system level design variables and constraints are created. Also, the subsystems have some control over the system level constraints. Thus, each subsystem may drive the system constraints to zero, yet on return to the system, these constraints will be violated. The system level optimization must then overcome this constraint violation. Numerical experience has shown that this is a more theoretical than practical consideration and is implicitly dealt with as the optimization proceeds.

Considering the size and complexity of the design problems we wish to solve, it is clear that large scale optimization will require immense computing power. From Figure 2, it is seen that distributed computing will also play a major role. Depending on the computer architecture, this may be done on parallel processors on a single machine, on multiple machines, or on a combination of the two. As our ability to analyze larger and more complex systems grows, the use of numerical optimization to integrate this analysis into a formal design capability will become increasingly attractive. Perhaps more importantly, with the increasing sophistication of engineered systems, coupled with a lack of experience in designing such systems, optimization promises to provide the competitive edge.

DESIGN EXAMPLES

Examples are offered here to demonstrate the methods discussed above. The first is a relatively simple planar truss, demonstrating the state of the art in discrete element shape optimization, while the second is the three dimensional continuum shape optimization of a connecting rod. The third example is a simple two bay planar frame solved by the multilevel method described above.

Figures 3a and 3b show the design of a planar tower designed to support three separate loading conditions [12]. Symmetry was imposed and there are a total of 27 independent area (sizing) design variables and 17 coordinate design variables. Figure 3a gives the initial configuration and Figure 3b shows the final configuration. Constraints included stress limits and Euler buckling limits in the members. This design required a total of 11 analysis/gradient combinations with the associated solution of the approximate optimization problem, an order of magnitude improvement over previous methods for truss shape optimization. This example is computationally trivial for the supercomputer, but indicates the general state of the art in truss configuration optimization. It is however not hard to extrapolate this simple example to the design of a large space antenna, which would overwhelm the computational power of smaller machines.

Figures 4, and 5 show the design of an engine connecting rod [13]. This structure was modeled with 20-node solid elements and a 1/4th model used just over 2000 displacement degrees of freedom. Eight design variables were used to define the geometry and over 900 stress constraints were considered. As with the first example, this is a result of recent research wherein the member forces are approximated. Figure 4 shows the finite element model for the 1/4th model. The optimum was achieved after six design iterations. The initial and final shape, as well as the design variable definition are shown in Figure 5. This design required just under one hour on a super-computer. Some modifications in the optimization algorithm were made which reduced this time to about 40 minutes. The research code used to solve this problem is based on the SAP IV program [14] and uses finite difference gradient calculations. The code is presently being modified to utilize parallel processors. Also, during the approximate optimization phase, numerous calculations are done which can make efficient use of parallel computations. It is anticipated that making these relatively minor modifications will reduce the computational time to the order of ten minutes or less on a parallel process-ing machine. For the real design of a connecting rod, several loading conditions must be considered and more geometric detail is needed. Also, the proper analysis of the interface between the connecting rod and the crankshaft requires nonlinearities be considered. Thus, even for this "component" design problem, the computational needs will tax the power of our best super-computers.

The two bay frame shown in Figure 6 was designed for mini-mum material using the proposed method [17]. The system level constraints are shown in the figure, as well as the loading conditions. The subsystem constraints included stress, local buckling, and sizing limits. Symmetry was used so the system is comprised of four subsystems, being the vertical members of each bay and the floor members of each bay. Each subsystem consists of six design variables for a total of twenty four in-dependent design variables. The iteration history is shown in Figure 7, where a single iteration consists of a set of subsys-tem level optimizations followed by a system level optimiza-tion.

SUMMARY

Numerical optimization provides a powerful tool for en-gineering design. For structural design in particular, the state of the art is becoming relatively well developed and op-timization is now included or is being added to most major com-mercial finite element programs. Much of the progress in this field has come from improved optimization algorithms and care-ful formulation of the problem based on approximation tech-niques. However, the technology that has made the application of these methods, where finite element analysis is now treated as a subroutine in a larger framework, has come from advances in computer technology.

The purpose here has been to provide a brief introduction to numerical optimization and identify some of the recent developments in structural optimization that are driving the technology. Relative to the size and complexity of the problems we expect to solve in the foreseeable future, the examples presented here are considered small and simple. The key to solving the larger problems lies in the availability of software and, most of all, the continued advances in supercomputing technology. It is clear that as we move to the multi-level and especially the multidiscipline design environment, all of the common issues relative to computing become important. These include not just raw speed, but database systems tuned to the hardware, parallel processing, distributed computing and a variety of others.

Reference 19 offers a state of the art review of structural optimization up to the 1980 timeframe. There it was noted that, while it is difficult to predict the future, it was easy to state that at least structural optimization has a future. Today a part of that future is past and it can be said with some assurance that on the one hand structural optimization is a mature or at least maturing technology and on the other that past successes have only served to identify the need for continued research as well as the need for immense computational power. The key to success in such a computationally intensive area lies in the hands of the supercomputer.

REFERENCES

1.   Schmit, L. A., "Structural Design by Systematic Synthesis," Proc. 2nd Conference on Electronic Computation, American Society of Civil Engineers, New York, 1960, pp. 1249-1263.

2.   Schmit, L. A., "Structural Synthesis - Its Genesis and Development," AIAA Journal, Vol. 19, No. 10, 1981.

3.   Vanderplaats, G. N., Numerical Optimization Techniques for Engineering Design; with Applications, McGraw-Hill, 1984.

4.   Haftka, R. T. and Kamat M. P., Elements of Structural Optimization, Martinus Nijhoff, The Hague, 1985.

5.   Houg, E. J., Choi, K. K. and Komkov, V., Design Sensitivity Analysis of Structural Systems, Academic Press, 1985.

6.   Schmit, L. A. and Miura, H., "Approximation Concepts for Efficient Structural Synthesis," NASA CR-2552, March 976.

7.   Fleury, C. and Braibant, V., "Structural Optimization - A New Dual Method Using Mixed Variables," LTAS Report SA-115, University of Liege, Liege, Belgium, March, 1984.

8.   Vanderplaats, G. N. and Salajegheh, E., "A New Approximation Method for Stress Constraints in Structural Synthesis," (Accepted for publication, AIAA Journal)

9. Vanderplaats, G. N. and Salajegheh, E., "An Efficient Approximation Technique for Frequency Constraints in Frame Optimization," Int. Journal for Numerical Methods, Vol. 26, 1988, pp 1057-1069..

10. Kodiyalam, S., "Shape Optimization of Three Dimensional Continuum Structures Using Efficient Approximation Techniques," Ph.D. Dissertation, University of California, Santa Barbara, CA, June 1988.

11. Hansen, S. R., "An Efficient Method of Truss Configuration Optimization Subject to Constraints on Stress and Euler Buckling," Master of Science Thesis, University of California, Santa Barbara, CA, December 1987.

12. Hansen, S. R. and Vanderplaats, G. N., "An Approximation Method for Configuration Optimization of Trusses," Proc. 1988 AIAA/ASME/ASCE/AHS Structures, Structural Dynamics and Materials Conference, Williamsburg, Virginia.

13. Kodiyalam, S. and Vanderplaats, G. N., "Shape Optimization of 3D Continuum Structures Via Force Approximation Technique," Proc. 1988 AIAA/ASME/ASCE/AHS Structures, Structural Dynamics and Materials Conference, Williamsburg, Virginia.

14. Sobieszczanski-Sobieski, J., "A linear Decomposition Method for Large Optimization Problems - Blueprint for Development," NASA Technical Memorandum 83,248, February 1982.

15. Sobieszczanski-Sobieski, J., James, B., and Dovi, A., "Structural Optimization by Multilevel Decomposition," AIAA Paper No. 83-0832, Proc. AIAA/ASME/ASCE/AHS 24th Structures, Structural Dynamics and Materials Conference, Lake Tahoe, Nevada, May 2-4, 1983.

16. Beers, M. and Vanderplaats, G. N., "A Linearization Method for Multilevel Optimization," Proc. NUMETA 87, 2nd Int. Conference on Advances in Numerical Methods in Engineering, Theory and Applications, University College, Swansea, U. K., July 6-10, 1987.

17. Vanderplaats, G. N., Yang, Y. J. and Kim, D. S., "An Efficient Multilevel Optimization Method for Engineering Design, Proc. 1988 AIAA/ASME/ASCE/AHS Structures, Structural Dynamics and Materials Conference, Williamsburg, VA.

18. Bath, K. J., Wilson, E. L. and Peterson, F. E., "SAP-IV: A Structural Analysis Program for Static and Dynamic Response of Linear Systems," EERC, University of California, Berkeley, CA.

19. Vanderplaats, G. N., "Structural Optimization - Past, Present and Future," AIAA Journal, Vol. 20, No. 7, July 1982.

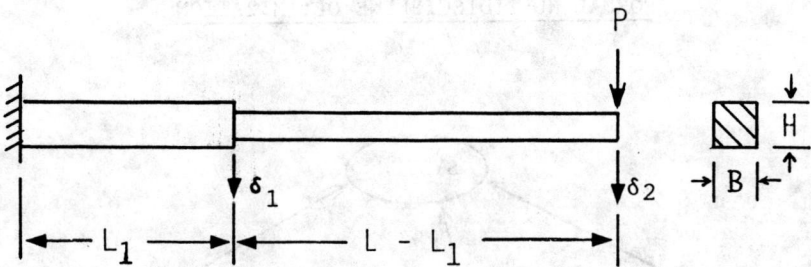

Figure 1: **Cantilevered Beam**

FORMAL MULTIDISCIPLINE OPTIMIZATION

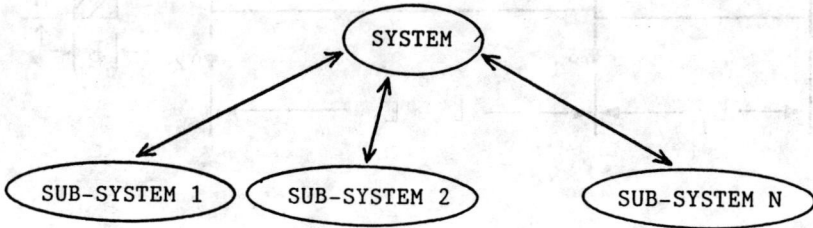

ADDITIONAL LEVELS OF SUB-SYSTEMS MAY EXIST

Figure 2:   General Multilevel Program Structure

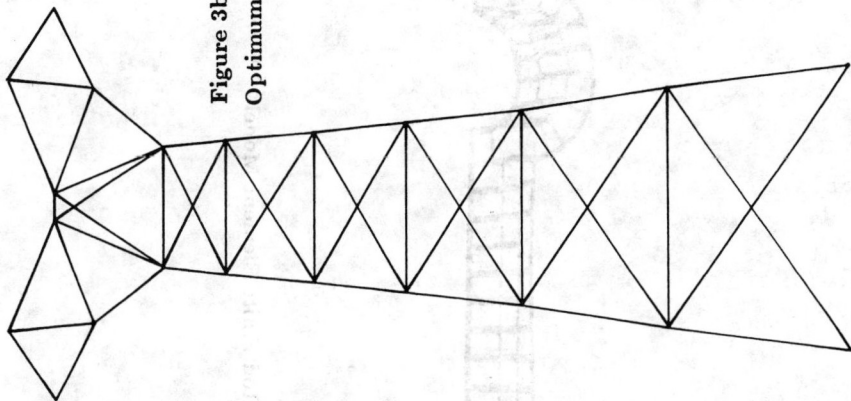

Figure 3b: Planar Tower:
Optimum Configuration

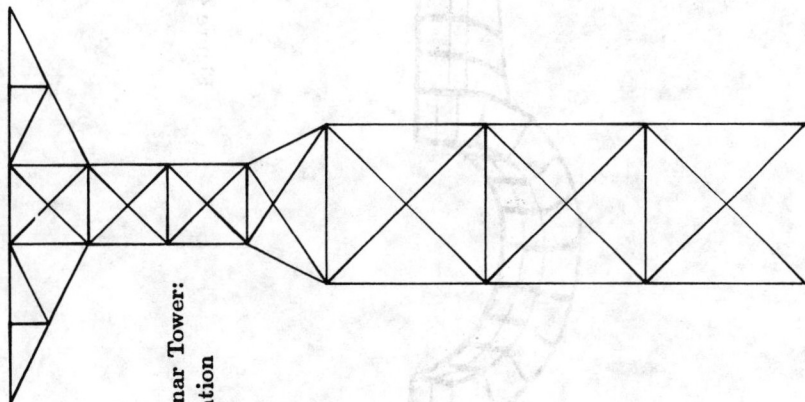

Figure 3a: Planar Tower:
Initial Configuration

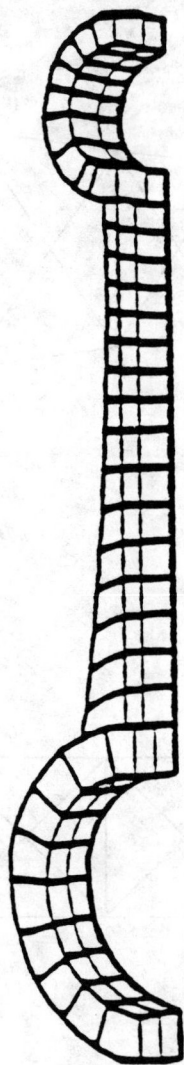

Figure 4: Connecting Rod Finite Element Model

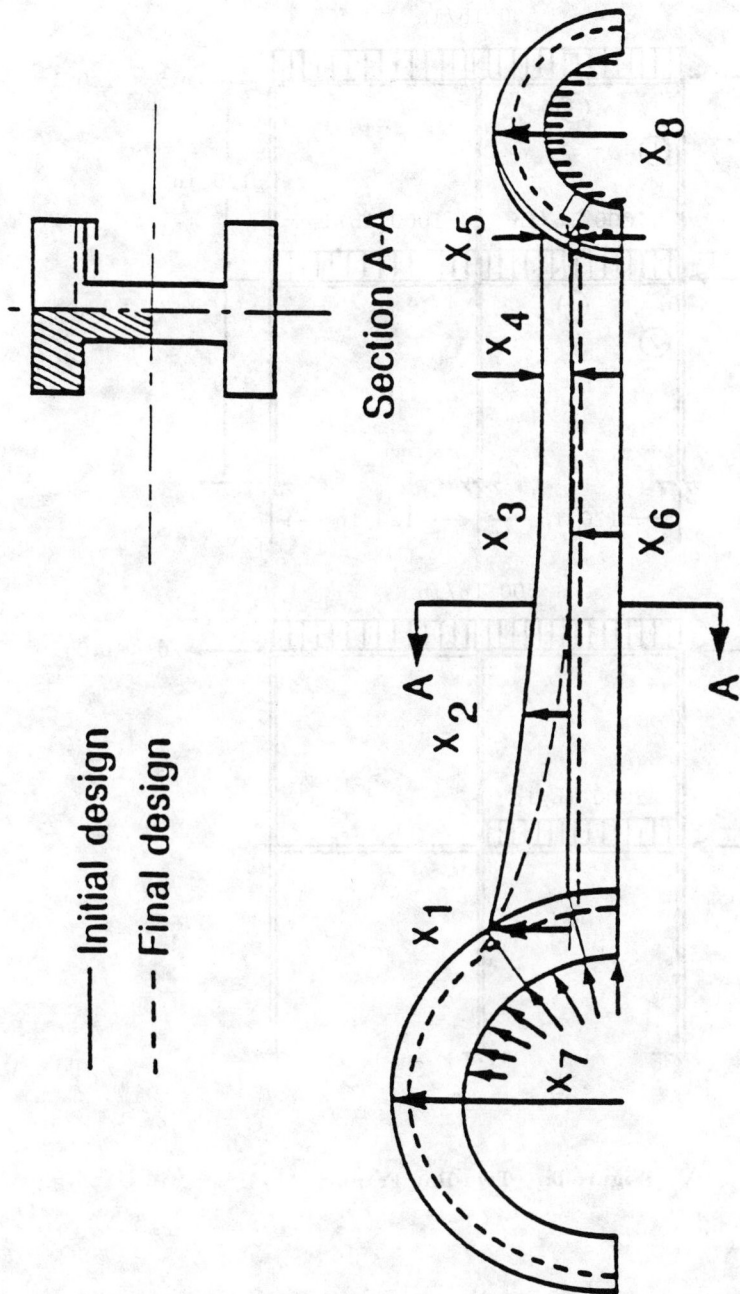

Note: x's are design variables

Figure 5: Connecting Rod Initial and Final Shape

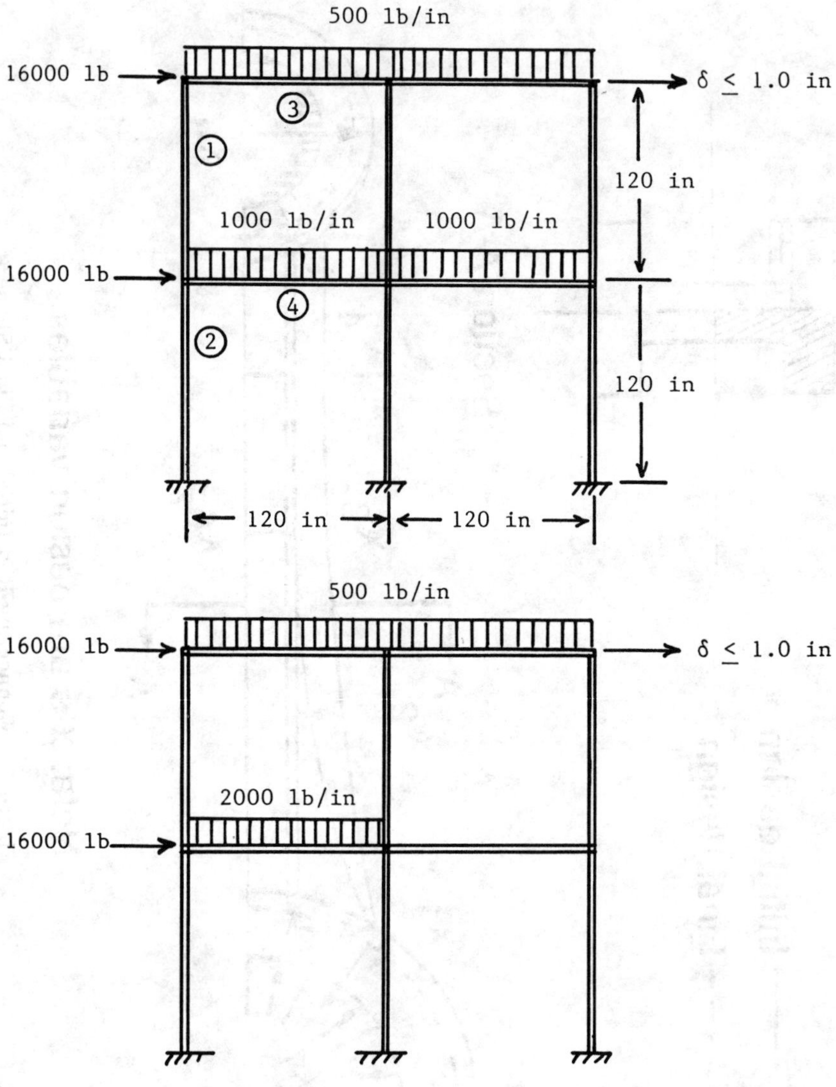

Figure 6: Two-Bay Frame

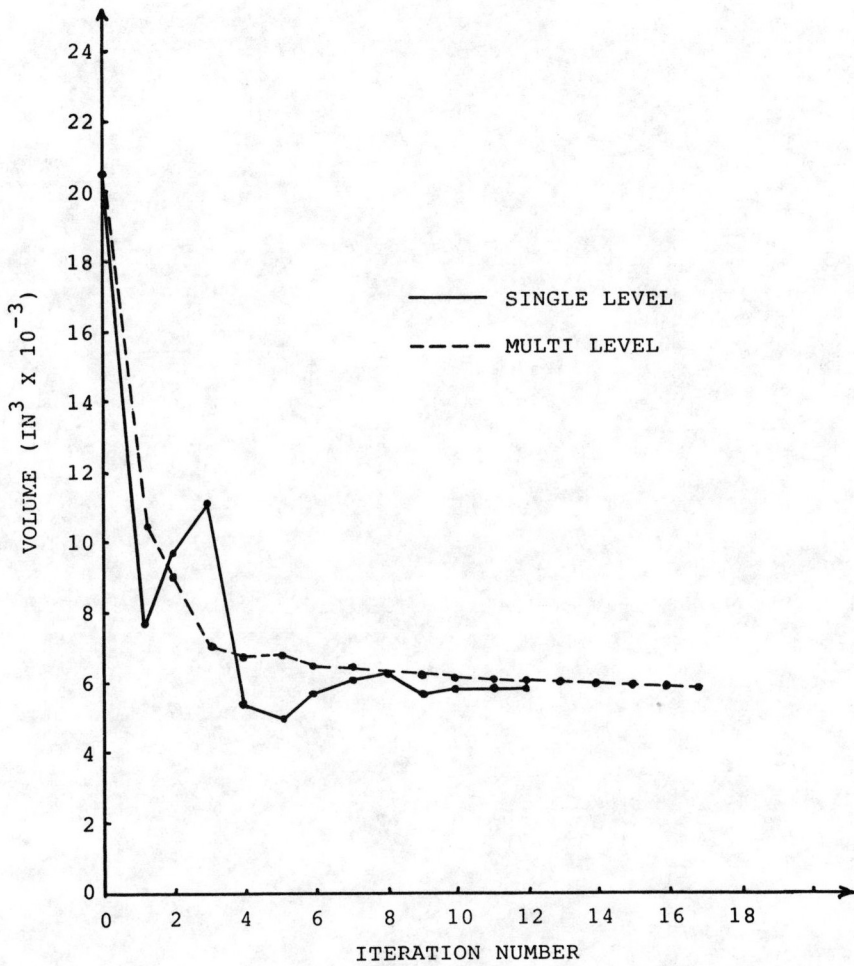

Figure 7: Iteration History for Two-Bay Frame